THE MICROBE 1984
Part II Prokaryotes and Eukaryotes

SYMPOSIA OF THE
SOCIETY FOR GENERAL MICROBIOLOGY*

* Published by the Cambridge University Press, except for the first Symposium, which was published by Blackwell Scientific Publications Limited.

THE MICROBE 1984

Part II Prokaryotes and Eukaryotes

EDITED BY

D. P. KELLY AND N. G. CARR

THIRTY-SIXTH SYMPOSIUM OF

THE SOCIETY FOR GENERAL MICROBIOLOGY

HELD AT

THE UNIVERSITY OF WARWICK

APRIL 1984

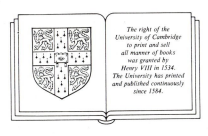

*The right of the
University of Cambridge
to print and sell
all manner of books
was granted by
Henry VIII in 1534.
The University has printed
and published continuously
since 1584.*

Published for the Society for General Microbiology

CAMBRIDGE UNIVERSITY PRESS

CAMBRIDGE

LONDON   NEW YORK   NEW ROCHELLE

MELBOURNE   SYDNEY

Published by the Press Syndicate of the University of Cambridge
The Pitt Building, Trumpington Street, Cambridge CB2 1RP
32 East 57th Street, New York, NY 10022, USA
296 Beaconsfield Parade, Middle Park, Melbourne 3206, Australia

First published 1984

Printed in Great Britain at The Pitman Press, Bath

Library of Congress catalogue card number: 83–19004

British Library Cataloguing in Publication Data

Symposium of the Society for General Microbiology (36th: 1984: Warwick)
The Microbe 1984. – (Symposia of the Society for
General Microbiology; 36)
Pt. II: Prokaryotes and eukaryotes
1. Micro-organisms – Congresses
I. Title   II. Kelly, D. P.   III. Carr, N. G.
IV. Series
576   QR41.2
ISBN 0 521 26057 4

CONTRIBUTORS

Arbuthnott, J. P. Dept of Microbiology, Moyne Institute, Trinity College, University of Dublin, Dublin 2, Ireland

Dijkhuizen, L. Dept of Microbiology, University of Groningen, Kerklaan 30, 9751 NN Haren, The Netherlands

Goodfellow, M. Dept of Microbiology, University of Newcastle upon Tyne, Newcastle upon Tyne NE1 7RU, UK

Harder, W. Dept of Microbiology, University of Groningen, Kerklaan 30, 9751 NN Haren, The Netherlands

Hopwood, D. A. John Innes Institute, Colney Lane, Norwich NR4 7UH, UK

Jannasch, H. W. Dept of Biology, Woods Hole Oceanographic Institute, Woods Hole, MA 02543, USA

Johnston, A. W. B. John Innes Institute, Colney Lane, Norwich NR4 7UH, UK

Morris, J. G. Dept of Botany and Microbiology, University College of Wales, Aberystwyth SY23 3DA, UK

Pfennig, N. Fakultät für Biologie, Universität Konstanz, D-7750 Konstanz, Federal Republic of Germany

Postgate, J. ARC Unit of Nitrogen Fixation, University of Sussex, Brighton BN1 9RQ, UK

Reznikoff, W. S. Dept of Biochemistry, College of Agricultural and Life Sciences, University of Wisconsin, Madison, WI 53706, USA

Schlegel, H. G. Institut für Mikrobiologie der Georg-August-Universität Göttingen, Grisebachstrasse 8, D-3400 Göttingen, Federal Republic of Germany

SHAPIRO, J. A. Dept of Microbiology, University of Chicago, 920 East 58th Street, Chicago, IL 60637, USA

THAUER, R. K. Fachbereich Biologie, Mikrobiologie, Universität Marburg, Lahnberge, D-3550 Marburg, Federal Republic of Germany

VELDKAMP, H. Dept of Microbiology, University of Groningen, Kerklaan 30, 9751 NN Haren, The Netherlands

VICKERS, J. C. Dept of Botany, University of Liverpool, Liverpool L69 3BX, UK

WILLIAMS, S. T. Dept of Botany, University of Liverpool, Liverpool L69 3BX, UK

CONTENTS

Contents: Part I Viruses

EDITORS' PREFACE

to a time when truth exists and what is done cannot be undone
(Orwell, *Nineteen Eighty-Four*)

It is common practice for Editors of this series to look back and comment upon the period of time that has elapsed since a Symposium based on a related topic took place at a Meeting of the Society. If the interval has been long, comment is made on the urgent necessity to conduct a reappraisal; if the interval be short, why, the vitality of the subject is self-evident! We have had no such requirement. To celebrate the One-hundredth Meeting of the Society for General Microbiology if was decided to hold a Symposium that would present some kind of overview of microbiology – looking back to some relatively immediate past achievements, and forwards to where and how the subject could develop. Since a comprehensive survey was obviously impossible, the Editors were left with the enviable task of making some personal choices, aided by valuable, though sometimes conflicting suggestions from other members of the Council.

Nineteen eighty-four is a date that has been imprinted over the past thirty-five years into the literary heritage and folklore of the mind of the English-speaking world by the doom-laden thesis of George Orwell – a date which has been approached with increasing trepidation year by year, and beyond which new worlds and aspirations must lie. It is, therefore, not surprising that a Committee of a learned Society should have been unable to resist the temptation to use this evocative date as part of the title of a Centenary Symposium Volume appearing in this year. To be honest, we make no apology for this perhaps less than imaginative title, that hardly does justice to the erudite and inspiring writing of our contributors. Suffice it to say that it is an occasion to review our subject in general terms.

It is perhaps well to recall that the role of the Society, in a life of less than forty years, notwithstanding its one hundred meetings, is of relatively recent significance, whereas the history of microbiology may reasonably be recognized as dating from about a century ago. Microbiology emerged as a coherent discipline in the late nineteenth and early twentieth centuries, born from a multidisciplinary coalescence of studies on the nature of disease, the very existence of the microbe, the 'battle of the chemists' over the nature of fermentation, into the recognition of microbes as the heritage and future of all the basic processes of physiology, ecology and genetics. Nevertheless, the Society has always attempted to present a general and integrated view of the micro-organism *per se*, rather than simply promoting the use of microbes merely as tools for the pursuit of other endeavours in biology. Thus one may fairly say that microbiology has become a microcosm of virtually all of biology and we hope that in each of the diverse chapters of this volume there will be matter of interest to all its readers. The essays our authors have been provoked to produce are as diverse as the subject: some long, some short, all relevant, some generalist, some minutely detailed, some linking the scientist and philosopher. To all we are indebted for making the work of the Editors such a pleasure. Certainly the quality of the contributors' command of language will slow the incursion into the microbiological literature of the 'doubleplusungood' of 'Newspeak' ('the only language in the world whose vocabulary gets smaller every year': Orwell, *Nineteen Eighty-Four*). We hope that this volume will bring enjoyment and pleasure as well as knowledge to all its readers.

D.P.K.
N.G.C.

GLOBAL IMPACTS OF PROKARYOTES AND EUKARYOTES

HANS G. SCHLEGEL

Institut für Mikrobiologie der Georg-August-Universität Göttingen, Grisebachstrasse 8, D-3400 Göttingen, Federal Republic of Germany

The central questions which kept biological research moving over the last fifty years concerned photosynthesis, genetics and evolution. With micro-organisms as model systems, fundamental discoveries followed each other in rapid succession. Kluyver's concept of unity in biochemistry originating from hydrogen transfer reactions (Kluyver & Donker, 1925), which emphasized the uniformity of biochemical reactions in all organisms (Kluyver & Donker, 1926), was able to be expanded to the universality of ATP as the elementary quantum of biological energy, of electrochemical transport mechanisms, of degradative and biosynthetic pathways, of DNA structure and of the genetic code. We started to understand the essential regulatory mechanisms that guarantee the harmony of cellular metabolism and can now even enjoy observing the first steps being taken to unravel the complexity of multicellular organisms.

Another revolutionary change in our view on the diversity of organisms and biological evolution came about with the 'concept of a bacterium' (Stanier & van Niel, 1962), the distinction between prokaryotes and eukaryotes. These two groups of organisms represent two different levels of cellular organization. Although this differentiation does not violate the principle of unity the differences between the groups are enormous. I would like to concentrate here on the impact of each of these groups on the environment in the course of biological evolution and on interactions between the groups.

THE EVOLUTION OF PROKARYOTES

Sixty years ago the earliest dominant organisms were considered to have been the autotrophic bacteria.

Als in den Urtagen des Lebens nach Abkühlung der Erde die ursprünglich als Wolken in der Luft hängenden Ozeane vollständig in flüssiger Form niedergeschlagen waren, konnten sich die primären Lebensformen zum ersten Mal die Sonnenenergie nutzbar machen und jenen 'normalen' Stoffwechsel herausbilden, den wir bei den grünen Pflanzen als Assimilation zu bezeichnen pflegen.*

These moving words, used by W. Bavendamm (1936) in a review article on the physiology of the sulphur and non-sulphur purple bacteria, reflected the prevailing hypothesis that the earliest dominant organisms had synthesized all their organic constituents from carbon dioxide, reduced compounds and light. In contrast, A. Oparin (1957) proposed in 1936 that the first organisms were heterotrophs and derived organic compounds from the environment. Oparin's basic thoughts remained unquestioned. They implied that the primeval forms of life were fermentative bacteria and that all other metabolic types, which derive energy from respiration or photosynthesis, originated from these strict anaerobes. Metabolic evolution was considered to have been a long process characterized by a slow, gradual development of metabolic types and to have culminated in the appearance of oxygenic photosynthesis (Broda, 1975). It was expected that early microfossil findings and geochemical analysis would reflect this sequence of events. The interpretation of isotopic data and fossil structures in old sediments is not in line with these assumptions (Schidlowski, 1982). As reliable data are still scarce and the survey of anaerobic bacteria is still fragmentary, the expansion of analytical studies (Brock, 1980) and investigation of anoxic ecosystems will be a rewarding challenge.

As the dogma implies, the original primitive atmosphere of the earth was anoxic and consisted largely of carbon dioxide, methane, nitrogen, ammonia, hydrogen and water. There is agreement among the experts concerning the absence of oxygen; mainly because otherwise the life span of the organic molecules which were formed abiotically and were required for synthesis of complex molecules, would have been too short, especially in the presence of ultraviolet light. However, the proposed compositions of the early atmosphere range from strongly reducing (methane, ammonia, hydrogen, water) to non-reducing (carbon dioxide, nitrogen, water) (Ponnamperuma, 1983). Experimental evidence indicates that orga-

---

* In the primeval days at the Dawn of Life, after the cooling of the Earth, when the Oceans hanging as clouds in the sky had reconstituted themselves into liquid form, then was it possible for the primordial life forms to make use of the Sun's energy for the first time, and to develop 'normal' metabolism, which we now term 'assimilation' in the green plants.

nic compounds can be synthesized abiotically under both conditions, but at a considerably higher yield in the more reducing
atmospheres. The time period taken for life to arise is completely
undefined:

Periods of 10^9 years are so far removed from our experience that we have
no feeling or judgement as to what is likely or unlikely in them. If the
origin of life took only 10^6 years, I would not be surprised. It cannot be
proved that 10^4 years is too short a period. (Miller, 1982)

There is almost complete agreement that the first living cells
consisted of self-reproducing polymers enclosed within a semi-
permeable membrane. While relying on abiotically formed organic
material, these cells obtained energy by substrate-level phosphorylation and acquired membrane proteins for osmoregulation, gaining
advantage from the efflux of protons and ion pumping devices.
When the rate of consumption of organic energy sources exceeded
the replenishment rate, the purely fermentative type of metabolism
had to be supplemented. This energy crisis may have given rise to
the development of the lithoautotrophic mode of life, possibly on
the basis of molecular hydrogen, carbon dioxide and 'left-overs' not
used by the fermentative bacteria. At least four divergent developmental paths can be considered, namely via: (i) the acetogenic
bacteria which use hydrogen and carbon dioxide to form acetate and
are metabolically closely related to the fermentative bacteria; (ii)
the sulphate-respiring bacteria, which can utilize hydrogen, carbon
dioxide and acetate with sulphate as electron acceptor; (iii) the
methanogenic bacteria, which utilize hydrogen, carbon dioxide and
acetic acid and contain coenzymes (M, F_{420}, F_{430}) and characteristic
membrane lipids not encountered in any other bacterial group; and
(iv) the phototrophic bacteria which exploit light as an energy
source and hydrogen, hydrogen sulphide (anoxygenic photosynthesis) or water (oxygenic photosynthesis) as reductants.
This picture coincides with a polyphyletic evolution of metabolic
types from the fermentative bacteria onwards. It indicates that large
amounts of cellular carbon can already have been formed in the
absence of oxygenic photosynthesis. It is in accordance with
thoughts on the evolution of membrane-bound bioenergetic systems
(Wilson & Lin, 1980; Garland, 1981) and includes the recent studies
on the archaebacterial type of metabolism and cell structure (Woese
& Fox, 1977; Stackebrandt & Woese, 1981; Kandler, 1982) and on
the sulphate- and sulphur-respiring bacteria (Widdel & Pfennig,

1981, 1982). The gaps are obvious; they seem, however, to be accessible to the experimental approach.

THE TIME SCALE OF METABOLIC EVOLUTION

With respect to the time scale of metabolic evolution one was inclined to assume a slow, gradual development of metabolic types lasting one or two Ga (1 Ga = 1 gigayear = 10^9 years) and culminating in the advent of oxygenic photosynthesis and aerobic respiration. Since the earth is 4.5 Ga old, and the earliest microfossils are 3.5 Ga old, cellular life must have existed very early. The beginning of the phanerozoic æon, when the land was colonized by plants and animals, has been set at 0.7 Ga ago. The prephanerozoan or Precambrian time period comprises four-fifths of the period that life has existed. In this time the prokaryotes were the dominant organisms on earth. The succession of metabolic types during this long time span and the time of appearance of the possibly continuous evolutionary line of anoxygenic photosynthesis, oxygenic photosynthesis, respiration, denitrification, nitrogen fixation and other metabolic traits is a matter of complete speculation.

The discussion on the time scale (Holland & Schidlowski, 1982) was recently stimulated by data derived from isotope (^{13}C) measurements which indicate a remarkably constant deposition of reduced organic carbon in the sediments, starting from the time at which the first banded iron formation (BIF) (in the Isua area of West Greenland) and stromatolites appeared, i.e. from 3.5 Ga onwards (Schidlowski, 1982). Assuming that all of the reduced carbon has been derived from living organisms, lithoautotrophic biomass production must have started very early. The discovery of microfossils and stromatolites in the 3.5 Ga-old rocks (Warrawoona Group, Western Australia) has been assumed to indicate the existence of cyanobacteria (Awramik, 1982).

The identification of these filamentous microfossils was based on the shape and size of these inclusions and on the lack of further examples allowing alternative interpretations. Since filamentous forms have been discovered among the sulphate-reducing bacteria (*Desulfonema*: Pfennig, Widdel & Trüper, 1981), the methanogenic bacteria (*Methanothrix soehngenii*: Huser, Wuhrmann & Zehnder, 1982) and the phototrophic sulphur bacteria (*Chloroflexus aurantiacus*: Pierson & Castenholz, 1974), the existence of these and other, now extinct forms – even among the purple sulphur bacteria – can easily be imagined. The size of filamentous bacteria may vary within wide limits, as known, for example, within the genus *Beggiatoa* (1–55 μm width). They may even have formed a kind of stromatolite, which may be any organo-sedimentary structure produced by trapping of carbonate and silicate minerals as a result of microbial growth (Krumbein *et al.*, 1979). These arguments in favour of the opinion that the oldest microfossils may have originated from filamentous bacteria other than the cyanobacteria are in accordance with a recent review (Schopf & Walter, 1982); this considers the presumptive evidence for the presence of cyanobacteria 2.8–2.5 Ga ago as convincing, but their existence 3.5 Ga ago as very doubtful.

The time of appearance of free oxygen in the atmosphere is completely unknown. The deposition of the BIFs in the course of the early Precambrian ages (James & Trendall, 1982) indicates that dissolved oxygen must already have been present in the upper water layers of lagoons or the continental belts 2.5 Ga ago while the depths of the oceans were still anoxic.

Furthermore, the appearance of heterocystous cyanobacteria in stromatolitic cherts of the Gunflint iron formation, which are about 2.0 Ga old, is indicative of at least local accumulation of dissolved oxygen. Other arguments for the availability of free oxygen during prephanerozoic ages are also indirect. The environment selecting for eukaryotes, which are – with very few exceptions – strict aerobes, must have contained free oxygen.

The temporal sequence of events involved in the evolution of prokaryotic metabolic types and the absolute time scale are, of course, very uncertain. However, speculation and experimental work will be rewarding. More details of the evolution of prokaryotes wait to be revealed. An optimistic view is supported by the following facts and availability of means: (i) The ecological habitats of the most important early organisms, the anaerobes, have endured on earth. In anoxic ecosystems the early organisms had a chance to survive. The rate at which new bacteria and metabolic types are now being discovered promises a fast expansion of our knowledge on prokaryotes. (ii) Relationships between groups of bacteria are being explored by ribosome analysis, by protein, DNA and RNA sequencing, and by comparison of cytochromes, ferredoxins, hydrogenases, cell walls, membrane lipids and other structures (Dayhoff & Schwartz, 1980). (iii) As even very early Precambrian sediments contain large amounts of reduced organic carbon, extractable and non-extractable (kerogen) matter, there is the hope of revealing the presence of various bacterial groups by the study of biological marker compounds such as derivatives and decomposition products of steroids, terpenoids, hopanoids, porphyrins and others. A vast literature on such chemofossils formed during the phanerozoic era exists (Mackenzie *et al.*, 1982; McKirdy & Hahn, 1982; Ourisson, Albrecht & Rohmer, 1982; Chappe, Albrecht & Michaelis, 1982). The application of the wide variety of modern analytical tools to carefully selected prephanerozoic samples (Brock, 1980) should provide valuable information about the evolution of prokaryotes.

THE GEOCHEMICAL EFFECTS OF PROKARYOTES IN THE PRECAMBRIAN

Precambrian life forms may have been involved in the mobilization and sedimentation of minerals in various ways: directly by the formation of acids and slime and by oxidation–reduction reactions, and indirectly as producers of accumulated organic carbon which acts as a reducing agent itself and can as such exert a mobilizing effect on various minerals.

One of the most uniform sediments, the BIFs, which contain 90% of the recoverable iron deposits of the earth, probably formed as a result of the precipitation of iron from water rich in ferrous iron. The model (Holland & Cloud, cited in Holland & Schidlowski, 1982) involves a stratified circulation system in relatively shallow basins with the water rich in ferrous iron at the bottom and oxygenated water on top, the mixing of which gave rise to the oxidation and precipitation of iron. It is assumed that the world-wide appearance of BIFs 2–2.6 Ga ago was due to the development of continental belts and to the biological production of oxygen by cyanobacterial plankton.

THE EVOLUTION OF EUKARYOTES

It is logical to suppose that the present-day bacteria evolved from early prokaryotes which are now extinct. Although the prokaryotes have the basic biochemistry and systems for catalytic, regulatory and genetic functions in common they have diversified with respect to metabolism and structure. The line of the archaebacterial cell-type may have diverged rather early from the evolutionary line of all other bacteria (Woese & Fox, 1977; van Valen & Maliorana, 1980; Stackebrandt & Woese, 1981). The prokaryotes learnt to exploit many ecological niches in both anoxic and oxic environments. Their metabolic evolution may have drawn benefit from highly evolved regulatory systems, a relatively simple cell structure, high growth rates and various means of gene transfer. However, the prokaryotic cell structure soon met insurmountable constraints prohibiting further evolution. These constraints may concern the genome size, its basically haploid nature and the small size of the cellular entity.

New habitats and environments required more advanced cellular functions and in consequence an increase in genetic material. Repeated successive doublings of the DNA may have been the response to the progressively increasing requirement for storage of genetic information. With the average genome size of a molecular mass of about 5×10^9, the upper size limit for a single-piece genome may have been reached. The replication and spatial separation of larger pieces may have encountered difficulties. Therefore, the expansion of genetic information required new principles of segregating duplicated DNA in the dividing cell.

The haploid nature of the prokaryotic genome represents another constraint. Evolution requires mutations. Most of these single mutations are disadvantageous or lethal to the cell. Favourable changes may occur when mutations are accumulated in a 'silent' state and then become dominant. As two sets of genetic information are required to conserve silent altered information, diploidy confers a decisive advantage for the manifestation of new complex properties. Further selective advantages are based on frequent recombination of DNA during meiosis, a process which appeared at the level of the eukaryotic cell.

A further obstacle to the evolution of the prokaryotic cell may have been its small size. High efficiency and success in a competitive population required more complexity, a high diversity of proteins and structural differentiation. To cope with a new style of life and to open up the surface of the land as a biotope required a new cellular design.

The protocyte and eucyte differ from each other in many aspects of structure and function. These differentiating characters are impressive and inspiring. They call for future research and deserve reiteration although they have recently been presented in an excellent review by Cavalier-Smith (1981). The characteristics of the eukaryotic cell are: (1) separation of the DNA from the metabolically active compartments by a nuclear membrane; (2) related to that, uncoupling of transcription and translation; (3) synthesis of exoproteins intracellularly and export by exocytosis; (4) intracytoplasmic transport systems based on actin and tubulin and on directed transport of vesicles (lysosomes, peroxisomes and others) and chromosomes during mitosis, meiosis and nuclear pairing; (5) multiple linear chromosomes with a plurality of replicons, limitation of DNA replication to an interphase and separation of daughter chromosomes by mitosis; (6) association of DNA with histones to form nucleosomes; (7) split genes and RNA splicing; (8) sexuality, chromosome pairing and meiosis to make sexual combination, new combinations of genes and alteration between haplophase and diplophase possible; (9) endocytosis (phagocytosis and pinocytosis) and the capability for harbouring cellular endosymbionts; (10) organelles (mitochondria and chloroplasts) for energy (ATP) regeneration; (11) 9 + 2 cilia for locomotion.

Although a few eukaryotes exist which lack one or other of these characters, practically no primitive organisms survived which allow us to recognize the temporal sequence in which the new characters appeared. Obviously each single step was accompanied by only a small selective advantage, at least compared with the better-equipped existing organisms. Intermediate members were outgrown by their successors; they did not survive and their vulnerable structures were not conserved as fossils in an analysable state. There are only a few living organisms which can be assumed to originate from intermediate forms. The possibility of recognizing the sequence of events is small, and my personal view is pessimistic: *credo ignorabimus*.

However, this statement should not discourage young scientists from studying the structure and metabolism of present-day eukaryotic micro-organisms, as well as symbiotic relationships and exchange of genetic information between these eukaryotes and prokaryotes. The early evolution of eukaryotic cells may have produced various models of cellular organization before multicellular organisms evolved which could cope with changing environments, strong radiation, dryness and unevenly distributed nutrients. At least a few unicellular eukaryotes may reflect their origin from intermediate forms or alternative models. Comparative studies on mitochondrial DNA revealed an astonishing diversity of sizes, structures and function.

The global impact of eukaryotes during the major part of their evolution in aquatic ecosystems (from 1.6 to 0.6 Ga) is completely unknown. We can reconstruct their influence only from the time at which they colonized the land. The eukaryotes specialized into photosynthesis and an aerobic way of life and left to the prokaryotes anaerobic ecological niches as well as some other ecological functions of vital significance such as nitrogen fixation, nitrification, denitrification, sulphate respiration, sulphur and metal ion oxidation, and methane formation and its utilization.

BIOGEOCHEMICAL CYCLES

The origin of organic carbon and molecular oxygen

Since life first existed organisms produced biomass part of which was conserved in the sediments. Sediments contain about 20 times more carbon than magmatic rocks. On the basis of $^{13}C/^{12}C$ isotope fractionation data it is assumed that the major proportion of the reduced carbon is of biogenic origin (Welte, 1967; Schwarcz, Hoefs & Welte, 1969; Böger, 1980; Schidlowski, 1982). Potential sources of the organic carbon (recognizable by its ^{13}C values) are all organisms able to fix carbon dioxide from the environment, such as anoxygenic phototrophic bacteria, oxygenic prokaryotes and eukaryotes, as well as methanogenic (Fuchs *et al.*, 1979) and acetogenic bacteria (Leigh, Mayer & Wolfe, 1981; Braun & Gottschalk, 1982). In the course of the last 0.5 Ga the highest productivities were reached on the land. However, almost all organic matter produced is ultimately recycled to carbon dioxide either through mineralization by micro-organisms or physico-chemical destruction. Only a very minor part of the biomass is trapped in sediments. This 'leak' in the carbon cycle has resulted, over the ages, in the deposition of organic carbon (10^{16} tonnes) which exceeds the amount of carbon present in the living biomass by at least 4 orders of magnitude.

The most important accumulation of organic carbon occurred during the past 0.5 Ga and resulted in the deposits of coal, petroleum and methane. The amount of recoverable carbon is, however, only a small fraction of the total carbon buried in the earth's crust. The biological contribution to fossilization and the formation of peat, coal, bitumen, kerogen, petroleum, methane and other components from plant matter is exclusively due to prokaryotes. The 'anaerobic food chain' provides an apt basis for speculation about the various processes, steps and their sequence. The microbial involvement deserves extensive investigation.

Atmospheric oxygen is the second product of oxygenic photosynthesis. The amount of atmospheric oxygen is, however, only a small fraction of the total oxygen produced; the major part, evolved early in the earth's history, is bound in iron oxides and sulphates. A tentative evaluation of the quantitative relationships between atmospheric oxygen, bound oxygen and carbon deposited supports this idea (Welte, 1970; Enos, 1973; Böger, 1980). The advent of free molecular oxygen had several consequences: it enabled the evolution and diversification of aerobically respiratory organisms and by the formation of ozone gave rise to a radiation shield in the upper atmosphere, thus making the fast development of plants on land possible, and allowed the functioning of the biological cycles of carbon, nitrogen and sulphur. Thus the present-day level of atmospheric oxygen and the deposits of recoverable carbon resources are the result of photosynthetic biomass production by eukaryotes.

The carbon cycle

In the carbon cycle, photosynthetic biomass production and the transient accumulation of organic materials is well balanced by mineralization. Photosynthates fall prey either directly or via the animal food chain to microbial degradation. About 99% of the carbon reaches the atmosphere in the form of carbon dioxide as a product mainly of aerobic processes. Only about 1% is released into the atmosphere as methane (Ehhalt, 1976) via anaerobic bacterial processes.

Aerobic degradation occurs in the oceans and on land. Most plant matter is decomposed *in situ* in the soil, which is inhabited by both prokaryotes and eukaryotes. The total microbial biomass is enormous. The total carbon content of mineral soils ranges from about 2400 to 70 000 kg ha^{-1}. (The weight of the 12.5 cm deep top layer of agricultural soil was assumed to be 1.5×10^6 kg ha^{-1}.) About 0.27–4.8% (average 2.5%) of total carbon is bound in the microbial biomass, that is 120–3500 kg biomass ha^{-1}. Old data on the ratio of bacterial to fungal biomass have been confirmed using new methods (Anderson & Domsch, 1980; Jenkinson & Ladd, 1981; Tate & Jenkinson, 1982). The contribution of bacteria to the total biomass ranges from 10 to 40% (average 25%). Similar values were obtained concerning the contributions to soil respiration (Anderson & Domsch, 1975). This means that the contribution of eukaryotes to the biomass and to the rate of destruction of organic matter is large even in the soil, and confirms the general conclusion that eukaryotes have replaced or pushed back the prokaryotes in all ecosystems which support aerobic organisms and with respect to all substrates which are accessible to them (such as proteins, nucleic acids, and polysaccharides, cellulose included).

Part of the organic matter is transported to anoxic environments before complete degradation, either directly or via the animal food chain as cellulosic detritus (Fenchel & Barker-Jørgensen, 1977). The anaerobic food chain, in which exclusively prokaryotes are involved, gives rise to methane which is released into the atmosphere. A remarkably high contribution to methane production originates from the rumen bacteria of the large number of ruminants (Ehhalt, 1976).

The annual cycles linked to the carbon cycle, such as the nitrogen and sulphur cycles, are either completely or predominantly the domain of prokaryotes (Fenchel & Blackburn, 1979; Schlegel, 1981). Nitrogen fixation is the privilege of bacteria, and some plants draw benefit from the nitrogen-fixing prokaryotes by intracellular, extracellular or associative symbiosis. Ammonium, the mineralization product of nitrogenous compounds, is almost exclusively oxidized to nitrite and nitrate by lithoautotrophic bacteria. In the vital function of denitrification also, bacteria alone are involved (Payne, 1981). Only prokaryotes are involved in the oxidation of hydrogen sulphide and sulphur, either aerobically or anaerobically, and in dissimilatory sulphate reduction (Trüper, 1982; Trudinger & Williams, 1982). The oxidation of iron, hydrogen and carbon monoxide as sole energy sources is also confined to bacteria. Prokaryotes alone could keep the cycles of matter running and maintain the biosphere while eukaryotes alone could not.

Extreme environments other than the anoxic ecosystems are also the domain of the prokaryotes – though of course there are exceptions (Kushner, 1978; Shilo, 1979; Gould & Corry, 1980; Schlegel & Jannasch, 1981). Aquatic ecosystems with temperatures up to the normal boiling point of water (Brock, 1978; Castenholz, 1979; Aragno, 1981) or even above (Stetter, 1982; Baross & Deming, 1983), with pH values lower than 2 (Langworthy, 1978) or higher than 10 (Horikoshi & Akiba, 1982) or with extremely high salt concentrations (Brock, 1980) are almost solely the preserve of prokaryotes. Tolerance to high salinity, up to 2 M sodium chloride, is the property of the halobacteria and some other halophilic bacteria. Heat resistance of bacterial endospores, which exceeds by far that of any vegetative cell or of fungal and plant spores, has an enormous impact on mankind by causing high energy consumption in the food industry.

ASSOCIATIONS BETWEEN PROKARYOTES AND EUKARYOTES

Associations between different species representing different metabolic types of prokaryotes are usually relatively loose. Many bacteria are linked to each other by food chains such as interspecies hydrogen transfer, syntrophism and exchange of growth factors without, however, permanent contact between the interdependent cells. Only in rare cases are two different bacteria closely associated with each other, as in the case of *Chlorochromatium aggregatum* or *Pelochromatium* in which photosynthetic green bacteria are attached to non-photosynthetic bacteria. As tissue formation and endocytosis do not occur among prokaryotes, closer symbiotic relationships between prokaryotic partners are lacking.

While the prokaryotes were able to develop for at least 2 Ga in splendid isolation, eukaryotes never had this chance. The evolution of eukaryotes occurred in an environment in which almost all the prokaryotic metabolic types known to-day were already present. Eukaryotes had to struggle continuously with prokaryotes. The developing eukaryotes offered new ecological niches, shelters and prey to the prokaryotes. Their inner and outer surfaces, intestinal tracts and skin lent themselves to colonization by prokaryotes. On the other hand, prokaryotes must, by their aggressiveness, have been the major causative agents of the highly developed protective, defence and survival mechanisms of higher plants and especially higher animals. Compared with the diversity of these adaptations provoked by the prokaryotes in the eukaryotes, the adaptive changes in the prokaryotes are almost negligible.

The enormous diversification and differentiation of the eukaryotes, which occurred in the permanent presence of prokaryotes, may have been the reason that a multitude of close associations of many kinds originated between eukaryotes and prokaryotes. Prokaryotes are the main ectosymbionts and endosymbionts of the eukaryotes, and recruiting cyanobacteria and aerobic respiratory bacteria as endosymbionts for the purpose of energy conversion was the most important or even portentous accomplishment of eukaryotic cells. I may disregard the controversy about the endosymbiont hypothesis (Dayhoff & Schwartz, 1980; Gray & Doolittle, 1982). Fairly recently it has been found that many of the highly specialized metabolic abilities of the prokaryotes are exploited by eukaryotes. The symbiotic relationships may be associative or, in various kinds of tissues, extracellular or intracellular.

Associative symbiosis may be as loose as in the case of the normal flora of the human skin (Noble & Pitcher, 1979) or the rhizosphere of plants. Symbionts in the intestinal tract of lower and higher animals, the gut of termites, wood-eating roaches and the rumen of cattle are all well-studied examples. However, recently described cases show that eukaryotes exploit not only the ability of bacteria to ferment cellulose and to fix nitrogen, but also their capacity to remove acetate and hydrogen and to utilize hydrogen sulphide for biomass production.

In many cases prokaryotes are sheltered extracellularly in plant tissues, as are the nitrogen fixing cyanobacteria in liverworts, *Azolla, Cycas* and *Gunnera*, the luminescent bacteria in fish and molluscs, and hydrogen-sulphide-oxidizing autotrophic bacteria in bivalves and other molluscs.

Endosymbiosis represents the highest level of symbiotic relationship between eukaryotes and prokaryotes. It is due to the ability of eukaryotic cells to engulf and at least transiently maintain foreign cells in their cytoplasm, and allows several of the metabolic activities developed at the level of the prokaryotic cell to be exploited by eukaryotes. The classic examples concern nitrogen fixation, oxygenic photosynthesis and luminescence. Nitrogen fixation in the root nodules of leguminous and some non-leguminous plants (*Parasponia*) by rhizobia and in the rhizothamnia of *Alnus, Casuarina, Shepherdia, Hippophaë, Ceanothus* and others by the actinomycete *Frankia* are outstanding examples of endosymbiosis. The presence of leghaemoglobin in most root nodules and its high homology with the globins of animals (Østergaard Jensen *et al.*, 1981; Appleby, Tjepkema & Trinick, 1983) raise questions with respect to early horizontal DNA transfer from a third partner. Similarly well-known is the exploitation of oxygenic photosynthesis by cyanobacterial (cyanellae) or microalgal cells (zoochlorellae, zooxanthellae) as endosymbionts in protozoans, coelenterates, platyhelminths and molluscs (see Schwemmler & Schenk, 1980). Luminescent bacteria of the genus *Photobacterium* exist as ectosymbionts and in some cases as endosymbionts in molluscs and fish (Hastings & Nealson, 1980; Nealson *et al.*, 1981). In one system evidence has even been reported of gene transfer from the host to the bacterium (Martin & Fridovich, 1981).

There are two recently discovered examples of nutritional endosymbiosis. One concerns lithoautotrophic bacteria. Around the deep-sea hydrothermal vents of the Galapagos Rift and East Pacific Rise ocean spreading centres, ecosystems have been discovered which are based solely on aerobic lithoautotrophic production of biomass with hydrogen sulphide and carbon dioxide as energy and carbon sources (Jannasch & Wirsen, 1979, 1981; Cavanaugh *et al.*, 1981). Besides being free-living and loosely associated with clams, hydrogen-sulphide-utilizing bacteria were found to be harboured by tube worms (Cavanaugh, 1983). The tube worm *Riftia pachyptila*, belonging to the phylum Pogonophora, is an animal without a mouth or gut, the interior of which contains a large organ, the trophosome. This organ is densely filled with gram-negative bacteria; the presence of ribulose bisphosphate carboxylase activity indicated that these are autotrophs (see Jannasch, this volume). Parallel studies on bivalves living in other hydrogen-sulphide-rich habitats indicated that associations between hydrogen-sulphide-utilizing bacteria and animals are more common than anticipated and that at least in some cases the bacteria are endosymbionts (Cavanaugh, 1983). Lithoautotrophy on the basis of hydrogen sulphide may therefore be added to those prokaryotic activities which are exploited by animals via associative, extracellular or even intracellular symbiosis.

The other example of a nutritional relationship involving endosymbiotic bacteria concerns interspecies substrate transfer in sapropels and in the rumen. Ectosymbiotic associations of sapropelic protozoans with methanogenic bacteria attached to the surface of the ciliate cell have been repeatedly described (Vogels, Hoppe & Stumm, 1980). Recently the endosymbiotic bacteria in the giant amoeba *Pelomyxa palustris*, the sapropelic ciliate *Metopus striatus* and various other anaerobic protozoans examined were identified as methanogenic bacteria (van Bruggen, Stumm & Vogels, 1983). Although the basic metabolism of these protozoans has not yet been studied in detail one may speculate that this type of endosymbiosis allows to channelling of fermentation products of the host, such as acetate and hydrogen, into methane and biomass production. A lucrative mechanism of waste disposal!

The number of endosymbiotic relationships between bacteria and host eukaryotes is high. Although the ground was well prepared by the cytological studies of Buchner (1953) the symbiosis of insects with micro-organisms has not received as much attention as it deserves.

DEVELOPMENT OF PROTECTIVE, DEFENCE AND SURVIVAL MECHANISMS

The acceptance by a eukaryote of a prokaryote as symbiont has been a rare exception rather than the rule. Symbiosis, which in essence leads to reciprocal exploitation of the partner's resources, is the result of a delicate balance between the offensive and defensive activities of the partners. For every organism the best source of nutrients is another organism, so eukaryotes must have been welcome victims for the existing prokaryotes, whose metabolic versatility and inventiveness made them the most important parasites of eukaryotes. The majority of bacteria have the potential to affect multicellular organisms for good or ill: one or a few mutations in a soil bacterium are sufficient to transform it into a plant pathogen (Stolp, 1961); many pathogenic bacteria are closely related to harmless saprophytic soil bacteria; and virulence can even be transferred by plasmids (Elwell & Shipley, 1980).

The defence mechanisms of higher organisms are numerous, and their efficiency is extraordinary. Many physical and chemical arrangements as well as special proteins (lysozyme, betalysin, peroxidase, transferrins, immunoglobulins) and specialized cells (various types of leucocytes) act synergistically to prevent the invasion, multiplication and destructive actions of parasites. This becomes obvious when the defence mechanisms of the host have been weakened, for instance by therapeutic means such as surgery, immunosuppression, antibiotic treatment, or just illness. Such compromised hosts often become victims of opportunistic bacteria which normally do not affect the healthy animal (von Graevenitz, 1977). The sequence of events which resulted in the highly complex defence systems of higher animals on the one hand and in the offensive mechanisms of the pathogenic and parasitic organisms on the other, poses an evolutionary problem that needs studying.

Prokaryotes adapted to eukaryotes as ecological niches to a comparatively small extent. The changes concern either a drastic reduction of envelope structures and metabolic routes in obligate cell parasites such as the rickettsiae (Hackstadt & Williams, 1981; Weiss, 1982) and chlamydiae (Becker, 1978), or simply point mutations affecting single enzymes in various biosynthetic pathways as among the lactic acid bacteria (Morishita *et al.*, 1981).

GROWTH KINETICS

The growth of a population (where a population is defined as a group of individuals of the same species) is principally nutrient-limited. The population density is linearly proportional to the nutrient concentration. However, at high substrate concentrations and cell densities growth rates become drastically reduced. This may either be due to the accumulation of fermentation products (e.g. ethanol, n-butanol or organic acids) exerting concentration-dependent growth-inhibitory effects, or to other constraints related to diffusion rates, shearing forces or physical stress. This has been shown in studies on the growth of bacteria, microfungi and microalgae in both batch and continuous culture. In many cases the factors limiting growth yields are not known. In nature, where organisms are members of food chains and subject to predation, the limits of growth are seldom reached.

Macro-organismic populations are subject to essentially the same rules of growth kinetics and population dynamics. Nutrient limitation, competition, predation and parasites determine the equilibrium within ecosystems. Selected examples and the fundamentals of general and microbial ecology can be found in the standard textbooks (Alexander, 1971; Odum, 1977). The strategies of evolution have been given some thought; while the prokaryotes and some other micro-organisms seem to have adopted the principle of *r*-selection, the success of the eukaryotic macro-organisms is ascribed to *K*-selection (Pianka, 1970; Carlile, 1980, 1982). These general principles even pertain to the human population. Its size expanded stepwise when, as a result of increased food production – from 'hunting and gathering' via primitive agriculture to support by machines and fertilizers and finally 'green revolutions' – nutrients became available in excess. The almost exponential growth of mankind was for a long time retarded by epidemic infectious diseases; rats, lice, protozoans, bacteria and viruses balanced population growth and in this way competed with food supply as the growth-retarding factor. Thus prokaryotes and other single-celled organisms could be said to have done their best to maintain the biosphere in an equilibrium state.

With the development of modern medicine, hygiene, chemotherapeutics and antibiotics, diseases have been almost abolished as regulating factors, and mankind faces essentially two kinds of constraints. One concerns nutrient limitation; if there is an over-shoot of population growth, which cannot be compensated for by increased food production, starvation and death will result. The other, more serious consequence concerns the biosphere as a whole. The activities of man to accommodate and supply the rapidly increasing human population will severely damage and even destroy the biosphere by the overproduction of carbon dioxide, sulphur dioxide, nitrous oxide, toxic fumes and vapours, and by diminishing the area of arable soil because of erosional and other changes. Some of these consequences can be avoided, for instance by switching from fossil carbon to nuclear energy as a main energy source and subsequently from agricultural to industrial food production on the basis of hydrogen, carbon dioxide (Schlegel, 1969), methanol and other then easily accessible substrates (Litchfield, 1983). Other consequences, originating from highly intensive agricultural production are probably unavoidable, such as increased release of nitrous oxide from previously unexpected sources (Blackmer, Bremner & Schmidt, 1980; Seiler & Conrad, 1981).

Without question, man as the dominant eukaryotic organism is causing the most drastic and severe changes on our planet and endangers the biosphere as a whole. Not less but more research is necessary (Perutz, 1982), followed by efficient distribution of information to the consumers. Finally intelligence, the most recent accomplishment in the evolution of eukaryotes, should not only evaluate the requirements of man in terms of kilocalories and kilowatt-hours but also consider Aurelio Peccei's (1981) proposals for a human world: 'The individuum should not be identified with its basic requirements; in addition it has further needs, as it is a reasonable, intellectual, and artistic creature which likes to dream, is creative and enjoys himself.' A minimum of impacts on the biosphere by the dominant eukaryote would meet this definition best.

REFERENCES

ALEXANDER, M. (1971). *Microbial Ecology*. New York: Wiley.

ANDERSON, J. P. E. & DOMSCH, K. H. (1975). Measurement of bacterial and fungal contributions to respiration of selected agricultural and forest soils. *Canadian Journal of Microbiology*, **21**, 314–22.

ANDERSON, J. P. E. & DOMSCH, K. H. (1980). Quantities of plant nutrients in the microbial biomass of selected soils. *Soil Science*, **130**, 211–16.

APPLEBY, C. A., TJEPKEMA, J. D. & TRINICK, M. J. (1983). Hemoglobin in a nonleguminous plant, *Parasponia*: possible genetic origin and function in nitrogen fixation. *Science*, **220**, 951–3.

ARAGNO, M. (1981). Responses of microorganisms to temperature. In *Physiological Plant Ecology I*, vol. 12A, ed. O. L. Lange, P. S. Nobel, C. B. Osmond & H. Ziegler, pp. 339–69. Berlin, Heidelberg & New York: Springer-Verlag.

AWRAMIK, S. M. (1982). The pre-phanerozoic fossil record. In *Mineral Deposits and the Evolution of the Biosphere*, ed. H. D. Holland & M. Schidlowski, *Dahlem Konferenzen*, pp. 67–82. Berlin, Heidelberg & New York: Springer-Verlag.

BAROSS, J. A. & DEMING, J. W. (1983). Growth of 'black smoker' bacteria at temperatures of at least 250 °C. *Nature, London*, **303**, 423–6.

BAVENDAMM, W. (1936). Physiologie der schwefelspeichernden und schwefelfreien Purpurbakterien. *Ergebnisse der Biologie*, **13**, 1–53.

BECKER, Y. (1978). The chlamydia: molecular biology of procaryotic obligate parasites of eucaryocytes. *Microbiological Reviews*, **42**, 274–306.

BLACKMER, A. M., BREMNER, J. M. & SCHMIDT, E. L. (1980). Production of nitrous oxide by ammonia-oxidizing chemoautotrophic microorganisms in soil. *Applied and Environmental Microbiology*, **40**, 1060–6.

BÖGER, P. (1980). The O_2/CO_2 cycle: development and atmospheric consequences. In *Biochemical and Photosynthetic Aspects of Energy Production*, ed. A. San Pietro, pp. 175–90. New York & London: Academic Press.

BRAUN, M. & GOTTSCHALK, G. (1982). *Acetobacterium wieringae* sp. nov., a new species producing acetic acid from molecular hydrogen and carbon dioxide. *Zentralblatt für Bakteriologie, Parasitenkunde, Infektionskrankheiten und Hygiene I. Abteilung Originale Reihe C*, **3**, 368–76.

BROCK, T. D. (1978). *Thermophilic Microorganisms and Life at High Temperatures*. New York, Heidelberg & Berlin: Springer-Verlag.

BROCK, T. D. (1980). Precambrian evolution. *Nature, London*, **288**, 214–15.

BRODA, E. (1975). *The Evolution of the Bioenergetic Processes*. Oxford: Pergamon Press.

VAN BRUGGEN, J. J. A., STUMM, C. K. & VOGELS, G. D. (1983). Symbiosis of methanogenic bacteria and sapropelic protozoa. *Archives of Microbiology*, **136**, 89–95.

BUCHNER, P. (1953). *Endosymbiose der Tiere mit pflanzlichen Mikroorganismen*. Basel: Birkhäuser Verlag.

CARLILE, M. J. (1980). From prokaryote to eukaryote: gains and losses. In *The Eukaryotic Microbial Cell*, ed. G. W. Gooday, D. Lloyd & A. P. J. Trinci, *Symposium of the Society for General Microbiology 30*, pp. 1–40. Cambridge University Press.

CARLILE, M. (1982). Prokaryotes and eukaryotes: strategies and success. *Trends in Biochemical Sciences*, **7**, 128–30.

CASTENHOLZ, R. W. (1979). Evolution and ecology of thermophilic microorganisms. In *Strategies of Life in Extreme Environments*, ed. M. Shilo, *Dahlem Konferenzen*, pp. 373–92. Weinheim: Verlag Chemie.

CAVALIER-SMITH, T. (1981). The origin and early evolution of the eukaryotic cell. In *Molecular and Cellular Aspects of Microbial Evolution*, ed. M. J. Carlile, J. F. Collins & B. E. B. Moseley, *Symposium of the Society for General Microbiology 32*, pp. 33–84. Cambridge University Press.

CAVANAUGH, C. M. (1983). Symbiotic chemoautotrophic bacteria in marine invertebrates from sulphide-rich habitats. *Nature, London*, **302**, 58–61.

CAVANAUGH, C. M., GARDINER, S. L., JONES, M. L., JANNASCH, H. W. & WATERBURY, J. B. (1981). Prokaryotic cells in the hydrothermal vent tube worm *Riftia pachyptila* Jones: possible chemoautotrophic symbionts. *Science*, **213**, 340–2.

CHAPPE, B., ALBRECHT, P. & MICHAELIS, W. (1982). Polar lipids of archaebacteria in sediments and petroleums. *Science*, **217**, 65–6.

DAYHOFF, M. O. & SCHWARTZ, R. M. (1980). Early biological evolution derived from chemical structures. In *Biological Chemistry of Organelle Formation*, ed. Th. Bücher, W. Sebald & H. Weiss, *Mosbach Colloquium 31*, pp. 71–86. Berlin, Heidelberg & New York: Springer-Verlag.

EHHALT, D. H. (1976). The atmospheric cycle of methane. In *Microbial Production and Utilization of Gases*, ed. H. G. Schlegel, G. Gottschalk & N. Pfennig, pp. 13–22. Göttingen: Verlag E. Goltze.

ELWELL, L. P. & SHIPLEY, P. L. (1980). Plasmid-mediated factors associated with virulence of bacteria to animals. In *Annual Review of Microbiology*, **34**, 465–96.

ENOS, P. (1973). Photosynthesis and atmospheric oxygen. *Science*, **180**, 515–16.

FENCHEL, T. M. & BARKER-JØRGENSEN, B. (1977). Detritus food chains in aquatic ecosystems: the role of bacteria. *Advances in Microbial Ecology*, **1**, 1–58.

FENCHEL, T. M. & BLACKBURN, T. H. (1979). *Bacteria and Mineral Cycling*. New York & London: Academic Press.

FUCHS, G., THAUER, R., ZIEGLER, H. & STICHLER, W. (1979). Carbon isotope fractionation by *Methanobacterium thermoautotrophicum*. *Archives of Micro-biology*, **120**, 135–9.

GARLAND, P. B. (1981). The evolution of membrane-bound bioenergetic systems: the development of vectorial oxidoreductions. In *Molecular and Cellular Aspects of Microbial Evolution*, ed. M. J. Carlile, J. F. Collins & B. E. B. Moseley, *Symposium of the Society for General Microbiology 32*, pp. 273–83. Cambridge University Press.

GOULD, G. W. & CORRY, J. A. L. (1980). *Microbial Growth and Survival in Extremes of Environment*. New York & London: Academic Press.

GRAEVENITZ, A. VON (1977). The role of opportunistic bacteria in human disease. *Annual Review of Microbiology*, **31**, 447–77.

GRAY, M. W. & DOOLITTLE, W. F. (1982). Has the endosymbiont hypothesis been proven? *Microbiological Reviews*, **46**, 1–42.

HACKSTADT, T. & WILLIAMS, J. C. (1981). Biochemical stratagem for obligate parasitism of eukaryotic cells by *Coxiella burnetii*. *Proceedings of the National Academy of Sciences, USA*, **78**, 3240–4.

HASTINGS, J. W. & NEALSON, K. H. (1980). Exosymbiotic luminous bacteria occurring in luminous organs of higher animals. In *Endocytobiology*, vol. 1, ed. E. Schwemmler & H. E. A. Schenk, pp. 467–71. Berlin: Walter de Gruyter.

HOLLAND, H. D. & SCHIDLOWSKI, M. (1982). *Mineral Deposits and the Evolution of the Biosphere. Dahlem Konferenzen*. Berlin, Heidelberg & New York: Springer-Verlag.

HORIKOSHI, K. & AKIBA, T. (1982). *Alkalophilic Microorganisms: A New Microbial World*. Tokyo/Berlin, Heidelberg & New York: Japan Scientific Societies Press/Springer-Verlag.

HUSER, B. A., WUHRMANN, K. & ZEHNDER, A. J. B. (1982). *Methanothrix soehngenii* gen. nov. sp. nov., a new acetotrophic non-hydrogen-oxidizing methane bacterium. *Archives of Microbiology*, **132**, 1–9.

JAMES, H. L. & TRENDALL, A. F. (1982). Banded iron formation: distribution in time and paleoenvironmental significance. In *Mineral Deposits and the Evolution of the Biosphere*, ed. H. D. Holland & M. Schidlowski, *Dahlem Konferenzen*, pp. 199–218. Berlin, Heidelberg & New York: Springer-Verlag.

JANNASCH, H. W. & WIRSEN, C. O. (1979). Chemosynthetic primary production at East Pacific sea floor spreading centers. *BioScience*, **29**, 592–8.

JANNASCH, H. W. & WIRSEN, C. O. (1981). Morphological survey of microbial mats near deep-sea thermal vents. *Applied and Environmental Microbiology*, **41**, 528–38.

JENKINSON, D. S. & LADD, J. N. (1981). Microbial biomass in soil: measurement and turnover. In *Soil Biochemistry*, vol. 5, ed. E. A. Paul & J. N. Ladd, pp. 415–71. New York & Basle: Marcel Dekker.

KANDLER, O. (1982). Cell wall structures and their phylogenetic implications. *Zentralblatt für Bakteriologie, Parasitenkunde, Infektionskrankheiten und Hygiene, I Abteilung, Originale Reihe C*, **3**, 149–60.

KLUYVER, A. J. & DONKER, H. J. L. (1925). The unity in the chemistry of the fermentative sugar dissimilation processes of microbes. *Proceedings of the Royal Academy of Amsterdam*, **28**, 297–313.

KLUYVER, A. J. & DONKER, H. J. L. (1926). Die Einheit in der Biochemie. *Chemie der Zelle und Gewebe*, **13**, 134–90.

KRUMBEIN, W. E., BUCHHOLZ, H., FRANKE, P., GIANI, D. C. & WONNEBERGER, K. (1979). O_2 and H_2S coexistence in stromatolites. A model for the origin of mineralogical lamination in stromatolites and banded iron formations. *Naturwissenschaften*, **66**, 381–9.

KUSHNER, D. J. (1978). *Microbial Life in Extreme Environments*. New York & London: Academic Press.

LANGWORTHY, T. A. (1978). Microbial life in extreme pH values. In *Microbial Life in Extreme Environments*, ed. D. J. Kushner, pp. 279–315. New York & London: Academic Press.

LEIGH, J. A., MAYER, F. & WOLFE, R. S. (1981). *Acetogenium kivui*, a new thermophilic hydrogen-oxidizing acetogenic bacterium. *Archives of Microbiology*, **129**, 275–80.

LITCHFIELD, J. H. (1983). Single-cell proteins. *Science*, **219**, 740–6.

MACKENZIE, A. S., BRASSEL, S. C., EGLINTON, G. & MAXWELL, J. R. (1982). Chemical fossils: the geological fate of steroids. *Science*, **217**, 491–504.

McKIRDY, D. M. & HAHN, H. (1982). The composition of kerogen and hydrocarbons in Precambrian rocks. In *Mineral Deposits and the Evolution of the Biosphere*, ed. H. D. Holland & M. Schidlowski, *Dahlem Konferenzen*, pp. 123–54. Berlin, Heidelberg & New York: Springer-Verlag.

MARTIN, J. P., JR. & FRIDOVICH, I. (1981). Evidence for a natural gene transfer from the ponyfish to its bioluminescent bacterial symbiont *Photobacter leiognathi*. *Journal of Biological Chemistry*, **256**, 6080–9.

MILLER, S. L. (1982). Prebiotic synthesis of organic compounds. In *Mineral Deposits and the Evolution of the Biosphere*, ed. H. D. Holland & M. Schidlowski, *Dahlem Konferenzen*, pp. 155–76. Berlin, Heidelberg & New York: Springer-Verlag.

MORISHITA, P., DEGUCHI, Y., YAJIMA, M., SAKURAI, T. & YURA, T. (1981). Multiple nutritional requirements of Lactobacilli: genetic lesions affecting amino acid biosynthetic pathways. *Journal of Bacteriology*, **148**, 64–71.

NEALSON, K. H., COHN, D., LEISMAN, G. & TEBO, B. (1981). Co-evolution of luminescent bacteria and their eukaryotic hosts. *Annals of the New York Academy of Sciences*, **361**, 76–91.

NOBLE, W. C. & PITCHER, D. G. (1979). Microbial ecology of the human skin. *Advances in Microbial Ecology*, **2**, 245–89.

ODUM, E. P. (1977). *Ecology: The Link between the Natural and the Social Sciences*, 2nd edn. London & New York: Holt, Rinehart & Winston.

OPARIN, A. I. (1957). *Die Entstehung des Lebens auf der Erde*, 3rd edn. Berlin: VEB Deutscher Verlag der Wissenschaften.

ØSTERGAARD JENSEN, E., PALUDAN, K., HYLDIG-NIELSEN, J. J., JØRGENSEN, P. & MARCKER, K. A. (1981). The structure of a chromosomal leghaemoglobin gene from soybean. *Nature, London*, **291**, 677–9.

OURISSON, G., ALBRECHT, P. & ROHMER, M. (1982). Predictive microbial biochemistry from molecular fossils to procaryotic membranes. *Trends in Biochemical Sciences*, **7**, 239–9.

PAYNE, W. J. (1981). *Denitrification*. New York: Wiley.

PECCEI, A. (1981). *Cent pages pour l'avenir*. Paris: Economica. (German edition: *Die Zukunft in unserer Hand*. Vienna & Munich: Molden.)

PERUTZ, M. F. (1982). *Ging's ohne Forschung Besser? Der Einfluss der Naturwissenschaften auf die Gesellschaft*. Stuttgart: Wissenschaftliche Verlagsgesellschaft.

PFENNIG, N., WIDDEL, F. & TRÜPER, H. G. (1981). The dissimilatory sulfate-reducing bacteria. In *The Prokaryotes. A Handbook on Habitats, Isolation and Identification of Bacteria*, vol. 1, ed. M. P. Starr, H. Stolp, H. G. Trüper, A. Balows & H. G. Schlegel, pp. 926–40. Berlin, Heidelberg & New York: Springer-Verlag.

PIANKA, E. R. (1970). On r- and K-selection. *American Naturalist*, **104**, 592–7.

PIERSON, B. K. & CASTENHOLZ, R. W. (1974). A phototrophic gliding filamentous bacterium of hot springs, *Chloroflexus aurantiacus*, gen. and sp. nov. *Archives of Microbiology*, **100**, 5–24.

PONNAMPERUMA, C. (1983). *Cosmochemistry and the Origin of Life*. Dordrecht, The Netherlands: Reidel.

SCHIDLOWSKI, M. (1982). Content and isotopic composition of reduced carbon in sediments. In *Mineral Deposits and the Evolution of the Biosphere*, ed. H. D. Holland & M. Schidlowski, *Dahlem Konferenzen*, pp. 103–22. Berlin, Heidelberg & New York: Springer-Verlag.

SCHLEGEL, H. G. (1969). From electricity via water electrolysis to food. In *Fermentation Advances*, ed. D. Perlman, pp. 807–32. New York & London: Academic Press.

SCHLEGEL, H. G. (1981). Microorganisms involved in the nitrogen and sulfur cycles. In *Biology of Inorganic Nitrogen and Sulfur*, ed. H. Bothe & A. Trebst, pp. 3–12. Berlin, Heidelberg & New York: Springer-Verlag.

SCHLEGEL, H. G. & JANNASCH, H. W. (1981). Prokaryotes and their habitats. In *The Prokaryotes. A Handbook on Habitats, Isolation and Identification of Bacteria*, vol. 1, ed. M. P. Starr, H. Stolp, H. G. Trüper, A. Balows & H. G. Schlegel, pp. 43–82. Berlin, Heidelberg & New York: Springer-Verlag.

SCHOPF, J. W. & WALTER, M. R. (1982). Origin and early evolution of cyanobacteria: the geological evidence. In *The Biology of Cyanobacteria*, ed. N. G. Carr & B. A. Whitton, pp. 543–64. Oxford: Blackwell Scientific.

SCHWARCZ, H. P., HOEFS, J. & WELTE, D. (1969). Carbon. In *Handbook of Geochemistry*, vol. II/1. Berlin, Heidelberg & New York: Springer-Verlag.

SCHWEMMLER, W. & SCHENK, H. E. A. (1980). *Endocytobiology*. Berlin: Walter de Gruyter.

SEILER, W. & CONRAD, R. (1981). Mikrobielle Bildung von N_2O (Distickstoffoxid) aus Mineraldüngern – ein Umweltproblem. *Forum Mikrobiologie*, **6**, 323–30.

SHILO, M. (1979). *Strategies of Microbial Life in Extreme Environments. Dahlem Konferenzen*. Weinheim: Verlag Chemie.

STACKEBRANDT, E. & WOESE, C. R. (1981). The evolution of prokaryotes. In *Molecular and Cellular Aspects of Microbial Evolution*, ed. M. J. Carlile, J. F. Collins & B. E. B. Moseley, *Symposium of the Society for General Microbiology 32*, pp. 1–31. Cambridge University Press.

STANIER, R. Y. & VAN NIEL, C. B. (1962). The concept of a bacterium. *Archiv für Mikrobiologie*, **42**, 17–35.

STETTER, K. O. (1982). Ultrathin mycelia-forming organisms from submarine volcanic areas having an optimum growth temperature of 105 °C. *Nature, London*, **300**, 258–60.

STOLP, H. (1961). Neue Erkenntnisse über phytopathogene Bakterien und die von ihnen verursachten Krankheiten. I. Verwandtschaftsbeziehungen zwischen phytopathogenen Pseudomonas-Arten und saprophytischen Fluoreszenten auf der Grundlage von Phagenreaktionen. *Phytopathologische Zeitschrift*, **42**, 197–262.

TATE, K. R. & JENKINSON, D. S. (1982). Adenosine triphosphate measurement in soil: an improved method. *Soil Biology and Biochemistry*, **14**, 331–5.

TRUDINGER, P. A. & WILLIAMS, N. (1982). Stratified sulfide deposition in modern and ancient environments. In *Mineral Deposits and the Evolution of the Biosphere*, ed. H. D. Holland & M. Schidlowski, *Dahlem Konferenzen*, pp. 177–98. Berlin, Heidelberg & New York: Springer-Verlag.

TRÜPER, H. G. (1982). Microbial processes in the sulfur cycle through time. In *Mineral Deposits and the Evolution of the Biosphere*, ed. H. D. Holland & M. Schidlowski, *Dahlem Konferenzen*, pp. 5–30. Berlin, Heidelberg & New York: Springer-Verlag.

VAN VALEN, L. M. & MALIORANA, V. C. (1980). The Archaebacteria and eukaryotic origins. *Nature, London*, **287**, 248–50.

VOGELS, G. D., HOPPE, W. F. & STUMM, C. K. (1980). Association of methanogenic bacteria with rumen ciliates. *Applied and Environmental Microbiology*, **40**, 608–12.

WEISS, E. (1982). The biology of rickettsiae. *Annual Review of Microbiology*, **36**, 345–70.

WELTE, D. H. (1967). Das Problem der frühesten organischen Lebensspuren. *Naturwissenschaften*, **54**, 325–9.

WELTE, D. H. (1970). Organischer Kohlenstoff und die Entwicklung der Photosynthese auf der Erde. *Naturwissenschaften*, **57**, 17–23.

WIDDEL, F. & PFENNIG, N. (1981). Studies on dissimilatory sulfate-reducing bacteria that decompose fatty acids. I. Isolation of new sulfate-reducing bacteria enriched with acetate from saline environments and description of *Desulfobacter postgatei* gen. nov., sp. nov. *Archives of Microbiology*, **129**, 395–400.

WIDDEL, F. & PFENNIG, N. (1982). Studies on dissimilatory sulfate-reducing bacteria that decompose fatty acids. II. Incomplete oxidation of propionate by *Desulfobulbus propionicus* gen. nov., sp. nov. *Archives of Microbiology*, **131**, 360–5.

WILSON, T. H. & LIN, E. C. C. (1980). Evolution of membrane bioenergetics. *Journal of Supramolecular Structure*, **13**, 421–46.

WOESE, C. R. & FOX, G. E. (1977). Phylogenetic structure of the prokaryotic domain: the primary kingdoms. *Proceedings of the National Academy of Sciences, USA*, **74**, 5088–90.

MICROBIAL BEHAVIOUR IN NATURAL ENVIRONMENTS

NORBERT PFENNIG

Fakultät für Biologie, Universität Konstanz, D-7750 Konstanz, Federal Republic of Germany

It must, however, be recognized that in nature the conditions are seldom simple. Hence we must learn to study more carefully the effects of complicating circumstances . . . This will require much imaginative work, and correlation of many kinds of observations.

C. B. van Niel (1955)

INTRODUCTION

It is certainly one of the aims of the life sciences to understand the real nature of an organism by the study of its functioning in the natural environment. In the case of the microbes this is a particularly difficult task because of their small size and the mostly indirect nature of the methods which have to be applied. The formulation of problems certainly stimulated the development of new methods. However, current concepts and methods were only gradually formulated and used in the course of the development of microbiology. As a result, different aspects of the life of microbes emerged step by step, often with a predominance of certain concepts at one time. At present one can be overwhelmed by the accumulated and detailed knowledge on certain microbes. It may, therefore, be appropriate to consider briefly here some of the guiding principles and methods that microbiologists have inherited, in order to become aware of their strengths and limitations in the search for understanding microbial behaviour in natural habitats.

Historical remarks and basic ideas

Usually microbes are hardly visible to us; it is the substrate in decomposition that we see. It was in the seventeenth century that Antonie van Leeuwenhoek first discovered the 'animalcules' (little animals) in all kinds of fluids he examined with his small microscope, and expressed his thoughts about their activities. But van Leeuwenhoek was so much ahead of his time with his discoveries that only in the nineteenth century would he have met congenial discussion partners.

The development of chemistry in the nineteenth century included the study of chemical transformations of organic substances both in nature and under experimental conditions. Interestingly, the chemical changes were not yet recognized to be directly dependent on the metabolic activity of the microbial cells that were observed microscopically in the decaying materials. F. Cohn (1872) was the first scientist to state clearly that the organic and inorganic substances remaining in dead plants and animals are degraded by the activities of microbes; he emphasized that this recycling activity sustains the conditions for life for all higher organisms on earth.

Cohn's idea greatly stimulated research on the chemical transformations of organic substances by microbes. At the end of the last century Winogradsky (1890) and Beijerinck (1895) introduced the most comprehensive experimental approach for this study – the elective culture method – which became one of the most important tools in general microbiology. In the second Marjory Stephenson memorial lecture, van Niel (1955) characterized the principle of elective culture as 'natural selection'. From the vast variety of different microbes present in the inoculum from nature, only those develop that are best adapted to thrive under the given experimental conditions. By the use of this method we have learnt and still learn what kinds of microbes carry out the degradation of a chosen substrate in competition and/or cooperation with other microbes under defined conditions. Even after nearly a hundred years the potential of this approach has by no means been exhausted or lost significance (Alexander, 1976).

The elective culture method gained its importance only by subsequent application of pure culture methods. Brefeld and Hansen introduced the dilution method in liquid media. The most important agar media were first developed and applied by Koch in 1876 for the pure culture isolation of pathogenic bacteria. Beijerinck combined elective culture with the isolation into pure culture of the microbes thus enriched. This offered the possibility for research to be pursued in different directions. On the one hand, enrichment and pure culture techniques became tools in experimental microbial ecology (van Niel, 1955); on the other, the available pure cultures were used for research away from the conditions in nature and towards a more detailed knowledge of the individual microbial species maintained in laboratory cultures. Microbial taxonomy and systematics evolved primarily on the basis of morphological, cytological and physiological studies of pure cultures (Orla-Jensen, 1909; Kluyver & van Niel, 1936). Fascinating details of the fine structure of all kinds of microbes were later revealed by electron microscopy.

The great variety of chemical transformations uncovered in microbial species obtained from enrichments by Beijerinck, his students and others led to an unintelligible descriptive science which demanded a basis for conceptual comprehension. This was achieved by Beijerinck's successor, A. J. Kluyver, who united the biochemical knowledge of his time and formulated the general principles applying to all biochemical transformations (Kluyver & Donker, 1926). Kluyver thereby actually founded comparative biochemistry (van Niel, 1949).

Biochemical research successfully progressed from the recognition of the chemical nature of the substrates and products of microbial metabolism to an understanding of the catabolic and anabolic intermediary metabolism of a great number of microbes. In 1945 M. Stephenson (Woods, 1953) defined five levels of equal importance at which research into microbial life in nature was carried out. The first level comprised the biochemical activities of mixed cultures in complex 'natural' media while the fifth level included the reactions of cell preparations, purified enzymes and defined substrates. It now appears justified to add a sixth level in order to accommodate the results of molecular biology and genetics. These fields of study provided evidence for the unity and monophyletic origin of life. Also, biochemical interpretations of heredity and hereditary changes were developed. This kind of research depended on mass pure cultures of microbes in which the requirements for carbon and energy sources, growth factors and trace elements were satisfied to enable maximum yields. The more research was extended in the characterized direction, the more it was realized that the accumulated knowledge of the biochemical details constituted only a part of the characteristics, capacities and activities that can potentially be expressed by microbes under various conditions in nature (Brock, 1971).

The pioneer of microbiological research directed towards the understanding of microbial behaviour in natural habitats was Winogradsky. He used direct microscopic observation of microbes in mud or soil samples and experiments with raw cultures or soils in addition to the study of enrichment and pure cultures in order to study development of individual species or biotypes in natural habitats. From his results Winogradsky recognized that under the heterogeneous and changing conditions (in both space and time) that obtain in the natural environment, microbes may exhibit special properties which allow them to survive or to compete successfully with other microbes. Such peculiarities may never be recognized in homogeneous mass cultures in the laboratory.

In his review of earlier work, Winogradsky (1949) postulated again that morphology and development of pure cultures of microbes should be studied in microcultures under conditions that closely correspond to those in nature. He referred to his studies of 1888 on the species and genera of the purple sulphur bacteria. The strains that were obtained in pure culture up to 1945 were rods and cocci of different sizes and, as Winogradsky put it 'conventional characters, deprived of all the *couleur locale*'. The special morphological and developmental details that these bacteria exhibit when studied in samples taken from natural sources were not seen.

In fact, when we compare the complex or synthetic media that were in use until about 30 years ago with the media that are used today, the excessive and unnecessarily high concentrations of organic nutrients and mineral salts in the original media become apparent. The use of substrates at growth-limiting concentrations and the determination of growth yields in the chemostat certainly favour a more conscious application of the necessary nutrients in culture media today (Herbert, Elsworth & Telling, 1956; Pirt, 1975). Also, much more is now known about the nutrient and mineral salt concentrations in various aquatic habitats. This has stimulated laboratory experiments using conditions more closely resembling those to which many microbes are adapted in nature (Brock, 1971; Rosswall, 1973; Poindexter, 1981a).

The subject of microbes in their natural environments was presented in an admirably detailed and fairly comprehensive way in the thirty-fourth Symposium of the Society for General Microbiology (Slater, Whittenbury & Wimpenny, 1983). And there have been several other symposia devoted to microbial ecology (e.g. Ellwood et al., 1980). In view of the enormous body of knowledge existing on every aspect of the subject, I have chosen to present only a few selected examples of the behaviour of prokaryotic microbes in natural environments. These may illustrate that studies of pure and mixed cultures in the laboratory as well as observations and measurements in natural habitats are required in order to increase our understanding of microbial life under conditions of competition in nature.

SELECTED EXAMPLES OF MICROBIAL BEHAVIOUR IN NATURAL ENVIRONMENTS

Microbiological research with the aim of understanding the development and survival of microbes in natural habitats has been carried out at different levels. On the one hand, the behaviour of individual microbial species or genera has been studied both in nature and in the laboratory. This has been possible in cases where the microbe could be unequivocally recognized in a sample from nature because of its particular morphological or cytological characteristics. Examples exist for, amongst others, the species *Thiospirillum jenense, Beggiatoa mirabilis, Gallionella ferruginea, Hyphomicrobium vulgare, Oscillatoria agardhii* and others. In other species a combination of morphological and physiological features of a microbe has been used for their identification in nature. On the other hand, the behaviour of groups or families sharing certain morphological or physiological features has been studied. This is the case, for, for example, the endospore-forming bacilli, the growth habits of mycelium-forming microbes, appendaged or prosthecate bacteria, the sulphate-reducing bacteria, the gas-vacuolate Cyanobacteria and others. In many cases microbes were not first detected in nature. Rather, they were discovered after enrichment and isolation in pure culture which allowed the exploitation of morphological and physiological characteristics and their taxonomic identification. Only on the basis of this knowledge did it subsequently become possible to recognize a particular microbe in its natural habitat. Field studies were more or less difficult depending on the size and nature of the microbe's environment. In soils or sediments the microbial habitats are usually of microscopic size. In aquatic environments or places with relatively uniform, highly selective conditions (e.g. hot springs, salt lakes), however, a habitat with populations of single or a few species may attain macroscopic dimensions. In this case the study of microbial behaviour *in situ* does not present too serious a problem.

Mycelial growth and soil

If we compare the microbial populations of the two basic environments – water and land – it is clear that the prevailing growth forms in open water are the unicellular coccoid, vibrioid, spirilloid and more or less rod-shaped microbes. In contrast, microbes with multicellular, filamentous or branching hyphal types of growth are found in water only in association with solid particles, plants and animals, or on the shore. Hyphal and mycelial growth appears to be an adaptation to life in inhomogeneous environments with solid surfaces, primarily soils.

This view is supported by the striking example of convergent evolution in the life styles of the prokaryotic actinomycetes and the eukaryotic fungi (Ensign, 1978). Both groups of microbes exhibit apical cell growth of branching hyphae which spread radially in different directions from a centre. When nutrients are limited (on agar plates in the laboratory, or in soils), both actinomycetes and fungi develop long, sparsely branched hyphae. With increasing amounts of nutrients available, the hyphae become shorter and more and more densely branched, forming a thick mycelium. In soil, hyphal growth over distances appears to be an advantage. Nutrient transport through hyphae allows the mycelium to spread over inorganic and organic particles as well as through substrates, pore water and air space in which aerial mycelium and desiccation-resistant spores may form. Insoluble organic substrates such as pieces of cellulose, chitin, starch or proteinaceous material are absorbed directly by hyphae after being locally dissolved by exoenzymes. In eukaryotic fungi, streaming protoplasm effects active transport of nutrients through the hyphae away from depleted areas to actively growing hyphal tips (Carlile, 1980). In this way fungi effect a kind of movement through the substratum by localized growth and translocation of protoplasm. Streaming of the cytoplasm does not exist in prokaryotic actinomycetes, although some nutrient transport certainly occurs. Growth of *Streptomyces* mycelium in soil was observed by the use of Cholodny's (1930) technique of buried glass microscope slides (Pfennig, 1958). Vegetative hyphae were seen to have grown over distances of more than 50–100 μm through nutrient-deficient areas. Also, growth of long aerial hyphae and spore formation at their ends cannot be understood in the absence of transport of nutrients through the hyphae.

Resting stages and survival

Stress-resistant stages

Every living organism gradually deprives its habitat of the necessary growth substrates by metabolic activity. If its nutritional requirements are not continuously or periodically re-established by the activity of other living organisms, it will locally die off or, perhaps, develop special dormant stages by which it survives starvation and deleterious environmental conditions. Only a few bacterial groups are able to form particular resting cells when the conditions become unsuitable for growth and metabolic activity (constitutive dormancy: Sussman & Halvorson, 1966). The bacteria of the genera *Bacillus* and *Clostridium* develop endospores many of which are extremely resistant to heat, desiccation, chemical agents or radiation. The mainly desiccation-resistant exospores and cysts are less resistant than endospores. Exospores are formed as buds from vegetative cells by the methane-oxidizing *Methylosinus trichosporium* and the phototrophic *Rhodomicrobium vannielii* (Dow & Whittenbury, 1980). Cysts are direct transformations of vegetative cells which developed a higher degree of structural rigidity and attained metabolic dormancy. Cysts are formed by species of the genus *Azotobacter*, by the methane-oxidizing *Methylocystis* (Whittenbury, Davies & Davey, 1970), and by the myxobacteria (Voelz & Dworkin, 1962).

The spores of streptomycetes are formed from aerial hyphae by septation at regular intervals. The inner cell wall is thickened and becomes the spore wall. These spores are dormant in the sense that they are metabolically much less active than vegetative cells. They may survive desiccation for several months or years, but are not heat-resistant (Ensign, 1978).

We may be inclined to think that endospore formation, and perhaps also exospore and cyst formation, might, from the evolutionary point of view, be the best solution to the challenge of survival in nutrient-depleted or otherwise adverse environments. Surprisingly, however, the spore-forming bacteria constitute only a small fraction of the microbial world. Most microbes in nature lack stress-resistant resting stages. This is certainly an indication of the effectiveness of wind, water movement and animals as carriers for the distribution of living microbes in the biosphere, and accounts for their ubiquitous presence. We have to accept it as an expression of microbial diversity and a characteristic of a particular species as to whether or not it forms special kinds of cells with different functions and survival values under particular conditions.

Survival of vegetative cells
Interestingly, the vegetative cells of different bacteria vary considerably in their stress resistance. While various degrees of starvation

resistance are widespread (exogenous dormancy: Sussman & Halvorson, 1966), resistance of vegetative cells to desiccation is rather exceptional. Because of their drastic changes in temperature and water content, soils are highly selective for microbes that are both starvation- and desiccation-resistant. It is common experience that desiccation of soil greatly increases the percentage of spore-forming bacteria. There are, however, some typical soil bacteria, the *Arthrobacter* species, that have been shown to be inherently resistant to starvation and desiccation as vegetative cells (Ensign, 1970). They were found to be the most numerous single group of bacteria in nutrient-poor soils. Robinson, Salonius & Chase (1965) compared the capacity of *Pseudomonas* and *Arthrobacter* species to survive drying in soils. After an initial 48 h drying time less than 1% of the *Pseudomonas* cells but 80% of the *Arthrobacter* cells survived. During a subsequent 4 week drying period about 85% of the *Pseudomonas* survivors died, while the number of living *Arthrobacter* cells remained unchanged.

Survival of long-term starvation by spherical and rod-shaped cells of *Arthrobacter crystallopoietes* in phosphate buffer was tested by Ensign (1970). Both cell forms remained 100% viable for 30 days and more than 60% of the cells survived 60 days of starvation. The 50% survival time of *Arthrobacter* was found to be at least 30 times longer than that of *Aerobacter aerogenes* or *Azotobacter agilis*. One of the most obvious reasons for the exceptional starvation resistance of arthrobacters is their capacity to respond rapidly to starvation conditions by reducing their endogenous respiration to an extremely low basal rate; this rate was found to be more than 10 times lower than in other bacteria. Under good growth conditions the doubling time of arthrobacters is about 3 h and, therefore, considerably longer than that of *Pseudomonas* strains in the same habitat (Ensign, 1970). The advantage that arthrobacters have in nutrient-poor soils is their particular starvation and desiccation resistance as vegetative cells.

There is another unusual microbe, the genus *Caulobacter*, which is likewise capable of surviving starvation as vegetative cells (Poindexter, 1981a, b). Unlike the arthrobacters, though, *Caulobacter* cannot withstand drying. This is understandable in view of the fact that the caulobacters are typical aquatic bacteria which are commonly found in all nutrient-poor aquatic environments as well as in soils of high moisture content and low organic matter content (Krasil'nikov & Belyaev, 1967). Their growth is inhibited by nutrient concentrations adequate for other chemotrophic bacteria. Poindexter expressed for the caulobacters what applies equally well to the arthrobacters in soils, namely that their competitive advantage (relative to *Pseudomonas* species) lies in not multiplying faster, but dying slower. Together with the less common genera *Asticcacaulis* and *Hyphomicrobium* the caulobacters are prosthecate bacteria with particular morphological and functional traits that can be understood as adaptations to growth and survival in low-nutrient aquatic habitats. Present knowledge as well as the unanswered questions in this field are discussed in detail by Poindexter (1981*a*, *b*). She characterizes the caulobacters as the ultimate scavengers and mineralizers of the low amounts of organic substances that are diluted in the oceans and freshwater environments which occupy the greater part of the earth's biosphere. *Caulobacter* is adapted to degrading the organic compounds that are released or left over by the other, copiotrophic microbes more directly concerned with the decomposition of plant and animal residues.

Movement and taxes

The different kinds of movements of microbial cells are ecologically significant when they are guided by the perception and sensory transduction of certain stimuli in the environment. Depending on whether such stimuli indicate the presence of favourable or unfavourable conditions to the cell, the movements become positive or negative responses to the stimuli. Prerequisites for such reactions are, therefore, physicochemical inhomogeneities or gradients in the habitat. In order to cause a suitable orientation of the microbe, the dimensions of such gradients must be in a particular ratio to the size of the microbe and to the speed of its movement.

Carlile (1980) pointed out that the highly polarized growth of fungal hyphae in soil has some attributes of active movement. From this point of view, the gliding movement of filamentous bacteria can be considered as a more independent method of translocation which is, however, still dependent on particle surfaces. In the swimming of flagellate bacteria free oriented movement in the liquid phase is realized.

Filamentous gliding bacteria and sediments

The gliding bacteria are typical inhabitants of soils and marine or freshwater sediments. Only in these environments with all kinds of particulate solid surfaces, can these bacteria display their oriented creeping movement. Well-known examples are the filamentous gliding *Beggiatoa* species that are common in the upper few centimetres of sandy and muddy sediments. Jørgensen (1977*a*) observed *Beggiatoa alba* with 3–5 μm wide filaments in sandy-silty sediments with small interstitial spaces. In coarse-grained sand or sediments composed of sand and pellet mud, *Beggiatoa* species with 10–20 μm wide filaments dominated. The largest forms (> 23 μm) were found in pure pellet mud with a relatively loose structure.

In marine and brackish-water sediments even the sulphate-reducing bacteria are represented, by, among other species, the filamentous gliding species *Desulfonema limicola* and *D. magnum* (Widdel, Kohring & Mayer, 1983). Since these bacteria are strictly anaerobic they are likely to adapt themselves to permanently anoxic sediment layers below *Beggiatoa*. Ecological studies comparable to those on *Beggiatoa* are so far lacking.

The top layer of sediments in marine lagoons or marshes is often formed by phototrophic filamentous gliding cyanobacteria. Species of the genera *Microcoleus*, *Lyngbya*, *Oscillatoria* and *Pgormidium* form laminated algal mats which are important agents of sediment stabilization (Javor & Castenholz, 1980; Jørgensen, 1982). Sandy sediments containing detrital organic particles, and the laminated algal mats with precipitated minerals, inhibit water mixing and rapid exchange of interstitial water. These aquatic habitats are therefore characterized by steep gradients of the various environmental parameters. The gliding microbes use their chemo- or phototactic responses to attain positions in these gradients that provide them with suitable conditions for growth (Jørgensen, 1982).

Nelson & Castenholz (1982) showed that the apparent negative response of *Beggiatoa* towards light (which had already been noted by Winogradsky, 1887) is in fact a direct photophobic response to blue light (maximum at 426 nm) and not a secondary tactic reaction. In the natural habitat *Beggiatoa* filaments move down into the sediments by early morning and reappear shortly after sunset on the sediment surface, where they form whitish tufts and layers. These diurnal vertical migrations are of adaptive significance for these catalase-negative bacteria which thrive in the narrow region where both sulphide and dissolved oxygen occur simultaneously. Since filamentous cyanobacteria and diatoms at the sediment surface produce oxygen in daylight, oxygen concentrations in the upper sediment layers are significantly higher during the day than at night.

The opposite is true for sulphide, which reacts with the oxygen. The upward movements of the *Beggiatoa* filaments at night can thus be understood as a true positive aerotaxis of these essentially oxygen-dependent bacteria. The vertical position of *Beggiatoa* in sediments is, therefore, regulated by the filaments' responses to dissolved oxygen and light. *Beggiatoa* is the first known non-phototrophic bacterium in which the photophobic response is of apparent ecological significance (Nelson & Castenholz, 1982).

Microbial zonation in a marine sulphuretum
Jørgensen (1982) gave an exciting report of the diurnal cycle of oxygen and sulphide distribution and the interplay between three different microbes in the upper millimetres of the sediment in a marine sulphuretum near Aarhus, Denmark. In addition to the filamentous *Oscillatoria* spp. and *Beggiatoa*, small polarly flagellated purple sulphur bacteria of the *Chromatium* type were the visibly dominant species. *Oscillatoria* and *Beggiatoa* behaved in the same way as described above. During the day *Oscillatoria* formed the dark-green top layer with *Beggiatoa* below it at the oxygen/hydrogen sulphide interface (redox potential, E_h, between 0 and +100 mV); *Chromatium* was abundant in colonies in the hydrogen sulphide zone below *Beggiatoa*. Soon after sunset *Beggiatoa* moved through the *Oscillatoria* layer and formed the whitish surface layer. Concomitant with the consumption of oxygen and the increase in sulphide in and above the *Beggiatoa* layer (around midnight), swarming *Chromatium* cells appeared and changed the colour of the top layer from white to purple-red. Shortly after sunrise, when *Oscillatoria* started to form oxygen, *Chromatium* retreated into the sediment, soon followed by *Beggiatoa*. With increasing daylight the sediment surface reassumed its dark-green colour.

Chemotaxis
It is apparent from the preceding section that microbes with mycelial growth or gliding motility will successfully compete with other microbes particularly under environmental conditions which not only satisfy their necessary nutrient requirements but also provide the physical conditions in which their morphological characteristics and capacities for oriented movement may be expressed positively.

Experience with enrichment cultures has shown that low-nutrient stationary liquid cultures inoculated from soil preferentially yield fairly uniform populations of highly motile microbes such as pseudo-monads, vibrios or spirilla, often followed by ciliates feeding on the bacteria (van Niel, 1955). However, when agar plate cultures containing the same nutrients and inoculated from the same soil sample are studied, a great variety of moulds, actinomycetes, bacteria and protozoans may be encountered, which appear unable to compete successfully with motile bacteria in entirely aqueous media.

On a comparison of a motile and a non-motile strain of *Pseudomonas fluorescens*, Smith & Doetsch (1969) showed that the positively aerotactic motile strain was favoured only in stationary liquid cultures while in shaken cultures the original ratio between the two populations remained unchanged. The competition between a chemotactically active strain and an insensitive mutant strain of *Proteus mirabilis* was studied in liquid and semi-solid media by Pilgram & Williams (1976). Both strains grew equally well in shaken liquid medium. However, in the semi-solid medium with a stable gradient of nutrients, the chemotactically active strain was able to outgrow the mutant strain. This result confirmed that sensory responses are only of advantage in stagnant aqueous habitats and microhabitats that allow free movement and in which gradients of growth conditions occur. In several cases, the significance of tactic responses is directly apparent: for example, the positive chemotaxis of symbiotic nitrogen-fixing *Rhizobium* species to chemoattractants on the root hairs of their legume host plants (Ames & Bergman, 1981), or the positive chemotactic response of plant pathogenic *Pseudomonas* species to plant exudates (Chet & Mitchell, 1976).

With respect to the structure and function of the bacterial flagellum as well as the alternation of swimming and tumbling, the most detailed studies have been carried out on *Escherichia coli* and *Salmonella typhimurium*. The entire molecular basis of movement, from stimulus perception to sensory transduction and motor response, is being elucidated in these bacteria (Macnab, 1979). Positive chemotactic responses were observed towards sugars, amino acids and oxygen. Negative responses were elicited by non-physiological pH values and high concentrations of certain amino acids and fatty acids. By means of their chemotactic reactions these bacteria are directed towards nutrients and repelled from harmful conditions.

It is surprising to see how many permanently non-motile microbial species exist that seemingly thrive just as well as their motile companions. This may mean that we underestimate the effectiveness of the various forces in soils and sediments that cause the transport and distribution of non-motile cells as well as of nutrients (e.g. Brownian movement, capillary forces, water currents through the space between particles, or movements and mixing caused by the different kinds of motile protozoans and invertebrate animals).

Motile swarmers and sessile mother cells
In addition to the motile or permanently non-motile bacterial species, microbes exist that form both kinds of cells: e.g. *Caulobacter, Hyphomicrobium* and *Rhodomicrobium*. In all three genera flagellated motile swarmer cells are formed by polar growth and asymmetric division of the prosthecate non-motile mother cells. The swarmer cells are considered as the dispersal stage (Dow & Whittenbury, 1980), and in the caulobacters also as conjugal recipients or genetic vectors (Poindexter, 1981*a*). Although it is assumed that the swarmer cells should be capable of chemotactic responses, there is no experimental evidence for this. As long as the swarmer cells are motile, they are unable to grow and divide. As a first sign of further development the swarmers shed their flagella and develop an outgrowth of the cell envelope, a prostheca or stalk. The non-motile stalked cells of *Hyphomicrobium* and *Caulobacter* have holdfasts with which they attach themselves either to one another, to form free-floating rosettes, or to solid surfaces. In their natural habitats they were observed to be attached to inorganic or organic particles, but also to fungi or algae (Hirsch, 1974). Under favourable conditions the stalked cells grow and produce motile daughter cells. In *Rhodomicrobium* the stalked cells may grow into multicellular complexes which form peritrichously flagellated daughter cells under conditions of high carbon dioxide concentration and low light intensity (Dow & Whittenbury, 1980). It is assumed that the swarmer cell initiates the developmental sequence when it reaches suitable growth conditions. This behaviour appears to have an ecological meaning in view of the fact that the multicellular complexes were observed in the surface mud of pools and ponds where the conditions for swarmer cell formation could well exist. Under conditions of nutrient depletion, *Rhodomicrobium* cell complexes form exospores instead of swarmer cells.

Phototaxis and geotaxis
In contrast to soils and sediments, which are permanently dark, illuminated aquatic environments such as stagnant pools, ponds, lagoons, shallow bays or smaller lakes offer suitable living conditions

for phototrophic microbes. The degradation of plant and algal organic matter leads to the depletion of oxygen and the formation of hydrogen sulphide. Under these conditions the anaerobic phototrophic purple bacteria flourish. The flagellated species all exhibit scotophobic responses when crossing from a region of higher light intensity to lower light intensity or darkness (Clayton, 1959). This behaviour causes photoaccumulations of the cells in regions of optimal light intensities for growth, since the perception of the decrease in light intensity effects a simultaneous reversal of flagellar rotation which results in the reversal of the swimming direction (Buder, 1915). In the light the purple sulphur bacteria in addition show a negative aerotaxis (chemotactic response to dissolved oxygen) that causes them to return to anoxic sulphide-containing water. This explains the accumulation of large numbers of purple sulphur bacteria in pink to purple-red visible layers on the surface of sulphide-containing anoxic sediments exposed to daylight.

The large purple sulphur bacteria *Chromatium okenii*, *Chromatium weissei* and *Thiospirillum jenense* show in addition negative geotaxis, by means of which they swim upwards against gravity in their aquatic habitat or in liquid cultures. This capacity is shared by many free-swimming eukaryotic organisms such as ciliates and flagellates. When studied in horizontally mounted slide preparations, only random movement is seen. The negative geotaxis can be observed, however, when the microscope is turned horizontally with the slide in vertical position. Under these conditions the cells of *C. okenii*, *C. weissei* and *T. jenense* are seen to swim upwards and downwards, predominantly in a vertical direction with the flagella always upwards. In stagnant shallow aquatic habitats and in liquid cultures of sufficiently high cell density, the negative geotaxis causes characteristic bioconvection patterns (Platt, 1961; Plesset, Whipple & Winet, 1975). In nature, and under laboratory conditions of diurnal light and dark cycles, the capacity for negative geotaxis has been observed to confer a selective advantage upon *C. okenii*, *C. weissei* and *T. jenense* over the other species of purple sulphur bacteria (Pfennig, 1962).

In duckweed-covered ponds or shady parts of forest ponds, these three species may be found swarming in reddish clouds of cells that exhibit the typical bioconvection patterns. During the day, sulphide becomes exhausted by the photosynthetic activity of these bacteria and the clouds move closer and closer to the mud surface; the cells eventually disappear into the upper mud layers. At night, the sulphide concentration in the free water above the sediment increases again and the negatively geotactic purple sulphur bacteria rise until they reach water layers with dissolved oxygen. Such diurnal movements were also detected in stratified lakes. Sorokin (1970) reported on diurnal vertical movements over distances of up to 2 m for the purple-red layer of *Chromatium okenii* in the meromictic Lake Belovod.

For *Chromatium weissei* it was experimentally shown by van Gemerden (1974) that metabolic properties of the cells associated with the diurnal light and dark periods also contribute to its ability to compete successfully with the small *Chromatium vinosum*. The latter has a higher affinity for sulphide and a higher maximum growth rate, and therefore became dominant over *C. weissei* in continuously illuminated chemostat cultures. However, when light and dark periods alternated, *C. weissei* was able to coexist with *C. vinosum*. This was due to the fact that after a dark period *C. weissei* took up the accumulated sulphide 2.5 times faster than *C. vinosum* and formed globules of intracellular elemental sulphur from it. This reservoir of sulphur was excluded from the intraspecific competition for nutrients and allowed further growth of *C. weissei*. The other trait of selective advantage for *C. weissei*, its capacity for negative geotaxis, is effective only in stagnant water or stationary cultures, and did not come into play in van Gemerden's stirred cultures.

Vertical movements by buoyancy regulation
A quite unexpected kind of vertical movement is effected by the species of planktonic cyanobacteria and purple sulphur bacteria that contain gas vacuoles. Walsby (1971) was the first to determine the turgor pressure in prokaryotic cells, using gas-vacuolate *Anabaena flos aquae*. He also demonstrated convincingly that turgor pressure rises in the light and that this rise causes collapse of a proportion of the gas vesicles (Dinsdale & Walsby, 1972; Allison & Walsby, 1981). This type of response allows gas-vacuolate planktonic prokaryotic phototrophs to stay at an optimal depth on vertical light gradients in non-turbulent natural waters (Walsby, 1978). For a metalimnetic population of *Oscillatoria agardhii* at 4.8 m depth in a stratified lake, Walsby & Klemer (1974) showed experimentally that the position of the algae in the water column was dependent on the number of gas vesicles in the cells; this in turn was controlled by the turgor mechanism in response to the daily mean light intensity. The same kind of buoyancy regulation in response to light intensity was observed for colonies of *Aphanizomenon* in Lake Mendota during calm weather conditions (Konopka, Brock & Walsby, 1978).

As in the cases of chemotaxis and negative geotaxis mentioned earlier, the selective advantage of cells with gas vacuoles may become apparent only in undisturbed water or liquid culture. No difference in growth rate was observed between the wild-type and a gas-vacuoleless mutant of *Anabaena flos aquae* in stirred culture (Walsby, 1978). In a thermally stratified water column, however, the mutant strain sank to the bottom while the gas-vacuolate wild-type settled and developed at a depth with optimal growth conditions.

Many lakes in Britain which form an anaerobic hypolimnion in summer were found to have gas-vacuolate phototrophic and chemotrophic bacteria constituting a large part of the bacterial plankton (Walsby, 1978). This shows that gas vesicles apparently confer a selective advantage on microbes in non-turbulent aquatic habitats. Apart from the few impressive examples of vertical movements of gas-vacuolate cyanobacteria, experimental evidence for buoyancy regulation in phototrophic sulphur bacteria and anaerobic chemotrophic bacteria is so far lacking.

Some interactions of microbes involved in the anoxic degradation of organic matter

With the exception of lignin (Zeikus, 1981), the bulk of organic compounds of plant and animal origin is degradable under anaerobic conditions. Many different groups of microbes participate in this process and much is known about the initial decomposition steps which include the depolymerization of carbohydrates, pectin, chitin, lipids, proteins and nucleic acids as well as the fermentation of the intermediary monomeric substances thereby produced. Most of our knowledge on the fermentations of hexoses, pentoses, amino acids, purines and pyrimidines was obtained by pure culture studies. We know, however, that in pure cultures the quality and quantity of the fermentation products are more or less dependent on the fermentation conditions. This is even more the case under the complex conditions in natural habitats with their heterogeneous microbial populations. The fermentation products of one bacterium (e.g. ethanol, lactate or succinate) may become the substrates for another.

Sulphate-reducers and methanogens

Depending on the environmental conditions, further mineralization may essentially proceed in two different directions that are competitive but not necessarily mutually exclusive. In sulphate-sufficient marine and brackish-water sediments it is the sulphate-reducing bacteria that function predominantly as terminal oxidizers (Jørgensen & Fenchel, 1974; Jørgensen, 1977b; Laanbroek & Veldkamp, 1982). In sulphate-limited sediments, anoxic soils or digester sludge, the methanogenic bacteria in cooperation with special fermenting bacteria carry out the terminal degradation (Zehnder, 1978). For the not entirely sulphate-depleted sediment of Lake Mendota, Ingvorsen & Brock (1982) established by experiments *in situ* that the electron flow via sulphate reduction was about 25% of the electron flow via methanogenesis. Unlike the sulphate-reducers, the activity of the methanogens is never limited by their electron acceptor since carbon dioxide is abundant in anoxic environments.

The main function of these bacterial groups in their habitats is the terminal degradation of the two quantitatively most significant reduced products of the anaerobic degradation of organic substances: molecular hydrogen and acetate. In order to explain why sulphate-reducers outcompete methanogens in sulphate-sufficient environments, Winfrey & Zeikus (1977) suggested that the former may have higher affinities for acetate and hydrogen. Schönheit, Kristjansson & Thauer (1982) confirmed this assumption. The apparent K_s for acetate of *Desulfobacter postgatei* was lower than 0.2 mM, that of *Methanosarcina barkeri* around 3 mM, and that of *Methanothrix soehngenii* 0.5 mM (Zehnder, Ingvorsen & Marti, 1982). Also, the apparent K_s for hydrogen of *Desulfovibrio* was 1.0 μM, that of *Methanobrevibacter* 6.0 μM (Kristjansson *et al.*, 1982).

The high affinity of sulphate-reducers and methanogens for hydrogen allows its effective utilization in their different metabolic pathways, with far-reaching consequences for the entire degradation process in the habitat. Since the hydrogen partial pressure is kept low (3.5×10^{-3} to 4.7×10^{-5} atmospheres: Zehnder *et al.*, 1982) the liberation of hydrogen from reduced coenzymes via ferredoxins and hydrogenases becomes thermodynamically feasible (Wolin, 1976). In consequence, the fermentations of carbohydrates are shifted away from the generation of reduced end-products such as ethanol, lactate, propionate, butyrate or succinate towards the energetically more favourable formation of the more oxidized products acetate and carbon dioxide (Hungate, 1966; Tewes & Thauer, 1980). The reducing equivalents are evolved as hydrogen that is immediately consumed by sulphate-reducers or methanogens which thus restore the conditions for further hydrogen evolution.

During the past ten years considerable progress has been made in the recognition of the sulphate-reducing bacteria that carry out the terminal oxidation of anaerobic degradation products. The ecologically important oxidizers of hydrogen and formate are the classical *Desulfovibrio* species (Postgate, 1979; Brandis & Thauer, 1981). By the work of Widdel (1980), new genera and species of sulphate-reducers became known that completely oxidize acetate, lower and higher fatty acids, phenylsubstituted fatty acids and benzoate with sulphate as electron acceptor (Pfennig & Widdel, 1982). It appears, therefore, that most intermediates of anaerobic degradation are more or less directly oxidized by various species of sulphate-reducers. Future research should reveal in detail what compounds cannot be degraded in this way.

Formation of methanogenic substrates

In contrast to the sulphate-reducers, no known methanogenic bacteria degrade organic acids larger than acetate (Zeikus, 1977; Zehnder *et al.*, 1982). It is known, however, that many organic compounds are completely degradable to methane and carbon dioxide by bacterial populations in enrichment cultures (Tarvin & Buswell, 1934; Barker, 1956; Healy & Young, 1979). In many cases these degradations proceed in a food chain in which the fermentation products of one bacterial species are used by another one without significantly changing the ratio of the respective products. Recent examples are the complete fermentation of syringic acid and trimethoxybenzoic acid via gallic acid to methane and carbon dioxide by *Acetobacterium woodii*, *Pelobacter acidigallici* and *Methanosarcina* (Bache & Pfennig, 1981; Schink & Pfennig, 1982a; Kaiser & Hanselmann, 1982), and the complete degradation of choline via trimethylamine, ethanol and acetate to methane, carbon dioxide and ammonia by *Desulfovibrio* and *Methanosarcina* (Fiebig & Gottschalk, 1983).

In other cases the degradation becomes possible only in the presence of hydrogen-utilizing methanogens. Bryant postulated the existence of a particular group of bacteria capable of converting organic compounds in one or more steps into the methanogenic substrates hydrogen, carbon dioxide, one-carbon compounds and acetate (McInerney & Bryant, 1980).

The first experimental system that depended on interspecies hydrogen transfer was studied by Bryant *et al.* (1967). These authors showed that the fermentation of ethanol to acetate and hydrogen by one bacterium proceeded only in the presence of a methanogenic bacterium that directly consumed the hydrogen in methane formation. Later studies confirmed the view that the anoxic hydrogen-utilizing bacteria establish an ecological niche for a whole group of unusual hydrogen-forming, acetogenic bacteria. Three species were recently described: *Syntrophobacter wolinii* degrades propionate (Boone & Bryant, 1980), *Syntrophomonas wolfei* degrades longer-chain fatty acids (McInerney *et al.*, 1981), and an unnamed 'benzoate catabolizer' (Mountfort & Bryant, 1982) exclusively degrades benzoate. The interaction between hydrogen-producing and hydrogen-consuming bacteria represents an interesting example of metabolic symbiosis with unidirectional substrate supply from which mutual benefit is derived.

Our knowledge of the anaerobic breakdown of many organic acids, alcohols and hydroxylated aromatic compounds is rather incomplete. Also, we are just beginning to learn what types of compounds can only be degraded in syntrophy with hydrogen-utilizing bacteria and what types can be fermented in their absence (Laanbroek & Veldkamp, 1982; Schink & Pfennig, 1982a, b).

From natural habitats to experimental growth conditions

Nutritional studies in batch culture

Experience has shown that any one culture medium is somehow selective for the growth of certain microbes. This does not only pertain to the kind and concentration of the carbon and energy source supplied but also to the concentrations of mineral salts and trace elements. When hitherto unknown microbes are to be isolated from certain natural habitats it is advisable, therefore, not to rely too heavily on an otherwise approved culture medium. Laboratory media are usually designed to give high yields of certain microbes that are known to grow well in the medium; in consequence, the medium is selective. Good results can be obtained when the water of the natural habitat is used, supplemented with only small amounts of phosphate and a nitrogen source in addition to low concentrations of the supposed carbon and energy source (Widdel, 1983). During the isolation of a new microbe one does not require high yields. Once the organism is obtained in pure culture, the concentrations of the various constituents can gradually be increased according to the tolerance of the organism, and defined mineral media can be tested.

There is no single trace element solution that could be used equally well for all microbes. If the trace elements are provided without a chelating agent, the concentrations of one or another of them may be to some degree inhibitory for growth, or cause clumping of the cells. This was observed in the case of copper, selenium and molybdenum. Using chelating agents avoids inhibitory effects but the agents themselves may establish selective conditions. Some bacterial species were affected by ethylenediaminetetraacetate, others by nitrilotriacetate. There is no way out other than to test the different possibilities.

The situation with respect to mineral salts is similar. In the case of phototrophic bacteria, sulphur- and sulphate-reducing bacteria, the concentrations of phosphate, sodium chloride, calcium or magnesium were of significance for the success or failure to enrich and isolate certain species (Widdel & Pfennig, 1981; Widdel, Kohring & Mayer, 1983). Phosphate-buffered media often contained 10–50 mM phosphate; such abnormal concentrations were inhibitory for *Pelobacter acidigallici* (Schink & Pfennig, 1982a). Also, some sulphate-reducing bacteria should be cultivated with not more than 2–5 mM phosphate (Widdel *et al.*, 1983). Quite unexpectedly, sodium proved to be significant in the cultivation of several anerobic bacteria not only from marine but also from freshwater sources (Widdel & Pfennig, 1981; Schink & Pfennig, 1982a, b). Extremely low nutrient concentrations are required for enrichment and pure culture of aerobic heterotrophic caulobacters; the nutritional features and problems of low-nutrient-adapted bacteria are treated in detail by Poindexter (1981b).

Knowledge of the growth requirements of an organism in defined media is of considerable interest for ecological studies, since the organism's continued development and survival under conditions of competition in nature may depend on these characteristics. At present we are in the happy position of knowing many of the organic growth factors and trace elements required by microbes, and it should be possible to determine unknown requirements without much difficulty. Therefore, it will be an important task for future research to continue nutritional studies in order to achieve optimal growth conditions with defined media (Pirt, 1975).

Kinetic studies in continuous culture
In natural habitats such as lakes, poor soils or the open ocean, bacterial growth is largely limited by the concentration of the

dissolved carbon and energy sources. If the nutritional requirements of certain different microbes in such a habitat are very similar, their kinetic properties may become of primary importance for the outcome of competition. In low-nutrient environments, microbes that can grow slowly at very low substrate concentrations have a selective advantage over others that are adapted to growth at higher nutrient levels (Poindexter, 1981*b*). The latter may survive in low nutrient conditions only as dwarf cells in dormant stages (Jannasch, 1958, 1968). The selective isolation and characterization of bacteria with respect to their kinetic features is, therefore, a prerequisite to the recognition of their selective advantage in a natural population under certain environmental conditions. So far the most useful tool for the study of kinetic parameters, particularly at substrate concentrations approaching those in natural environments, is continuous culture in a chemostat (Jannasch 1967*a, b*; Jannasch & Mateles, 1974; Veldkamp, 1977; Matin & Veldkamp, 1978). These studies drew attention to the fundamental processes underlying the life of microbes and added a new dimension to our understanding of the diversity of microbial behaviour. From an ecological point of view, we have to realize that the chemostat is only one of various open-flow systems and experimental tools by means of which the kinetic phenomena of microbes in nature can be approached.

Chemostat enrichment of low-nutrient-adapted bacteria

Low dilution rate. At low dilution rate in a chemostat with growth limitation by the carbon source, the heterotrophic bacteria from a natural population that will establish a steadily growing population will be those that can remove the limiting substrate with the highest efficiency. This was shown experimentally by Jannasch (1967*a*) and Matin & Veldkamp (1978), who enriched cultures of marine and freshwater aerobic heterotrophic bacteria that had different kinetic properties. In all cases studied, the bacteria with high substrate affinity exhibited lower maximum growth rates than those that were selected for at high dilution rates. The former can be considered to belong to the significant catabolizers in the low-nutrient habitat while the latter are dependent on incidental increases in the nutrient level.

The kinetic properties of heterotrophic bacteria isolated by chemostat enrichment from low-nutrient habitats are not stable features. Jannasch (1968) showed that both the substrate concentration necessary for half-maximum growth rate and the maximum growth rate increased considerably during continued growth at higher substrate concentrations in batch cultures (e.g. 0.01 M lactate). Loss of the original growth characteristics of freshly chemostat-enriched seawater isolates was demonstrated for strains of the genera *Achromobacter, Pseudomonas, Vibrio* and *Spirillum*. These results strongly indicate that kinetic parameters of bacterial strains isolated from nutrient-limited environments are reliable only if the bacteria are continuously maintained under low-nutrient conditions.

Poindexter (1981*b*) drew attention to the fact that the chemostat as an open-flow system must be selective for a certain physiological type of low-nutrient-adapted microbes. Only those bacteria that are capable of continued growth and reproduction at the rate corresponding to the dilution rate chosen can establish steady-state populations. In the natural environment, however, continuous population dilution and wash-out does not occur, although grazing by protozoans may have similar effects. Consequently, in nature a type of low-nutrient bacterium may well persist in which physiological stages of gradual nutrient accumulation alternate with stages of reproduction. Such bacteria will be washed out from the chemostat.

Low population density. With the chemostat, a low dilution rate was not the only selective condition for the enrichment of low-nutrient-adapted aquatic bacteria. Jannasch (1967*b*) detected that the steady-state population density, which is experimentally adjusted by the concentration of the carbon source in the medium reservoir, will constitute an additional selective factor. At a fixed low dilution rate, bacteria with high substrate affinity and low maximum growth rate were found to be capable of continued growth at significantly lower population densities than others with lower substrate affinity and higher maximum growth rate. Initiation and continuation of growth of the latter type of bacteria was more or less dependent on the abolition of growth-inhibiting effects in the culture medium by the metabolic activity of the population. Such behaviour of different kinetic types of bacteria appears to be in agreement with conditions in different habitats. In low-nutrient environments which usually also have low population densities (around 10^5 bacteria per ml), the individual bacterial cells must most of all be capable of substrate assimilation and multiplication more or less independently of habitat conditioning by the metabolic activity of the population (Poindexter, 1981*b*).

At the lowest suitable dilution rate and substrate concentration, Jannasch (1967*a*) obtained a distribution of the bacteria in the chemostat that can only be compared to the distribution of bacteria in low-nutrient mountain streams. Ladd, Costerton & Geesey (1979) found the highest bacterial activity in the slimy surface layers that covered stones and gravel, while both numbers and activity of bacteria in the rushing water were negligible. Jannasch used the natural bacterial population of offshore seawater as inoculum and observed the establishment of such bacteria in a culture vessel. Slimy colonies developed that were attached to the glass walls. Of course, the growth rate of these adherent bacteria is independent of the dilution rate and they are thus unaffected by the selective stress of the chemostat. It was also most likely that, compared with freely suspended bacteria, they were favoured by the slightly increased substrate concentrations arising by adsorption at the glass–water and slime–water interfaces (Marshall, 1980). Under these conditions the chemostat changed to an undefined open-flow system.

In the laboratory, experimental conditions are arranged to be as simple, well-defined and reproducible as possible. This is the prerequisite to the formulation of clear concepts. In nature, however, the conditions are seldom simple; rather, they are enormously complex and often unpredictable. It was under these conditions that the microbes evolved. Hence, they reveal a far greater diversity and versatility in ecologically oriented studies than in routine laboratory work. To unravel nature's complexity must be the aim of microbial ecology. This may possibly be reached by two approaches. One is the increase in our knowledge of the special and complex conditions in a given natural habitat, by careful observations and measurements. The other approach is by experimental work in the laboratory under ecologically relevant conditions that are as well defined and reproducible as possible. The gradual advance in microbial ecology will essentially depend on the synthesis of these two approaches.

My thanks are due to Drs B. Schink and F. Widdel for stimulating discussions.

REFERENCES

ALEXANDER, M. (1976). Natural selection and the ecology of microbial adaptation in a biosphere. In *Extreme Environments*, ed. M. R. Heinrich, pp. 3–25. New York and London: Academic Press.

ALLISON, E. M. & WALSBY, A. E. (1981). The role of potassium in the control of turgor pressure in a gas-vacuolate blue-green alga. *Journal of Experimental Botany*, **32**, 241–9.

AMES, P. & BERGMAN, K. (1981). Competitive advantage provided by bacterial motility in the formation of nodules by *Rhizobium meliloti*. *Journal of Bacteriology*, **148**, 728–9.

BACHE, R. & PFENNIG, N. (1981). Selective isolation of *Acetobacterium woodii* on methoxylated aromatic acids and determination of growth yields. *Archives of Microbiology*, **130**, 255–61.

BARKER, H. A. (1956). *Bacterial Fermentations*. New York: Wiley.

BEIJERINCK, M. W. (1895). Ueber *Spirillum desulfuricans* als Ursache von Sulfatreduktion. *Centralblatt für Bakteriologie* (II Abteilung), **1**, 1–9, 49–59, 104–14.

BOONE, D. R. & BRYANT, M. P. (1980). Propionate-degrading bacterium, *Syntrophobacter wolinii* sp. nov., gen. nov., from methanogenic ecosystems. *Applied and Environmental Microbiology*, **40**, 626–32.

BRANDIS, A. & THAUER, R. K. (1981). Growth of *Desulfovibrio* species on hydrogen and sulphate as sole energy source. *Journal of General Microbiology*, **126**, 249–52.

BROCK, T. D. (1971). Microbial growth rates in nature. *Bacteriological Reviews*, **35**, 39–58.

BRYANT, M. P., WOLIN, E. A., WOLIN, M. J. & WOLFE, R. S. (1967). *Methanobacillus omelianskii*, a symbiotic association of two species of bacteria. *Archiv für Mikrobiologie*, **59**, 20–31.

BUDER, J. (1915). Zur Kenntnis des *Thiospirillum jenense* und seiner Reaktionen auf Lichtreize. *Jahrbücher für wissenschaftliche Botanik*, **56**, 529–84.

CARLILE, M. J. (1980). Positioning mechanisms – the role of motility, taxis and tropism in the life of microorganisms. In *Contemporary Microbial Ecology*, ed. D. C. Ellwood, J. N. Hedger, M. J. Latham, J. M. Lynch & J. H. Slater, pp. 55–74. New York & London: Academic Press.

CHET, J. & MITCHELL, R. (1976). Ecological aspects of microbial chemotactic behaviour. *Annual Review of Microbiology*, **30**, 221–39.

CHOLODNY, N. (1930). Ueber eine neue Methode zur Untersuchung der Bodenmikroflora. *Archiv für Mikrobiologie*, **1**, 620–52.

CLAYTON, R. K. (1959). Phototaxis of purple bacteria. In *Encyclopedia of Plant Physiology*, vol. 17/1, ed. W. Ruhland, pp. 371–87. Berlin: Springer Verlag.

COHN, F. (1872). *Ueber Bakterien, die kleinsten lebenden Wesen*, p. 165. Sammlung gemeinverständlicher wissenschaftlicher Vorträge, 7th Series. Berlin: C. Habel.

DINSDALE, M. T. & WALSBY, A. E. (1972). The interrelations of cell turgor pressure, gas vacuolation, and buoyancy in a blue-green alga. *Journal of Experimental Botany*, **23**, 561–70.

DOW, C. S. & WHITTENBURY, R. (1980). Prokaryotic form and function. In *Contemporary Microbial Ecology*, ed. D. C. Ellwood, J. N. Hedger, M. J. Latham, J. M. Lynch & J. H. Slater, pp. 391–417. New York & London: Academic Press.

ELLWOOD, D. C., HEDGER, J. N., LATHAM, M. J., LYNCH, J. M. & SLATER, J. H. (eds.) (1980). *Contemporary Microbial Ecology*. New York & London: Academic Press.

ENSIGN, J. C. (1970). Long-term starvation survival of rod and spherical cells of *Arthrobacter crystallopoietes*. *Journal of Bacteriology*, **103**, 569–77.

ENSIGN, J. C. (1978). Formation, properties, and germination of actinomycete spores. *Annual Review of Microbiology*, **32**, 185–219.

FIEBIG, K. & GOTTSCHALK, G. (1983). Methanogenesis from choline by a coculture of *Desulfovibrio* sp. and *Methanosarcina barkeri*. *Applied and Environmental Microbiology*, **45**, 161–8.

GEMERDEN, H. VAN (1974). Coexistence of organisms competing for the same substrate: an example among the purple sulphur bacteria. *Microbial Ecology*, **1**, 104–19.

HEALY, J. B. & YOUNG, L. Y. (1979). Anaerobic biodegradation of eleven aromatic compounds to methane. *Applied and Environmental Microbiology*, **38**, 84–9.

HERBERT, D., ELSWORTH, R. & TELLING, R. C. (1956). The continuous culture of bacteria; a theoretical and experimental study. *Journal of General Microbiology*, **14**, 601–23.

HIRSCH, P. (1974). Budding bacteria. *Annual Review of Microbiology*, **28**, 391–444.

HUNGATE, R. E. (1966). *The Rumen and its Microbes*. New York & London: Academic Press.

INGVORSEN, K. & BROCK, T. D. (1982). Electron flow via sulfate reduction and methanogenesis in the anaerobic hypolimnion of Lake Mendota. *Limnology and Oceanography*, **27**, 559–564.

JANNASCH, H. W. (1958). Studies on planktonic bacteria by means of a direct membrane filter method. *Journal of General Microbiology*, **18**, 609–20.

JANNASCH, H. W. (1967a). Enrichment of aquatic bacteria in continuous culture. *Archiv für Mikrobiologie*, **59**, 165–73.

JANNASCH, H. W. (1967b). Growth of marine bacteria at limiting concentrations of organic carbon in seawater. *Limnology and Oceanography*, **12**, 264–71.

JANNASCH, H. W. (1968). Growth characteristics of heterotrophic bacteria in seawater. *Journal of Bacteriology*, **95**, 722–3.

JANNASCH, H. W. & MATELES, R. J. (1974). Experimental bacterial ecology studied in continuous culture. *Advances in Microbial Physiology*, **11**, 165–212.

JAVOR, B. J. & CASTENHOLZ, R. W. (1980). Laminated microbial mats, Laguna Guerrero Negro, Mexico. *Geomicrobiology Journal*, **2**, 237–73.

JØRGENSEN, B. B. (1977a). Distribution of colorless sulfur bacteria (*Beggiatoa* spp.) in a coastal marine sediment. *Marine Biology*, **41**, 19–28.

JØRGENSEN, B. B. (1977b). The sulfur cycle of a coastal marine sediment (Limfjorden, Denmark). *Limnology and Oceanography*, **22**, 814–32.

JØRGENSEN, B. B. (1982). Ecology of the bacteria of the sulphur cycle with special reference to anoxic–oxic interface environments. In *Sulphur Bacteria*, ed. J. R. Postgate & D. P. Kelly, pp. 543–561. London: The Royal Society.

JØRGENSEN, B. B. & FENCHEL, T. (1974). The sulfur cycle of a marine sediment model system. *Marine Biology*, **24**, 189–201.

KAISER, J.-P. & HANSELMANN, K. W. (1982). Fermentative metabolism of substituted monoaromatic compounds by a bacterial community from anaerobic sediments. *Archives of Microbiology*, **133**, 185–94.

KLUYVER, A. J. & DONKER, H. J. L. (1926). Die Einheit in der Biochemie. *Chemie der Zelle und Gewebe*, **13**, 134–90.

KLUYVER, A. J. & VAN NIEL, C. B. (1936). Prospects for a natural system of classification of bacteria. *Zentralblatt für Bakteriologie und Parasitenkunde, Part I*, **133**, 301–32.

KONOPKA, A., BROCK, T. D. & WALSBY, A. E. (1978). Buoyancy regulation by planktonic blue-green algae in Lake Mendota, Wisconsin. *Archiv für Hydrobiologie*, **83**, 524–37.

KRASIL'NIKOV, N. A. & BELYAEV, S. S. (1967). Distribution of *Caulobacter* in certain soils. *Mikrobiologiya*, **39**, 1083–6.

KRISTJANSSON, J. K., SCHÖNHEIT, P. & THAUER, R. K. (1982). Different K_s-values for hydrogen of methanogenic bacteria and sulfate reducing bacteria: an explanation for the apparent inhibition of methanogenesis by sulfate. *Archives of Microbiology*, **131**, 278–82.

LAANBROEK, H. J. & VELDKAMP, H. (1982). Microbial interactions in sediment communities. *Philosophical Transactions of the Royal Society of London, Series B*, **297**, 533–50.

LADD, T. J., COSTERTON, J. W. & GEESEY, G. G. (1979). Determination of the heterotrophic activity of epilithic microbial populations. In *Native Aquatic Bacteria: Enumeration, Activity and Ecology*, ed. J. W. Costerton & R. R. Colwell, pp. 180–95. Philadelphia: American Society for Testing and Materials.

MCINERNEY, M. J. & BRYANT, M. P. (1980). Syntrophic associations of H_2-utilizing methanogenic bacteria and H_2-producing alcohol and fatty acid-degrading bacteria in anaerobic degradation of organic matter. In *Anaerobes and Anaerobic Infections*, ed. G. Gottschalk, N. Pfennig & H. Werner, pp. 117–26. Stuttgart: G. Fischer Verlag.

MCINERNEY, M. J., BRYANT, M. P., HESPELL, R. B. & COSTERTON, J. W. (1981). *Syntrophomonas wolfei* gen. nov. sp. nov., and anaerobic syntrophic fatty acid-oxidizing bacterium. *Applied and Environmental Microbiology*, **41**, 1029–39.

MACNAB, R. M. (1979). Chemotaxis in bacteria. In *Physiology of Movements*, ed. W. Haupt & M. E. Feinleib, pp. 310–34. Heidelberg: Springer Verlag.

MARSHALL, K. C. (1980). Reactions of microorganisms, ions and macromolecules at interfaces. In *Contemporary Microbial Ecology*, ed. D. C. Ellwood, J. N. Hedger, M. J. Latham, J. M. Lynch & J. H. Slater, pp. 93–106. London: Academic Press.

MATIN, A. & VELDKAMP, H. (1978). Physiological basis of the selective advantage of a *Spirillum* sp. in a carbon-limited environment. *Journal of General Microbiology*, **105**, 187–97.

MOUNTFORT, D. O. & BRYANT, M. P. (1982). Isolation and characterization of an anaerobic syntrophic benzoate-degrading bacterium from sewage sludge. *Archives of Microbiology*, **133**, 249–56.

NELSON, D. C. & CASTENHOLZ, R. W. (1982). Light responses of *Beggiatoa*. *Archives of Microbiology*, **131**, 146–55.

NIEL, C. B. VAN (1949). The 'Delft School' and the rise of general microbiology. *Bacteriological Reviews*, **13**, 161–74.

NIEL, C. B. VAN (1955). Natural selection in the microbial world. *Journal of General Microbiology*, **13**, 201–17.

ORLA-JENSEN, S. (1909). Die Hauptlinien des natürlichen Bakteriensystems nebst einer Übersicht der Gärungsphänomene. *Zentralblatt für Bakteriologie und Parasitenkunde* (II Abteilung), **22**, 305–46.

PFENNIG, N. (1958). Beobachtungen des Wachstumsverhaltens von Streptomyceten auf Rossi-Cholodny-Aufwuchsplatten im Boden. *Archiv für Mikrobiologie*, **31**, 206–16.

PFENNIG, N. (1962). Beobachtungen über das Schwärmen von *Chromatium okenii*. *Archiv für Mikrobiologie*, **42**, 90–5.

PFENNIG, N. & WIDDEL, F. (1982). The bacteria of the sulphur cycle. *Philosophical Transactions of the Royal Society of London, Series B*, **298**, 433–41.

PILGRAM, W. K. & WILLIAMS, F. D. (1976). Survival value of chemotaxis in mixed cultures. *Canadian Journal of Microbiology*, **22**, 1771–3.

PIRT, S. J. (1975). *Principles of Microbe and Cell Cultivation*. Oxford: Blackwell Scientific.

PLATT, J. R. (1961). Bioconvection patterns in cultures of free-swimming organisms. *Science*, **133**, 1766.

PLESSET, M. S., WHIPPLE, G. G. & WINET, H. (1975). Analysis of the steady state of bioconvection in swarms of swimming microorganisms. In *Swimming and Flying in Nature*, vol. 1, ed. T. Y. T. Wu, C. J. Brokaw & C. Brennan, pp. 339–60. New York: Plenum Press.

POINDEXTER, J. S. (1981*a*). The caulobacters: ubiquitous unusual bacteria. *Microbiological Reviews*, **45**, 123–79.

POINDEXTER, J. S. (1981*b*). Oligotrophy. Fast and famine existence. *Advances in Microbial Ecology*, **5**, 63–89.

POSTGATE, J. R. (1979). *The Sulphate-Reducing Bacteria*. Cambridge University Press.

ROBINSON, J. B., SALONIUS, P. O. & CHASE, F. E. (1965). A note on the differential response of *Arthrobacter* sp. and *Pseudomonas* spp. to drying in soil. *Canadian Journal of Microbiology*, **11**, 746–8.

ROSSWALL, TH. (1973). *Modern Methods in the Study of Microbial Ecology*. Stockholm: Swedish Natural Science Research Council.

SCHINK, B. & PFENNIG, N. (1982*a*). Fermentation of trihydroxybenzenes by *Pelobacter acidigallici* gen. nov. sp. nov., a new strictly anaerobic, non-sporeforming bacterium. *Archives of Microbiology*, **133**, 195–201.

SCHINK, B. & PFENNIG, N. (1982*b*). *Propionigenium modestum* gen. nov. sp. nov., a new strictly anaerobic non-sporing bacterium growing on succinate. *Archives of Microbiology*, **133**, 209–16.

SCHÖNHEIT, P., KRISTJANSSON, J. K. & THAUER, R. K. (1982). Kinetic mechanism for the ability of sulfate reducers to out-compete methanogens for acetate. *Archives of Microbiology*, **132**, 285–8.

SLATER, J. H., WHITTENBURY, R. & WIMPENNY, J. W. T. (eds.) (1983). *Microbes in their Natural Environments. Society for General Microbiology Symposium 34.* Cambridge University Press.

SMITH, J. L. & DOETSCH, R. N. (1969). Studies in negative chemotaxis and the survival value of motility in *Pseudomonas fluorescens*. *Journal of General Microbiology*, **55**, 379–91.

SOROKIN, YU. I. (1970). Interrelations between sulphur and carbon turnover in meromictic lakes. *Archiv für Hydrobiologie*, **66**, 391–446.

SUSSMAN, A. S. & HALVORSON, H. D. (1966). *Spores: Their Dormancy and Germination*. New York: Harper & Row.

TARVIN, D. & BUSWELL, A. M. (1934). The methane formation of organic acids and carbohydrates. *Journal of the American Chemical Society*, **56**, 1751–5.

TEWES, F. J. & THAUER, R. K. (1980). Regulation of ATP-synthesis in glucose fermenting bacteria involved in interspecies hydrogen transfer. In *Anaerobes and Anaerobic Infections*, ed. G. Gottschalk, N. Pfennig & H. Werner, pp. 97–104. Stuttgart: G. Fischer Verlag.

VELDKAMP, H. (1977). Ecological studies with the chemostat. *Advances in Microbial Ecology*, **1**, 59–94.

VOELZ, H. & DWORKIN, M. (1962). Fine structure of *Myxococcus xanthus* during morphogenesis. *Journal of Bacteriology*, **84**, 943–52.

WALSBY, A. E. (1971). The pressure relationships of gas vacuoles. *Proceedings of the Royal Society of London, Series B*, **178**, 301–26.

WALSBY, A. E. (1978). The gas vesicles of aquatic prokaryotes. In *Relations Between Structure and Function in the Prokaryotic Cell, Society for General Microbiology Symposium 28*, ed. R. Y. Stanier, H. J. Rogers & B. J. Ward, pp. 327–57. Cambridge University Press.

WALSBY, A. E. & KLEMER, A. R. (1974). The role of gas vacuoles in the microstratification of a population of *Oscillatoria agardhii* var. *isothrix* in Deming Lake, Minnesota. *Archiv für Hydrobiologie*, **74**, 375–92.

WHITTENBURY, R., DAVIES, S. L. & DAVEY, J. F. (1970). Exospores and cysts formed by methane-utilizing bacteria. *Journal of General Microbiology*, **61**, 219–26.

WIDDEL, F. (1980). Anaerober Abbau von Fettsäuren und Benzoesäure durch neu isolierte Arten Sulfat-reduzierender Bakterien. PhD. thesis, University of Göttingen.

WIDDEL, F. (1983). Methods for enrichment and pure culture isolation of filamentous gliding sulfate-reducing bacteria. *Archives of Microbiology*, **134**, 282–5.

WIDDEL, F., KOHRING, G. W. & MAYER, F. (1983). Studies on dissimilatory sulfate-reducing bacteria that decompose fatty acids. III. Characterization of the filamentous gliding *Desulfonema limicola* gen. nov. sp. nov., and *Desulfonema magnum* sp. nov. *Archives of Microbiology*, **134**, 286–94.

WIDDEL, F. & PFENNIG, N. (1981). Studies on dissimilatory sulfate-reducing bacteria that decompose fatty acids. I. Isolation of new sulfate-reducing bacteria enriched with acetate from saline environments. Description of *Desulfobacter postgatei* gen. nov., sp. nov. *Archives of Microbiology*, **129**, 395–400.

WINFREY, M. R. & ZEIKUS, J. G. (1977). Effect of sulfate on carbon and electron flow during microbial methanogenesis in freshwater sediments. *Applied and Environmental Microbiology*, **33**, 275–81.

WINOGRADSKY, S. (1887). Über Schwefelbakterien. *Botanische Zeitung*, **45**, 489–507, 513–610.

WINOGRADSKY, S. (1890). Recherches sur les organismes de la nitrification. *Annales de l'Institut Pasteur, Paris*, **4**, 257–75.

WINOGRADSKY, S. (1949). *Microbiologie du sol*. Oeuvres complètes. Paris: Masson et Cie.

WOLIN, M. J. (1976). Interactions between H_2-producing and methane-producing species. In *Microbial Formation and Utilization of Gases* (H_2, CH_4, CO), ed. H. G. Schlegel, G. Gottschalk & N. Pfennig, pp. 141–50. Göttingen: E. Goltze.

WOODS, D. D. (1953). The integration of research on the nutrition and metabolism of micro-organisms. *Journal of General Microbiology*, **9**, 151–73.

ZEHNDER, A. J. B. (1978). Ecology of methane formation. In *Water Pollution Microbiology*, vol. 2, ed. R. Mitchell, pp. 349–76. New York: Wiley.

ZEHNDER, A. B. J., INGVORSEN, K. & MARTI, T. (1982). Microbiology of methane bacteria. In *Anaerobic Digestion*, ed. D. E. Hughes, D. A. Stafford, B. J. Wheatley, W. Baader, G. Lettinger, E. N. Nyns, W. Verstrate & R. L. Wentworth, pp. 45–68. Amsterdam: Elsevier Biomedical Press.

ZEIKUS, J. G. (1977). The biology of methanogenic bacteria. *Bacteriological Reviews*, **41**, 514–41.

ZEIKUS, J. G. (1981). Lignin metabolism and the carbon cycle. *Advances in Microbial Ecology*, **5**, 211–43.

ENVIRONMENTAL REGULATION OF MICROBIAL METABOLISM

WIM HARDER, LUBBERT DIJKHUIZEN AND HANS VELDKAMP

Dept of Microbiology, University of Groningen, Kerklaan 30, 9751 NN Haren, The Netherlands

... it is a general property of [microbial] life to change its enzymatic equipment, and in doing so to meet nature's challenges.

(Kluyver, 1956)

Unlike higher organisms, microbes have only a very limited capacity for controlling their (micro)environment. In order for them to survive and compete successfully with other microbes under the heterogeneous and changing conditions that are so characteristic of natural environments, they have evolved the potential to respond to environmental change by changing themselves, both functionally and structurally. This potential for rapid adaptation is uniquely associated with the life style of the microbe and resides in its genome. This carries all the information necessary for an organism to become a functional self-reproducing unit and also confers upon it the ability to respond to changes in its environment. In principle adaptation to environmental change can occur in two ways, namely by changes in genetic constitution or by phenotypic adjustment. The former has given rise to various 'more permanent' forms of adaptation to environmental extremes such as high and low temperatures, pH values or salt concentrations and has also led to nutritional specialization. In this paper we shall only consider phenotypic responses to environmental change of micro-organisms of a given genotype.

Within a certain microbial genotype a variety of phenotypic responses may be encountered. It is evident that this potential to express different properties under various conditions must differ between genotypes. At one extreme we find exceedingly versatile creatures that are able to metabolize in excess of one hundred different substrates over a range of temperatures, oxygen partial pressures and pH values, whereas at the other extreme there are highly specialized organisms that can generate energy from one energy source only under otherwise strictly specified environmental conditions. There is evidence that at least the more versatile microbes generally do not express their entire genome under any one set of environmental conditions (Koch, 1976), but instead use only that part that enables them to become structurally and functionally adjusted to the prevailing set of conditions. This phenotypic variability may involve just one quantitative change in some cellular component or, more drastically, radical qualitative changes in cell structure and/or function. Consequently an organism of a given genotype is very much a product of its environment (Herbert, 1961; Tempest & Neijssel, 1978). This demonstrates that if our objective is to understand the way in which micro-organisms function in nature (in 1984 and beyond!), it is necessary not only to unravel basic mechanisms of metabolic control, but also to study the expression of the various characteristics, capacities and activities under conditions that prevail in nature.

A number of environmental parameters commonly influence the properties of microbial cells in nature. In addition to physical factors such as light availability, temperature and presence of various types of interfaces, chemical factors (i.e. pH, oxygen partial pressure, ionic strength and nutrient availability) are also of importance. Natural ecosystems are frequently depleted of one or more essential nutrients (see Konings & Veldkamp, 1983) as a consequence of the (potentially vigorous) metabolic activities of indigenous microbial populations. Nutrient insufficiency is, therefore, probably the most common environmental extreme to which microbes must accommodate themselves (Tempest & Neijssel, 1978); thus we shall consider not only how microbes respond to nutrient limitation, but also how they cope with periods of nutrient starvation. Since the chemical composition of many natural environments is probably more akin to a dilute soup containing a mixture of various potential nutrients than to laboratory media made up of a single carbon and energy source, we shall also discuss a few examples of how microbes respond to the presence of mixed substrates. No attempt will be made to provide a complete discussion of any of these topics. Instead we shall try to rationalize the observed responses in more general terms, because in this way we hope to do more justice to the functioning of the intact organism. For a more detailed discussion the reader is referred to a number of recent reviews (Tempest & Neijssel, 1978, 1981; Konings & Veldkamp, 1980, 1983; Harder & Dijkhuizen, 1982, 1983; Tempest, Neijssel & Zevenboom, 1983).

GENERAL MECHANISMS OF ADAPTATION

Micro-organisms possess a wide variety of mechanisms for adapting themselves both functionally and structurally to their environment. Not surprisingly, the main objectives of these mechanisms are to ensure the prolonged survival of the organism and to enable it to grow as fast as possible. The various mechanisms that have evolved in microbes become especially evident when the way in which they adapt themselves to a major change in one or more environmental parameters, for instance the availability of nutrients, is considered. Structurally, microbes may respond to such a situation by changing the macromolecular composition of the cells, their ultrastructure or their morphology.

An example of the first type of response is that of the RNA content of a micro-organism, which is an almost linear function of its growth rate (Herbert, 1961; Nierlich, 1978). Since most of this RNA is ribosomal, it has been suggested that the organism adapts itself to a change in the growth environment that enables an increase in the specific growth rate by increasing the cellular ribosome content approximately in proportion to the growth rate (Maaløe & Kjeldgaard, 1966). This is thought to be due to the fact that the rate of protein synthesis per ribosome is approximately constant.

Environment-induced changes in ultrastructure are particularly evident in yeasts. When these organisms are grown under conditions in which oxidases are involved in the metabolism of carbon and/or nitrogen sources, proliferation of peroxisomes occurs which drastically changes the ultrastructural appearance of these organisms (Veenhuis, van Dijken & Harder, 1983a).

Environment-induced morphological changes have also been observed frequently. Many bacteria respond to a decrease in the supply of nutrients by increasing their surface to volume ratio, thus enhancing the uptake capacity of the cells (Veldkamp, 1977; Matin, 1979). A striking example of this behaviour is the sphere–rod transition observed in *Arthrobacter* species (Luscombe & Gray, 1974; Clark, 1979). Under nutrient limitation the coccoid morphology dominates and this is the form thought to be most abundant in nature. Other bacteria respond to deprivation of a readily utilizable source of energy, nitrogen or phosphate by producing endospores, which allows the species to survive for a long time, albeit in a dormant state, in the virtual absence of nutrients (Hanson, 1979).

Finally, evidence has been presented recently (Dow, Whittenbury & Carr, 1983) to indicate that swarmer cells of the budding and prosthecate bacteria may enter a so-called shut-down state under conditions when the nutritional status of the environment is poor. Characteristically, these swarmer cells exhibit low endogenous metabolic activity and show virtually no rRNA or DNA synthesis. Thus, a relatively small part of the genome of these cells is expressed under conditions of nutrient constraints. This is thought to enable them to survive for a prolonged period in which they are not engaged in active net biosynthesis of cell material or reproduction. Differentiation of the swarmer cells into productive cells is triggered by environmental stimuli. In the case of the photosynthetic bacterium *Rhodomicrobium vannieli* for instance, the amount of light available per cell appears to be an important factor (Dow *et al.*, 1983). Conceivably, the presence of suitable energy sources may also initiate differentiation in this organism and other prosthecate bacteria.

In addition to the ability to change themselves structurally, microbes possess control mechanisms to adapt their intracellular metabolism, i.e. the levels and activities of enzymes present, in response to changes in their environment. The most important phenomena involved in metabolic regulation will be briefly surveyed here and illustrated with a few selected examples. For a discussion of the possible evolutionary origins of these control mechanisms and their selective advantages, the reader is referred to a recent review by Baumberg (1981). A detailed description of the molecular mechanisms for the control of protein synthesis is given elsewhere in this volume by Reznikoff.

In the presence of utilizable carbon or nitrogen substrates, an increased rate of synthesis of enzymes involved in their metabolism is generally observed. This induction phenomenon is a universal control mechanism for (peripheral) catabolic pathways and results from the interaction of a signal metabolite (inducer) with a repressor protein which, on its own, is able to block initiation of the transcription of relevant genes. In some cases it has been observed that the actual inducer is not the substrate of the first enzyme in a catabolic sequence of reactions, but rather one of the early products or even a side-product (Magasanik, 1976). This, of course, requires the presence not only of a specific transport system but also of early enzyme(s) in the reaction sequence, in low but sufficient levels under non-induced conditions. Induction of enzyme synthesis has also been observed for enzymes of the central pathways of intermediary metabolism. Synthesis of these enzymes in, for instance, *Escherichia coli* is, however, also strongly influenced by other environmental parameters such as the availability of oxygen or other respiratory electron acceptors (Smith & Neidhardt, 1983*a, b*).

The rate of synthesis of enzymes involved in the biosynthesis of cellular monomers or their precursors most commonly decreases in the presence of their end-products, irrespective of whether they are supplied externally or produced in the cell. This repression phenomenon, caused by a decrease in the frequency of initiation of transcription, has been widely observed in, for instance, biosynthetic pathways for amino acids (Umbarger, 1978). Especially in the case of branched pathways this may involve complex mechanisms in which the various end-products, often in concert, control the rate of synthesis of enzymes in that part of the sequence of reactions which they have in common.

Even for constitutive enzymes an adjustment of their rate of synthesis in response to environmental conditions may be produced. Recently, 3',5'-cyclic adenosine monophosphate (cyclic AMP), which is also the signal metabolite for carbon catabolite repression (see below) of numerous inducible enzymes in enterobacteria, and cyclic GMP have been implicated as mediators of this type of control in *E. coli* (Calcott, 1982). Although the precise controlling points remain to be established, evidence was obtained in chemostat cultures of this organism that the rate of synthesis of various constitutive enzymes was directly related to the concentration of these cyclic nucleotides.

Over the years it has become clear that other devices, in addition to the basic mechanisms outlined above, may be of importance in controlling the rate of enzyme synthesis in various organisms. These include autogenous regulation (Calhoun & Hatfield, 1975; Savageau, 1979), in which an enzyme controls its own synthesis, and attenuation (Crawford & Stauffer, 1980; Yanofsky, 1981), a mechanism by which transcription, once initiated, can be terminated before the regulated genes have been completely transcribed. The latter mechanism has been observed in a number of amino acid biosynthetic pathways and, in the case of tryptophan biosynthesis, functions in addition to end-product repression. Finally, an important role has been attributed to guanosine 5'-diphosphate 3'-diphosphate (ppGpp) in the regulation of expression of many genes. The action of this nucleotide becomes especially manifest during the

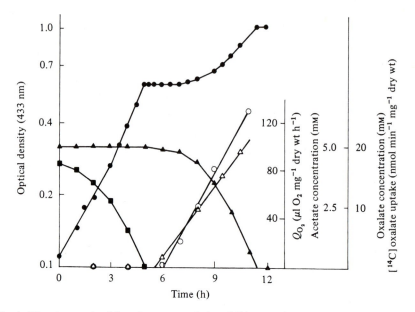

Fig. 1. Diauxic growth of *Pseudomonas oxalaticus* OX1 on a mixture of acetate (5 mM) and oxalate (20 mM). The inoculum was pregrown on acetate. ■, acetate concentration; ▲, oxalate concentration; ●, optical density at 433 nm; ○, Q_{O_2} for oxalate; △, rate of [^{14}C]oxalate transport in whole cells. (Reproduced from Dijkhuizen *et al.* (1980), with permission.)

so-called stringent response (Gallant, 1979) which is observed when certain strains of *E. coli* are deprived of an essential amino acid. This results in accumulation of ppGpp and a shift in transcriptional selectivity. Since an inverse relation between intracellular ppGpp concentration and growth rate has been observed (see Arbige & Chesbro, 1982), and because ppGpp inhibits transcription of genes coding for rRNA, ribosomal proteins and, possibly, other proteins of the protein-synthesizing machinery (Nierlich, 1978), it has been suggested that ppGpp is a negative effector of ribosome synthesis. It is evident that RNA polymerase is likely to be a prime target in the regulation of transcription. So far three systems have been identified that are involved in the control of the promoter selectivity of this enzyme *in vitro* (see Travers, Kari & Mace, 1981), namely a guanine nucleotide couple (ppGpp, GTP), an adenine nucleotide couple (ATP, AMP) and a system of macromolecular effectors (EF-Tu, EF-Ts, EF-G, IF-2, fMet-tRNA).

So far we have dealt with the response of microbes to the presence or absence of single nutrient factors. However, in many natural and man-made environments micro-organisms often encounter mixtures of nutrients that serve a similar physiological function (e.g. carbon source, energy source, nitrogen source). In principle, one might expect that control mechanisms have evolved that enable micro-organisms to utilize various carbon or nitrogen sources simultaneously. However, when a microbial species is placed under substrate-sufficient conditions in batch culture in a medium with, for instance, two utilizable carbon substrates, the cells often, although not invariably, adapt themselves in such a way that the substrate that supports the highest growth rate is used preferentially, while the synthesis of enzymes involved in the utilization of the second substrate remains repressed (Monod, 1942; Kornberg, 1966; Harder & Dijkhuizen, 1982). This results in sequential utilization of the two substrates and diauxic growth of the organism. Glucose is a preferentially used substrate in *E. coli*, but organic acids may be preferred in pseudomonads and some other bacteria (Ng & Dawes, 1973; Clarke & Ornston, 1975; Wood & Kelly, 1977). Diauxie is seen, for instance, during growth of *Pseudomonas oxalaticus* in batch culture on a mixture of the two organic acids oxalate and acetate (Fig. 1). In the first phase of growth the organism used acetate, and only after the complete exhaustion of acetate, in a lag phase of growth, was synthesis of the enzymes of oxalate metabolism started. This was then followed by a second, slower, growth phase on oxalate (Dijkhuizen, van der Werf & Harder, 1980).

Studies on mixed substrate utilization in various microbes have identified a number of cellular control mechanisms (Harder & Dijkhuizen, 1982). In various cases it has been shown that sequential substrate use was caused by interference of the preferred substrate with the uptake of the second substrate into the cell. Since this also prevents further induction of specific catabolic enzymes for the latter substrate, this phenomenon has been given the appropriate name of inducer exclusion. One of the best-studied examples is the effect of the presence of glucose on the uptake of a variety of substrates in *E. coli* and *Salmonella typhimurium*. In these organisms, glucose, mannose and fructose are transported across the cytoplasmic membrane via the phosphoenolpyruvate-dependent sugar phosphotransferase system (PTS). This so-called group translocation transport system (see Konings, Hellingwerf & Robillard, 1981) consists of two general proteins (Enzyme I, HPr) that occur in the cytoplasm of the cell, and membrane-bound proteins (Enzymes

II), which are sugar specific. At least one Enzyme II, namely that specific for glucose, requires an additional specific cytoplasmic protein, Enzyme III^{Glc} (E$_{III}^{Glc}$). The PTS couples the hydrolysis of phosphoenolpyruvate (PEP) via a series of phosphoryl-group transfer reactions to the transport of glucose and other hexoses (Postma & Roseman, 1976; Saier, 1977):

$$PEP + HPr \xrightleftharpoons{E_I, Mg^{2+}} P\text{--}HPr + Pyruvate$$

$$P\text{--}HPr + Sugar_{out} \xrightleftharpoons{E_{II}, E_{III}} HPr + Sugar\text{--}P_{in}$$

In the presence of glucose the utilization of many other substrates is blocked by a variety of mechanisms in which a component of the PTS plays a role. First of all glucose inhibits the uptake of other PTS sugars by successfully competing for the intracellular pool of P-HPr. In addition, glucose and other PTS sugars inhibit the uptake of several non-PTS substrates (such as melibiose, maltose, lactose and glycerol) via their specific transport systems. In the presence of PTS sugars, E$_{III}^{Glc}$ will become non-phosphorylated, and on the basis of genetic evidence obtained with mutants of *S. typhimurium* and measurements of the binding of E$_{III}^{Glc}$ to the lactose permease, it was concluded (see Postma, 1982) that this non-phosphorylated form of the protein inhibits transport of the compounds mentioned above. Since this model failed to explain the phenotype of certain mutants (*iex* and *gsr*) of *E. coli*, Parra, Jones-Mortimer & Kornberg (1983) postulated that in this organism not E$_{III}^{Glc}$ but a so far unidentified regulatory protein specified by the *iex* gene modulates the activities of transport systems for various non-PTS substrates, thus causing inducer exclusion.

In addition to inducer exclusion, further control mechanisms exist in *E. coli* and *S. typhimurium*. In these enteric bacteria cyclic AMP in a complex with a specific protein called catabolite-gene activator protein (CAP), must bind to the promoter of an operon before efficient binding of RNA polymerase and initiation of transcription is possible (Botsford, 1981). The metabolism of glucose, however, leads to a decreased intracellular concentration of cyclic AMP and, consequently results in inhibition of the synthesis of enzymes coded by various operons that require activation by the cAMP-CAP complex. Evidence is now accumulating that in this phenomenon, which has been termed (carbon) catabolite repression (Magasanik, 1961), a phosphorylated enzyme of the PTS plays an important role.

Most probably (see Postma, 1982) phosphorylated E_{III}^{Glc} is an activator of adenylate cyclase, the enzyme which catalyses the conversion of ATP into cyclic AMP. Thus, in the presence of glucose, which causes accumulation of non-phosphorylated E_{III}^{Glc}, both inducer exclusion and a lowered cyclic AMP concentration may prevent induction of catabolic enzymes for the use of other substrates. This results in preferential use of glucose and diauxic growth on mixtures of glucose and another sugar.

Whereas glucose is the carbon and energy source preferred over lactose or organic acids for growth of the enteric bacteria, ammonium sulphate is generally the preferred nitrogen source. During growth on mixtures of glucose with an organic nitrogen compound and ammonium sulphate a complex situation arises, in which a variety of regulatory mechanisms determines the sequence of substrate utilization. In *Klebsiella aerogenes*, for instance, the enzymes of histidine catabolism are subject to catabolite repression by glucose under these conditions, which results in the preferential use of glucose as the carbon source and ammonia as the nitrogen source (Magasanik, 1976, 1980). In the absence of ammonium sulphate, however, this catabolite repression can be partly overcome. This allows induction of the histidine enzymes and causes histidine utilization to occur at a rate sufficiently high to satisfy the cell's requirement for nitrogen. In general the use of organic nitrogen compounds is dependent not only on the energy status of the cells but also on the nitrogen status (nitrogen metabolite repression: Tyler, 1978; Magasanik, 1982). The identity and precise mechanism of the control system which allows relief of carbon catabolite repression under conditions of low nitrogen availability are still not completely clear. Recent evidence indicates that the products of genes closely linked to the structural gene for glutamine synthetase are involved in the regulation of various 'nitrogen-controlled' genes in *E. coli* and *S. typhimurium* (McFarland *et al.*, 1981; Magasanik, 1982; Pahel, Rothstein & Magasanik, 1982; Backman *et al.*, 1983).

In response to changes in their environment, micro-organisms may also adapt by controlling the activities of existing enzymes. This phenomenon, which may be considered as a 'quick action' control mechanism, plays an important role in regulating the flow of metabolites to accommodate the needs of the cells. Three types of control will be considered here, namely modulation, covalent modification and selective inactivation. Most of the enzymes sensitive to the modulation type of control are allosteric and their activities change in response to the presence of low-molecular-weight effector molecules (see Wyman, 1972; Baumberg, 1981). This is a fairly general control mechanism in biosynthetic pathways: for example, in amino acid synthesis (Umbarger, 1978) the end-product accumulates in the cell and often specifically inhibits the activity of the first enzyme involved in its formation (feedback inhibition). A second example of this type of regulation can be found among the enzymes involved in the central metabolism of the cell. In accordance with the amphibolic character of these pathways (glycolysis, tricarboxylic acid cycle) a number of their specific enzymes are sensitive to control by products of the cell's energy metabolism and/or precursors for biosynthesis. In *E. coli*, for instance, citrate synthase activity is inhibited by reduced nicotinamide adenine dinucleotide (NADH) and α-ketoglutarate. In contrast pyruvate kinase is activated by fructose-1,6-bisphosphate, while phosphofructokinase is inhibited by phosphoenolpyruvate and activated by adenosine 5'-diphosphate (Sanwal, 1970; see also Baumberg, 1981).

Examples of enzyme regulation by covalent modification and selective inactivation are less abundant in prokaryotes than in eukaryotes. Covalent modification characteristically interconverts two forms of an enzyme by a covalent substitution of an acetyl, adenyl or phosphoryl group. One example, namely that of glutamine synthetase, an enzyme which plays a most important role in the metabolism of nitrogen in the cell, may serve to illustrate this phenomenon. In various bacteria the activity of glutamine synthetase is regulated via reversible, enzyme-catalysed adenylylation and deadenylylation reactions which depend on the nitrogen status of the cell (Ginsburg & Stadtman, 1973; Tronick, Ciardi & Stadtman, 1973; Tyler, 1978). In the presence of low concentrations of ammonia in the cell, which may be caused by a low extracellular ammonia concentration or a low rate of utilization of a nitrogen source other than ammonia, the level of this enzyme in the cell increases and under these conditions it is largely in the active deadenylylated form. In cells growing with an excess of ammonia on the other hand, the enzyme is present in an adenylylated form which lacks the ability to catalyse the formation of glutamine. In addition, the two forms of the enzymes differ in their sensitivity to modulation by various effectors.

Selective inactivation of enzymes is frequently observed when an organism has to adapt itself to a major change in the availability of nutrients, such as a change in carbon or nitrogen sources, resulting in conditions where the continued activity of a particular enzyme is no longer required or even harmful (e.g. by creating a futile cycle). In this process the catalytic activity of an enzyme is lost rapidly and irreversibly, either by modification of the enzyme protein or by its degradation (Switzer, 1977). A classic example of this is the inactivation of several enzymes in *Saccharomyces cerevisiae* following transfer of cells from acetate to glucose, a phenomenon which has been termed catabolite inactivation (Holzer, 1976). Evidence for selective inactivation of enzymes has also been obtained in methylotrophic yeasts. During growth on methanol these organisms contain large amounts of alcohol oxidase and catalase. Both enzymes are involved in methanol metabolism and are localized in specific subcellular organelles called peroxisomes (Veenhuis *et al.*, 1983*a*). Exposure of methanol-grown cells of the yeast *Hansenula polymorpha* to an excess of glucose results in rapid inactivation of both these enzymes, a process that is paralleled by disappearance of alcohol oxidase and catalase protein (Bruinenberg *et al.*, 1982). Electron microscopical observations showed that this is accompanied by degradation of peroxisomes by means of an autophagic process in which the vacuole plays an important role by providing the enzymes required for proteolysis (Veenhuis *et al.*, 1983*b*). A rapid inactivation of alcohol oxidase is also observed after the addition of an excess of methanol to a methanol-limited chemostat culture of *H. polymorpha*, but this process is not associated with peroxisome degradation. Under these conditions an excess of methanol is potentially harmful to the cells because the alcohol oxidase activity present is very high compared with their capacity to oxidize methanol completely to carbon dioxide. Continued activity of alcohol oxidase will then result in accumulation and excretion of the toxic compounds formaldehyde and formate. An analysis of this phenomenon suggested that inactivation of alcohol oxidase observed under these conditions was due to the combined effect of a partial release of the prosthetic group (FAD) and a conformational change of the enzyme protein by excess formaldehyde (Bruinenberg *et al.*, 1982).

ADAPTATION TO CONDITIONS PREVAILING IN NATURE

In order to appreciate how environmental regulation of microbial metabolism may function in nature, we must have at least some basic information on the microbe's natural environment. This has been considered recently by Wimpenny, Lovitt & Coombs (1983) and Konings & Veldkamp (1983) and their analysis will not be reiterated here except for one or two major conclusions. Natural ecosystems are often heterogeneous and microbes thriving in such environments are thus exposed to physical and chemical gradients. This means that cells of one species living in close proximity to one another may express different phenotypic responses in order to accommodate to their specific microenvironment (Wimpenny, 1981; Wimpenny et al., 1983). In addition, the gradients may not be stable in time so that the microbes often find themselves in a condition of transient state. A further common characteristic of many natural ecosystems is that the available concentration of essential nutrients is very low (see Konings & Veldkamp, 1983), despite the fact that chemical analysis may indicate otherwise. This applies in particular to the carbon and energy sources, since these are the growth-limiting substrates most commonly encountered by chemo-organo-trophs, but also to the nitrogen and phosphorus sources required for growth.

Although Wimpenny et al. (1983) have argued that 'hetero-geneous models are needed to study [the] heterogeneous pheno-mena' that are so common in nature, most of our knowledge on the various ways in which microbes respond to ecologically significant conditions has come from studies of pure and mixed cultures grown in homogeneous liquid cultures, particularly in chemostats. In fact it is our firm belief that a better understanding of microbial behaviour in natural ecosystems must come from studies where the two approaches go hand in hand. Since our objective is to illustrate how microbial metabolism is influenced by the environment, we shall restrict ourselves mainly to strategies that have emerged from studies of pure cultures of organisms grown in homogeneous systems. This more classical approach to studies of microbial ecophysiology has led to at least some understanding of the various ways in which microbes respond to conditions that they may encounter in nature.

Nutrient limitation

Among the factors that may constrain microbial growth, such as temperature, light (in the case of phototrophs), pH, etc., it is the availability of essential nutrients (in particular carbon, nitrogen and phosphorus) that is probably the most common of environmental extremes (Tempest & Neijssel, 1978; Konings & Veldkamp, 1983). If it is accepted that nutrient-limited growth conditions have been and still are important in many natural environments, then it must be expected that these will have exerted a strong selective pressure in the course of evolution for organisms to evolve mechanisms to accommodate such restrictive conditions. The principal objective of these mechanisms must clearly be to enable an organism to grow as fast as possible at a given low environmental concentration of the limiting nutrient (Pardee, 1961; Veldkamp & Jannasch, 1972). This could be brought about if the organism were able to take up and metabolize that nutrient at the highest possible rate under conditions where its concentration outside the cell was very low and to produce cell material with a high yield factor (with respect to the limiting nutrient). Conceivably these mechanisms may include the ability to increase the rate of transport of a nutrient when its concentration becomes growth-limiting, the ability to increase the rate of initial metabolism of the nutrient that has accumulated inside the cell when its intracellular concentration is low, and the ability to rearrange the chemical composition of cellular structures by re-directing fluxes of metabolites containing the limiting chemical element (carbon, nitrogen or phosphorus). In the following we shall briefly consider these three mechanisms; for a more detailed discussion see Harder & Dijkhuizen (1983).

At the molecular level, the ability to increase the rate of transport of a growth-limiting nutrient may be brought about by synthesizing more of the existing carrier system (i.e. increase its V_{max}: see Tempest & Neijssel, 1978), by synthesizing a different 'high-affinity' uptake system, or by changing the kinetic properties of an existing uptake system. The last may involve changes in the binding affinity of the substrate, changes in the stoichiometry of the transport process or modulation of carrier activity. Unfortunately studies of solute transport systems are still mainly at a stage of revealing general principles and little detailed information is available with respect to the relative importance of the responses listed above. Nevertheless, evidence for the occurrence of all three mechanisms can be found in the literature and a few examples may serve to illustrate this.

Transport of most carbon, nitrogen and phosphorus substrates is facilitated by specific carrier proteins and driven by the proton-motive force or one of its components (Konings & Michels, 1980). In the case of sugar transport in a number of facultative and obligate anaerobes, a group translocation system (phosphoenolpyruvate-dependent sugar phosphotransferase system or PTS) involving several proteins has been demonstrated which chemically modifies the sugar during transport (Postma & Roseman, 1976). The activity of glucose-PTS in glucose-limited *Klebsiella aerogenes* cells was found to increase with decreasing dilution rates, indicating that this organism responded to decreasing concentrations of the limiting carbon source by increasing the V_{max} of the glucose uptake system (O'Brien, Neijssel & Tempest, 1980). Similarly, when a marine strain of *Rhodotorula rubra* was grown under phosphate limitation (Robertson & Button, 1979), the organism responded by synthesizing more of the already-existing carrier system by elaborating a phosphate transport system with an accumulation capacity between 12 and 35 times that required for growth under conditions of phosphate sufficiency.

Preferential synthesis of a high-affinity uptake system under conditions of carbon-substrate limitation has been reported in *Pseudomonas aeruginosa* (Dawes, Midgley & Whiting, 1976). During growth under glucose-sufficient conditions, the substrate was mainly metabolized via periplasmic glucose dehydrogenase (K_m for glucose approximately 10^{-13} M) and gluconate and 2-ketogluconate uptake systems. Under conditions of glucose-limitation, however, a high-affinity glucose uptake system (K_m for glucose 8×10^{-6} M) was elaborated, while the activities of enzymes and uptake mechanisms of the low-affinity system were very much reduced. Recently (Robillard & Konings, 1981; Konings & Robillard, 1982) evidence was provided for regulation of the activity of the *E. coli* membrane-bound protein (E_{II}) of the PTS and the carrier proteins of lactose and proline by the redox state of these carriers. It was found that the redox state of these proteins is determined by the redox potential of the cytoplasm and by the protonmotive force. The reduced form of the carriers had a high affinity for their substrates whereas the K_m increased when they became oxidized, and this could provide a mechanism by which the activity of substrate uptake systems is controlled. It remains to be established, however, whether such a mechanism has any significance under conditions of nutrient limitation.

The ability to increase the rate of initial metabolism of the limiting nutrient accumulated into the cell when its intracellular concentration is low, is a response frequently seen under conditions of carbon and nitrogen limitation. The significance of this response may be appreciated from the following considerations. The driving force for accumulation of most solutes is composed of components of the protonmotive force and of the solute gradient. When this driving force is maintained at a certain value by primary transport systems, the level of accumulation of a solute can be predicted from the translocation mechanism (Konings & Michels, 1980). For instance, for a neutral substrate that is accumulated in symport with one proton, the steady-state level of accumulation at a certain value of the protonmotive force may be 10^3. Thus, if the extracellular concentration of this substrate is in the nanomolar range, its concentration inside will not exceed micromolar levels. If the cell were able to metabolize nutrients rapidly that had accumulated to concentrations of this order, it would be at a competitive advantage. In order to acquire this potential, organisms may either synthesize more of the existing enzyme(s) (i.e. increase the V_{max}) involved in the initial metabolism of the substrate, or synthesize a different high-affinity enzyme system. Both these responses have been encountered, not only with respect to constraints imposed upon the cells by carbon limitation but also under conditions of nitrogen limitation. In fact, a wealth of information is available in the literature regarding the synthesis of microbial catabolic enzymes in response to decreasing concentrations of a carbon substrate in their environment (see Dean, 1972; Matin, 1979; Harder & Dijkhuizen, 1982, 1983).

An example may serve to illustrate the point. *Hansenula polymorpha* is able to utilize methanol as a sole source of carbon and energy. The substrate is oxidized by alcohol oxidase which catalyses the oxygen-dependent formation of formaldehyde and hydrogen peroxide. This enzyme is exclusively present in peroxisomes (Veenhuis *et al.*, 1976) of which the organism contains a considerable number under conditions of methanol limitation at low growth rates (Van Dijken, Otto & Harder, 1976). Organisms harvested from methanol-limited cultures grown at a dilution rate of $0.03\,h^{-1}$ possessed a twelve-fold higher capacity to oxidize excess methanol than cells grown at a dilution rate of $0.16\,h^{-1}$, this difference being caused by a considerable derepression of enzyme synthesis at low dilution rates. The basis for this vast oxidative overcapacity at low growth rates (i.e. low methanol concentrations) was thought to reside in the poor affinity of alcohol oxidase for methanol ($K_m = 1.3 \times 10^{-3}\,\text{M}$), which under these conditions apparently required that the cells devote approximately 20% of their cellular protein to this enzyme. When grown at low dilution rates the cells thus contain large numbers of peroxisomes, whose matrix exhibits a highly regular structure due to the presence of crystalline alcohol oxidase protein (Veenhuis *et al.*, 1981).

A similar increase in activity of an enzyme involved in ammonia assimilation in yeasts has been observed in response to ammonia limitation. Under these conditions most yeasts (see C. M. Brown, 1976) synthesize very high levels of NADPH-specific glutamate dehydrogenase, an enzyme that has only a low affinity for ammonia ($K_m = 1.5 \times 10^{-3}$ M). Synthesis of a different high-affinity enzyme system involved in the initial intracellular metabolism of a growth-limiting substrate has also been reported. This applies, for instance, to the metabolism of glycerol in *K. aerogenes* (Neijssel *et al.*, 1975) and to enzymes involved in the assimilation of ammonia in a variety of bacteria (Brown, MacDonald-Brown & Meers, 1974).

The ability to rearrange the chemical composition of the cell material by redirecting fluxes of metabolites containing the limiting element is a response that is particularly apparent in microbes exposed to conditions of nitrogen or phosphorus limitation. Nitrogen-limiting conditions generally lead to a significant accumulation of intracellular reserve materials (Dawes & Senior, 1973), whereas under conditions of phosphate limitation major changes in the chemical composition of the cell envelope of bacteria have been reported. These changes may involve both replacement of the phosphorus-containing glycerol teichoic acids by teichuronic acids in the walls of gram-positive bacteria placed under phosphorus limitation (Ellwood & Tempest, 1972), and the replacement of phospholipids by ornithine-containing lipids and acidic glycolipids as observed in a marine strain of *Pseudomonas fluorescens* (Minnikin & Abdolrahimzadeh, 1974).

Mixed substrate utilization

When micro-organisms are presented with more than one utilizable carbon source under the substrate-sufficient conditions of a batch culture, they usually metabolize the substrate that supports the highest growth rate, while enzymes for the metabolism of the other substrates remain repressed. This distinct preference for the utilization of only one compound to the exclusion of all others would be of little value in a nutritionally poor environment, where selective pressure would favour those organisms capable of using the differ-

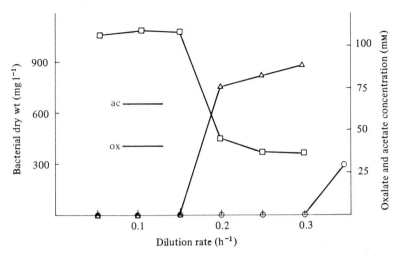

Fig. 2. Growth of *Pseudomonas oxalaticus* OX1 on a mixture of acetate ($S_R = 30$ mM) and oxalate ($S_R = 100$ mM) in a continuous culture with limited carbon and energy source. □, bacterial dry weight on the mixture; △, residual oxalate concentration; ○, residual acetate concentration. The bacterial dry weight values observed during growth of the organism on 30 mM acetate (ac) or 100 mM oxalate (ox), separately, are also given. (Reproduced from Harder & Dijkhuizen (1982), with permission.)

ent substrates simultaneously. Evidence is now accumulating (see Harder & Dijkhuizen, 1982) that many microbes do, in fact, simultaneously utilize a multiplicity of nutrients that serve a similar physiological function, provided the concentration of these nutrients is low.

A few examples may serve to illustrate the general response observed. As mentioned above, when *Pseudomonas oxalaticus* is grown under substrate-sufficient conditions in batch culture on a mixture of acetate plus oxalate, diauxic growth occurs with acetate being used first (see Fig. 1). Growth on this mixture in a continuous culture limited by carbon and energy source resulted, however, in simultaneous and complete utilization of the two organic acids, provided the dilution rate was kept below $0.15 \, h^{-1}$ (Fig. 2). At dilution rates above $0.15 \, h^{-1}$ an increasing proportion of the oxalate supplied to the culture remained unused, and this was paralleled by a significant drop in specific activity of oxalyl-CoA reductase, a key enzyme of oxalate assimilation. In contrast, acetate was completely consumed up to a dilution rate of $0.30 \, h^{-1}$, which is slightly below the maximum growth rate (μ_{max}) of the organism in media containing acetate alone. Thus, in a continuous culture limited by oxalate plus acetate, the organism is able to use the two carbon sources simultaneously provided their concentrations are low. This is the case at low growth rates. The ability of this organism and many other microbes to use a mixture of carbon substrates simultaneously during carbon-limited growth must be considered a consequence of the general tendency towards diminished catabolite repression of many catabolic enzymes under these conditions (Dean, 1972; Matin, 1979).

Another example relates to mixed electron acceptor utilization which was observed during studies with *Hyphomicrobium* X (Meiberg, Bruinenberg & Harder, 1980). This organism is able to grow in mineral media containing dimethylamine, both anaerobically with nitrate and aerobically. When grown in a dimethylamine-limited chemostat in the presence of nitrate at a dilution rate of $0.10\,h^{-1}$, synthesis of nitrate and nitrite reductases and nitrate consumption was observed at dissolved oxygen tensions (d.o.t.) below 15 mmHg (*c.* 2000 Pa). When the d.o.t. was reduced below this value, the activities of the two enzymes increased linearly with decreasing d.o.t. Meiberg *et al.* (1980) subsequently showed that the onset of denitrification in the presence of oxygen was dependent on the growth rate of the organism. Particularly at low growth rates $(0.01\,h^{-1})$, synthesis of the two enzymes (and utilization of nitrate) was already significant at relatively high d.o.t. values (up to *c.* 6600 Pa), whereas at relatively high growth rates $(>0.15\,h^{-1})$ activities of nitrate and nitrite reductases only appeared at d.o.t. values which were almost an order of magnitude lower. These enzymes were not detected when the organism was grown at the μ_{max} $(0.18\,h^{-1})$ unless the culture was brought to virtually anaerobic conditions. These results show that *Hyphomicrobium* X is able to carry out denitrification under 'partly aerobic' conditions, provided the growth rate is kept low. Evidence for 'aerobic' denitrification by the facultatively chemolithotrophic sulphur bacterium *Thiosphaera pantothropha* has also been presented (Robertson & Kuenen, 1983) and this, together with earlier reports, suggests that such a process may occur in nature and could thus account for the recent discovery of nitrate reductase activity in fully aerated soils (Tiedje *et al.*, 1982).

It is not generally known in which way and to what extent the utilization of a certain substrate influences the metabolism of another compound when they are used simultaneously. Under certain conditions the metabolism of each substrate may proceed completely independently as if the other substrate(s) were absent, but there is also evidence to indicate that the metabolism of a certain compound can be significantly influenced by the presence of another, even when both are utilized to completion (see Harder & Dijkhuizen, 1982). Although more information on these latter processes is urgently needed, particularly in view of the potential practical applications, the information currently available has disclosed the possible significance of mixed substrate utilization in natural ecosystems. In fact, evidence is now available to suggest that the ability of organisms to utilize various substrates simultaneously may confer a competitive advantage upon them (Smith & Kelly, 1979; Gottschal & Kuenen, 1980). This points to the possibility that there are in nature organisms that are specially adapted to life under mixed nutrient limitation and indicates that they could be of major importance in the cycling of nutrients in environments in which the turnover rates of substrates are comparable and low. However, the selection methods widely employed so far to isolate microbes from nature select *against* organisms of this specially adapted kind. We have therefore advocated (Harder & Dijkhuizen, 1982) the use of dual-nutrient-limited (or even multi-nutrient-limited) selection procedures to learn more about the possible significance of such organisms in nutrient cycling in nature.

Regulation of extracellular enzyme synthesis

Regulation of the synthesis and excretion of extracellular hydrolases by micro-organisms has intrigued microbiologists for many years, particularly with respect to the way in which an organism obtains signals that a polymer is present in its (micro)environment that could yield one or more useful carbon and energy sources for growth if hydrolysed by appropriate enzymes. In a number of micro-organisms able to break down polymeric compounds the synthesis of exoenzymes appears to be controlled as follows. The organism produces small quantities of hydrolases, particularly under conditions of growth limitation by the carbon and energy source. These hydrolases are excreted into the environment and act as 'scouting molecules'. If a substrate for such enzymes is available in the (micro)environment of the organism it will slowly be hydrolysed and the products of hydrolysis may diffuse back to the cell, thus signalling the presence of a utilizable polymeric substrate. These products of hydrolysis may thereby act as inducers for the synthesis of more enzyme(s). Synthesis of too much enzyme or production under conditions where it is not needed is prevented by (catabolite) repression, the enforcement of which may be linked to the energy status of the cell.

Although this scheme is a highly simplified account of the regulation of microbial extracellular enzyme production, there are many examples to be found in the literature that suggest it to be one of the major mechanisms in both gram-positive and gram-negative wild-type organisms (see Glenn, 1976; Burns, 1983). Two examples may serve to illustrate this in more detail. *Vibrio* SA1 is able to produce two extracellular proteolytic enzymes, namely an endopeptidase and an aminopeptidase (Wiersma *et al.*, 1978*a*), and it was shown that the concerted action of these two enzymes allows hydrolysis of protein molecules into fragments that the cell can take up and use as a carbon and energy source for growth. When the organism was grown in a lactate-limited chemostat in a medium containing 2 mM phenylalanine as an inducer (Wiersma & Harder, 1978), the rate of production of both enzymes was dependent on the dilution rate, while the shape of the production curves was compatible with that expected for enzymes whose synthesis is amplified by induction and decreased by catabolite repression (Clarke, Houldsworth & Lilly, 1968; Dean, 1972). Higher concentrations of amino acids in the environment also led to repression of enzyme synthesis. Repression of the synthesis of both exoenzymes could be overcome during iron-limited or oxygen-limited growth (Wiersma *et al.*, 1978*b*), which led us to postulate that this repression was controlled via the energy status of the cell. A similar conclusion was reached in studies on exoprotease production by *Pseudomonas aeruginosa* (Whooley, O'Callaghan & McLoughlin, 1983). In a subsequent paper Whooley & McLoughlin (1983) reported a negative correlation between the magnitude of the protonmotive force and the amount of exoprotease produced, in that a decrease in protonmotive force resulted in an increase in the rate of exoprotease production. These mechanisms allow, as Whooley *et al.* (1983) observed, a balance between the level of utilizable substrate, growth rate and amount of enzymes produced when growth is dependent on extracellular protein substrates.

The other example relates to the regulation of enzymes involved in the utilization of the algal polymer agarose by *Cytophaga flevensis* (van der Meulen & Harder, 1975, 1976*a, b*). The initial attack on this polymer is by an agarase. This enzyme has a low molecular weight (26 000 daltons), is partly truly exocellular and partly cell-bound, and hydrolyses agarose into a mixture of neoagaro-oligosaccharides. The most effective inducer for the enzyme was neoagarotetraose, although it was also induced by other components of agar; enzyme synthesis was further controlled by catabolite repression. The smallest oligosaccharide that this agarase can hydrolyse is neoagarohexaose (into neoagarotetraose and neoagarobiose), and therefore additional enzymes are required to hydrolyse the polymer into its constituent monomers. It was found that neoagarotetraose is hydrolysed by cleavage of the central α-galactosidic linkage, and the results obtained suggested that the site of hydrolysis of this tetramer was intracellular, and not periplasmic as in *Pseudomonas atlantica* (Young *et al.*, 1971). Neoagarobiose, produced either by action of the tetra-ase or agarase, is also hydrolysed intracellularly by cleavage of its β-galactosidic linkage into D-galactose and 3,6-anhydro-L-galactose. Both intracellular enzymes involved in the total hydrolysis of the agarose polymer were shown to be inducible by neoagaro-oligosaccharides (surprisingly neoagarooctaose was the best inducer for both the tetrasaccharidase and the disaccharidase), while in addition they were subject to catabolite repression by higher concentrations of various carbon sources, including neoagaro-oligosaccharides. Experiments conducted with resting cells of this organism suggested that these cells were able to form low levels of agarase constitutively under non-induced conditions. This may trigger the induction of all three enzymes required for the complete hydrolysis of agar when this polymer is present in the environment in which *C. flevensis* thrives.

In contrast to the situation in *Vibrio* SA1, part of the agarase of *C. flevensis* remained cell-bound. Such an association of hydrolases with the cell wall is a well-known characteristic of cytophagas and is generally assumed to be advantageous in the competition for products liberated by these enzymes from large molecules (Stanier, 1942).

Responses to changes in temperature, pH and ionic strength

Temperature
Growth of micro-organisms at barometric pressure is possible over the relatively large temperature range of -10 to $99\,°C$. However, for any bacterial species the temperature range that supports growth is much smaller and for most organisms is limited to a range of 35 to

40 °C. This has led to the recognition of various groups of organisms covering the whole of the biokinetic zone, namely the psychrophiles, mesophiles and thermophiles. Within each group, temperature has a profound effect on the physiology of an individual organism as may be exemplified by a study of its effect on the behaviour of an obligately marine psychrophilic *Pseudomonas* species (Harder & Veldkamp, 1967; see for a general discussion Morita, 1975). During growth of this organism in batch culture it exhibited a normal temperature profile with an optimum at 14 °C and a maximum between 19 and 20 °C, while the growth rate at −2 °C was still approximately 30% of that at the optimum growth temperature. The temperature characteristic as deduced from an Arrhenius plot was a typical $45.6\,\text{kJ}\,\text{mol}^{-1}$ ($10.9\,\text{kcal}\,\text{mol}^{-1}$). When *Pseudomonas* sp. was grown at the optimum growth temperature in a lactate-limited chemostat at different dilution rates, a linear relationship between cellular RNA content and dilution rate was found as has been reported for a variety of other organisms (see Harder & Dijkhuizen, 1982).

The question of how the cells adapted to lower and higher temperatures when grown at a constant submaximal rate was investigated as follows. A lactate-limited culture of the organism was grown at a constant dilution rate of $0.05\,\text{h}^{-1}$ at 14 °C and the effect of a temperature decrease or increase on the steady-state properties of the organism was studied (Fig. 3). The difference between the actual growth rate, μ (= dilution rate, $D = 0.05\,\text{h}^{-1}$ at steady state), and μ_{max} at each temperature was always negative and did not exceed $-0.04\,\text{h}^{-1}$. When the temperature of a culture growing at 14 °C ($D = 0.05\,\text{h}^{-1}$) was decreased, a marked increase in the cellular RNA content and of the respiratory potential of the cells at 14 °C (Q_{O_2} (14 °C)) was observed. Similar observations have been reported for the mesophilic bacterium *Klebsiella aerogenes* (Tempest & Hunter, 1965). This decrease in temperature below 14 °C naturally disturbed the steady state that was established at the optimum temperature. As a result μ became smaller than D and if this were a permanent change the culture would be washed out. However, since care was taken that at each lower temperature D was less than μ_{max}, the cells were able to offset the temperature-induced decrease in metabolic reaction rates by synthesizing RNA and enzymes involved in energy generation in greater quantity. Most RNA in the cells is associated with ribosomes (see below), so that this response was interpreted as an increase in ribosome

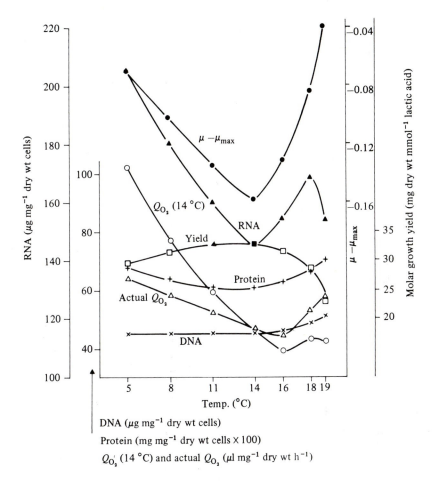

Fig. 3. Effect of temperature on some steady-state properties of lactate-limited *Pseudomonas* cells grown at a dilution rate of $0.05\,h^{-1}$. Q_{O_2} ($14\,°C$), oxygen uptake of cell suspensions at $14\,°C$; actual Q_{O_2}, oxygen uptake of cell suspensions at the growth temperature. (Reproduced from Harder & Veldkamp (1967), with permission.)

content of the cells. This was considered to result from the need to sustain the same rate of protein synthesis at the lower temperature as at $14\,°C$, which was required because the dilution rate was kept constant. The concomitant slight decrease in molar growth yield indicated that the energy needed per unit weight per unit time to sustain the growth rate increased with increasing deviation of the growth temperature from the optimum.

Increasing the temperature above the optimum also resulted in an increase in RNA content of the cells (Fig. 3). This obviously was a

Table 1. *RNA content and ratio of ribosomal RNA to total RNA of a Pseudomonas sp. grown at a constant dilution rate (D) at different temperatures and at a constant temperature at different dilution rates in a lactose-limited chemostat*

$D = 0.05\,h^{-1}$; temp. = 5–19 °C			Temp. = 14 °C; $D = 0.025$–$0.180\,h^{-1}$		
Temp. (°C)	Total RNA (μg mg^{-1} dry wt)	Ribosomal RNA (% of total RNA)	$D(h^{-1})$	Total RNA (μg mg^{-1} dry wt)	Ribosomal RNA (% of total RNA)
5	205	78	0.025	136	75
8	180	78	0.050	145	78
11	160	78	0.100	166	78
14	145	78	0.130	181	79
16	153	79	0.150	188	79
18	168	81	0.180	202	79
19	154	81			

From Harder & Veldkamp (1969).

response to a phenomenon entirely different from that observed on the lower side of the temperature optimum. The RNA increase at temperatures above 14 °C was considered to reflect an effort by the cells to compensate an impairment in the protein-synthesizing machinery at superoptimal temperatures. Concomitantly a decrease in molar growth yield and an increase in maximal rate of oxygen uptake at the growth temperature (actual Q_{O_2}) was observed. As soon as thermal damage (at temperatures above 18 °C) prevented a further increase in cellular RNA content, the growth rate decreased dramatically upon a small increase in temperature and the culture was washed out at temperatures between 19 and 20 °C. This behaviour could not be explained by denaturation of respiratory enzymes, since the optimum temperature for respiration was 23 °C (Harder & Veldkamp, 1968).

The above phenomena were further studied by analysing the nature of the RNA present in the cell at the various temperatures (Harder & Veldkamp, 1969). It appeared that the ratio of ribosomal RNA over total RNA was virtually constant and independent of growth rate or temperature at which this RNA was synthesized (Table 1). The observed increase in RNA in cells grown at the submaximal rate in the temperature range 14–18 °C therefore indicated a decreased rate of protein synthesis per ribosome. Direct evidence for this decreased efficiency was obtained in subsequent experiments (Harder & Veldkamp, 1969) which demonstrated that the protein-synthesizing machinery of the cells grown at 18 °C was impaired. Sedimentation analysis of ribosomes obtained from cells grown at 14 or 18 °C did not reveal any differences, so that it is possible that one or more of the aminoacyl-tRNA synthetases in this organism are subject to thermal inactivation at superoptimal temperatures, as has been reported for *Micrococcus cryophilus* (Malcolm, 1968). The above results could also be accounted for by assuming that a high proportion of the cells were not viable at temperatures above the optimum. However, in the above experiments it appeared that more than 95% of the cells were viable at temperatures between 5 and 19 °C.

The response of psychrophilic *Pseudomonas* cells to changes in the temperature of their environment described above probably also holds for a large variety of other organisms, although special effects such as the development of a requirement for growth factors or divalent cations at superoptimal temperatures may play an additional role. If the behaviour of the *Pseudomonas* sp. is generalized to encompass other organisms also, it can be concluded that microbes possess mechanisms that enable them (within limits) to offset potentially negative effects of temperature, provided their growth rate remains suboptimal over the temperature range in which they thrive. This is probably the case in many natural environments (Veldkamp & Jannasch, 1972). At the same time the generalized picture developed above enables a tentative explanation of the μ_{max}–temperature profiles commonly encountered among microorganisms. Assuming that the ribosome content of a bacterial cell cannot exceed a certain fixed maximum (Tempest & Hunter, 1965), this maximum may be expected to occur in cells growing at the maximal rate in any particular medium at the optimum temperature. When the temperature is decreased below the optimum, the cells cannot compensate for the temperature-induced slow-down of the rate of protein synthesis by an increased concentration of ribosomes (as is the case in cells growing at submaximal rate) and thus the maximum specific growth rate must decrease. Similarly, any inefficiency of protein synthesis at temperatures above the optimum cannot be offset by an increase in ribosome concentration, which results in a progressive decrease of the maximal rate of growth as the maximum growth temperature is approached.

Hydrogen ion concentration

Bacteria have been found in environments with pH values from approximately 1 to up to 11 but, as is the case with temperature, no single species is known to be able to grow over this whole pH range. The internal pH of acidophiles is poised between 6 and 7 and that of alkalophiles between 9 and 10, while neutrophiles maintain the internal pH around 7.5 (see Padan, Zilberstein & Schuldiner, 1981), so the question arises as to how micro-organisms maintain their internal pH around neutrality despite wide differences in the pH of their environment. This has been studied extensively and there is now convincing evidence to suggest that most bacteria, including alkalophiles and acidophiles, possess various mechanisms that are involved in the control of cytoplasmic pH. These mechanisms have been reviewed recently (Padan *et al.*, 1981) and need not be reiterated here, except for the basic mechanisms involved. These include: (i) primary proton pumps that cause extrusion of protons from the cytoplasm and thus act to increase the internal pH; (ii) sodium/hydrogen antiporter systems that extrude sodium ions and return protons inside the cell; and (iii) potassium/hydrogen antiporters that extrude potassium ions in exchange for protons. These mechanisms are also involved in preventing major internal pH fluctuations when the external pH changes drastically, for instance as a result of biological activity. It is, however, far from clear in which way the intracellular pH relative to that of the external environment is monitored by the cells and how this information is used to trigger any of the above mechanisms into action. It is equally obscure what the contribution of each of the three mechanisms is to the regulation of the internal pH in different bacteria.

Ionic strength

As with temperature and pH, different micro-organisms are able to grow in environments of widely different osmolarity. In bacteria this ability has apparently been acquired through adjustments of the internal osmolarity so that it always exceeds that of the environment. Microbes may not need to regulate their internal osmolarity with great precision, because they are enclosed by a cell wall that is capable of withstanding a considerable osmotic pressure. However, maintenance of a relatively high and constant ionic strength within the cell is probably of critical importance, because the stability and behaviour of enzymes and other macromolecules are strongly dependent on this factor. There is ample evidence that an organism's response to an increased osmolarity in its environment depends on whether this increase is brought about by higher salt levels or higher non-electrolyte levels (see Reid, 1980). In the following we shall consider only the former of these responses, since most natural environments of high osmolarity contain high concentrations of salts (in particular sodium chloride). Exposure to increased salt concentrations probably occurs frequently in nature: for instance when micro-organisms are transported down a freshwater river into the sea.

For most freshwater or soil organisms there appears to be no requirement for sodium chloride for growth. Nevertheless, the presence of increasing concentrations of sodium chloride in their environment does have profound effects, both on the growth rate and on the physiology of these organisms. Consider an organism that is suddenly transferred from an environment without salt to one with a high salt concentration. The initial response is likely to be a loss of water from the cytoplasm to the environment. This may continue until the original difference in osmolarity between the cytoplasm and the environment is restored. Alternatively the initial response may be followed by either of two longer-term responses in which the organism's metabolism is involved. The organism may either transfer external solutes (in particular cations) across the cell membrane into the cytoplasm in order to restore the initial difference in osmolarity, or the cell may synthesize internal (organic) solutes which have essentially the same effect. These osmoregulatory mechanisms may involve one, or several, of five main groups of compounds, namely inorganic ions, amino acids, disaccharides, polyols and betaines (Stewart, 1983). Inorganic ions, in particular potassium, appear to be important osmotica in marine and halophilic bacteria (Larsen, 1967; Dundas, 1977), while in freshwater microbes potassium and organic compounds are of major significance (Tempest, Meers & Brown, 1970; Stanley & Brown, 1974; Measures, 1975).

A study of the adaptive response of *Klebsiella aerogenes* cells growing in continuous culture to increasing sodium chloride concentration was made by Tempest *et al.* (1970). A three-fold increase in cellular potassium content was observed following an increase in sodium chloride concentration in the reservoir (from 0 to 4% w/v), along with a marked increase in the intracellular glutamate concentration. In subsequent experiments evidence was obtained that the glutamate pool increased rapidly (2–5 min) following addition of 2% (w/v) sodium chloride to a steady-state magnesium-limited culture of this organism and that the glutamate arose by *de novo* synthesis.

In an attempt to unravel the mechanism responsible for this rapid accumulation of glutamic acid in *K. aerogenes*, Tempest *et al.* (1970) noted that addition of sodium chloride to washed suspensions of organisms caused an immediate efflux of protons. This efflux, which is in accordance with a later mechanistic interpretation by Konings & Veldkamp (1980), conceivably could have caused activation of cytoplasmic glutamate dehydrogenase, an enzyme that has a sharp pH optimum around 8, which in turn may have led to an increased rate of glutamic acid synthesis. Since a continued presence of sodium chloride in the organism's environment probably required a low but steady-state proton efflux due to the operation of a sodium/hydrogen antiport system, the internal hydrogen ion concentration eventually returned to a new equilibrium value in which the rate of glutamate synthesis was decreased to a new steady-state value to ensure the corresponding level of glutamate in the pool.

In prokaryotic phototrophs, organic compounds appear to serve as major osmotica. In *Ectothiorhodospira halochloris* betaine accumulates (Galinski & Trüper, 1982) while in marine cyanobacteria glucopyranosylglycerol is a major osmoticum. Eukaryotic microalgae also regulate their osmolarity via the accumulation of organic compounds. Among the compounds implicated are sugars such as sucrose, and the polyols glycerol, mannitol and glycosylglycerol (see Stewart, 1983). This latter group of compounds also plays a role in osmoregulation in yeasts. The nature of the biochemical sensors that monitor the osmolarity of the environment relative to that of the cytoplasm is quite unresolved and in most cases the mechanism(s) that control the synthesis of organic osmotica in response to the ionic strength of the environment are not understood at all. This is clearly an area where results from further studies are eagerly awaited.

Organic osmotica, in addition to their role in osmoregulation, are probably also important as compatible solutes in the cell. At high intracellular concentrations they have hardly any inhibitory effect on enzymatic reactions, yet protect enzymes from inhibition by high concentrations of inorganic ions that may (temporarily) accumulate in the cytoplasm (A. D. Brown, 1976).

STRATEGIES FOR SURVIVAL UNDER STARVATION CONDITIONS

So far we have only considered the response of micro-organisms to environmental constraints under conditions in which they were assumed to be able to grow continuously. It must, however, be expected that in nature periods of (relatively rapid) growth are followed by periods during which the rate of supply of one or more essential nutrient(s) is virtually zero and hence many microbes must have evolved strategies to enable them to cope with prolonged periods of starvation. Under these conditions the cellular integrity must be maintained and those organisms that are not specially equipped for survival must be able to maintain a certain (albeit low) energy flux. This could be derived from the metabolism of endogenous reserves or cellular polymers (Dawes, 1976; Strange, 1976), and the nature and extent of this endogenous metabolism largely determines the survival characteristics of organisms. The general picture appears to be that energy-reserve polymers only enhance survival if they are degraded comparatively slowly (Dawes, 1976). Obviously further environmental constraints such as water activity, temperature, radiation, etc., are also of importance.

Many studies on microbial starvation have appeared, in particular on the relationship between endogenous metabolism and survival (for reviews see Dawes & Ribbons, 1964; Dawes, 1976; Strange, 1976). These studies have shown that there is a marked difference in survival characteristics between different organisms (see for instance Mink, Patterson & Hespell, 1982). Some, like *Staphylococcus epidermidis*, rapidly lose viability: when washed suspensions of this organism were incubated under anaerobic conditions more than 90% of the cells became non-viable within 12 h (Horan, Midgley & Dawes, 1981). Other organisms survive much longer periods of starvation. For instance the half-life of *Arthrobacter crystallopoites* was estimated to be approximately 100 days (Boylen & Ensign, 1970). Some organisms possess special survival mechanisms. This applies, for instance, to the endospore-formers (Hanson, 1979), but equally to organisms that are able to form exospores (Whittenbury & Dow, 1977), which often have survival properties akin to those of *Bacillus* endospores.

A special adaptation for survival occurs among organisms that undergo complex morphogenetic cell cycles. These organisms, including rhodomicrobia, hyphomicrobia and caulobacters, form so-called swarmer cells during asymmetric division. These differ from their parent cells in that they must first differentiate into a reproductive unit before they can themselves give rise to progeny and have therefore been called 'growth precursor' cells (Dow *et al.*, 1983). Most importantly the swarmer cells do not immediately embark upon differentiation after their formation, but do so only in response to an environmental stimulus. This may be light, as in the case of *Rhodomicrobium vannielii*, or an organic energy source, as in *Hyphomicrobium* swarmer cells. In a thought-provoking paper Dow *et al.* (1983) recently discussed the possible significance of the 'growth precursor' cell concept as a mechanism for survival of microbes in a potentially hostile environment. Their reasoning is mainly based on a number of specific properties of the swarmer cells and inspection of these properties reveals that these cells are, at least in principle, well-equipped for survival. Unfortunately data on long-term starvation experiments with swarmer cells are not yet available to substantiate the argument. Dow *et al.* (1983) also suggested that the 'growth precursor' concept as a survival mechanism might apply not only to those organisms that give rise to swarmer cells as an obligate and transient stage in their cell cycles (i.e. the budding and prosthecate bacteria including *Rhodopseudomonas palustris* and *Nitrobacter* species), but also to *Bdellovibrio* species, *Sphaerotilus* species, cyanobacteria and possibly even to more conventional forms of bacteria such as rods and cocci.

An alternative strategy for enhancing the chances of survival of a species would be to increase the number of individual cells in response to starvation conditions. This would be particularly successful if the newly formed cells were in addition resistant to starvation. This possibility was encountered by Novitsky & Morita (1978) in a study of the starvation survival of a marine psychrophilic *Vibrio* species. When log-phase cells of this organism were suspended in a starvation medium, the total biomass remained approximately constant, while the rod-shaped cells divided to form small cocci of 0.4 μm diameter; after 70 weeks approximately 15 times the original number of cells were still viable. Upon addition of small amounts of nutrients the cocci increased in size and changed shape to resemble the rod-shaped cells characteristic of normal laboratory cultures.

The ability of starved cells (or 'growth precursor' cells) to respond to environmental conditions that are favourable for growth, a property that may be referred to as metabolic reactivity, depends primarily on the ability of the cell to take up nutrients. Since the electrochemical proton gradient ($\Delta\bar{\mu}_{H^+}$) plays an important role in the uptake of many potential energy sources (Konings & Michels, 1980), it must be expected that the reactivity of starved cells is determined, at least in part, by their ability to sustain a certain measure of the membrane potential. Unfortunately, studies designed to unravel the relationship between $\Delta\bar{\mu}_{H^+}$ and reactivity of cells have so far been very few and those that have been reported do not allow an unequivocal interpretation of this relationship.

An informative study of the response of starved $E.$ $coli$ cells to a fresh supply of nutrients has been reported by Koch (1979). His results demonstrated that starved cells were able to respond quickly to a fresh supply of the energy source glucose. Even after 48 h of starvation the cells were capable of rapid biosynthesis which was shown in 'shift-up' experiments. Koch (1979) also obtained evidence for a distinct overcapacity for glucose transport in starved $E.$ $coli$ cells, indicating that this energy source can be taken up rapidly by such cells, provided the organism can still energize its translocation across the membrane. Previously Chapman, Fall & Atkinson (1971) had shown that during starvation of $E.$ $coli$ the energy charge fell rapidly from about 0.8 to 0.5–0.6, but was maintained at this level for 60–80 h. Since the total adenine nucleotide pool declined during starvation along with the phosphate potential, it must be expected that the electrochemical proton gradient had also decreased. Evidence that this may have been the case has been reported by Kashket (1981a). In view of this it is of interest that the intracellular amino acid pool in $E.$ $coli$ remained virtually unchanged upon starvation (Dawes & Ribbons, 1965). This could mean either that this organism is able to sustain a certain minimum value of the $\Delta\bar{\mu}_{H^+}$ for the duration of the starvation experiments or that the amino acid pool is not in equilibrium with $\Delta\bar{\mu}_{H^+}$. The latter conclusion was also reached in studies on the response of $Streptococcus$ $cremoris$ to starvation (R. Otto, unpublished).

The survival of lactic streptococci following starvation has been extensively investigated. Since the fate of $\Delta\bar{\mu}_{H^+}$, the intracellular ATP pool, the phosphate potential and pools of intermediary metabolites have been investigated in these studies, we shall briefly discuss the current views, particularly since we feel that these studies are exemplary of the type needed to obtain a more complete understanding of the various factors involved in starvation survival and metabolic reactivity of starved cells. Lactic streptococci do not synthesize reserve polymers and RNA appears to be the only one of the cellular polymers whose amount decreases significantly during starvation (Thomas & Batt, 1969a). This RNA degradation may not serve any significant role in providing the cell with metabolic energy because the products of its hydrolysis cannot be used by these organisms as energy sources. As is the case in other organisms provision of an external energy source such as arginine results in prolonged survival of *Streptococcus lactis*, particularly at low pH (Thomas & Batt, 1969b). Surprisingly, arginine fermentation in this organism does not lead to the generation of a pH gradient across the cytoplasmic membrane (Barker, 1977). This is unexpected since the metabolism of arginine does result in the production of ATP (Abdelal, 1979), which in principle could be used by the proton-translocating membrane-bound ATPase (Kashket, 1981b). Prolonged survival of *S. lactis* in the presence of arginine may therefore be due to an increase in the cellular ATP pool, which is possibly only slight.

A study of the various parameters that together determine the cellular energy status during growth and starvation of *S. cremoris* was recently made by Otto *et al.* (1983 and unpublished). The organism was grown in a lactose-limited chemostat which enabled an investigation of the early events following the onset of lactose-starvation imposed by stopping lactose supply to the culture. In the first hour after the onset of starvation, a rapid decay of the internal ATP pool was observed, while a concomitant decrease in the phosphate potential and the energy charge (from 0.85 to 0.15–0.20) occurred (Fig. 4). In contrast the internal AMP concentration increased, initially via ATP hydrolysis and myokinase activity and later because of RNA degradation, as was indicated by the total adenine nucleotide pool. During the first minutes after the lactose was exhausted the internal phosphoenolpyruvate (PEP) level increased, as was observed in *S. lactis* (Thompson, 1978). This increase in internal PEP level is most probably due to inhibition of pyruvate kinase resulting from a significant drop in the intracellular concentration of early glycolytic intermediates (in particular fructose-1,6-bisphosphate) which are known positive effectors of this enzyme (Thomas, 1976). The advantage of a response in which an increased internal PEP level is built up during early starvation of streptococci is clear. This compound is required by the organism for the transport of lactose and other carbohydrates, which is mediated by the PEP-dependent sugar phosphotransferase system, and an increased level of PEP would ensure rapid uptake of sugars when they again became available. The period during which the internal PEP level remained high was limited to only 10 min following the onset of starvation; after 30 min it had dropped to a level of <0.25 mM where it remained for many more hours. Despite this, *S. cremoris* cells rapidly respond to the presence of lactose, even

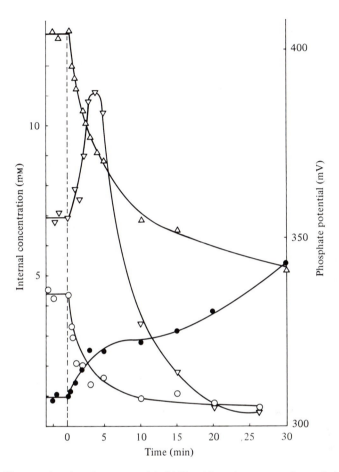

Fig. 4. Changes in phosphate potential ($\Delta G'_p$, $\triangle$) and concentrations of ATP ($\bigcirc$), phosphoenolpyruvate ($\triangledown$) and AMP ($\bullet$) in *Streptococcus cremoris* cells following a switch-off of the medium supply at zero time. The organism was grown in a lactose-limited chemostat at a dilution rate of $0.1\,h^{-1}$, at pH 6.8 and 30 °C, and a steady state was established before switching off the supply of fresh medium. (Unpublished data of R. Otto.)

after they have been starved for 24 h, by synthesizing ATP rapidly and restoring the energy charge (Fig. 5; R. Otto & J. Vije, unpublished; see also Konings & Veldkamp, 1983). Probably the pool of PEP remaining after 2 h is sufficient for the cells to take up at least some substrate molecules which in turn generate energy for the uptake of further molecules of the energy source.

In lactic streptococci ATP hydrolysis by the membrane-bound Ca^{2+}–Mg^{2+}-stimulated ATPase may lead to the formation of an electrochemical proton gradient ($\Delta\bar{\mu}_{H^+}$) across the membrane

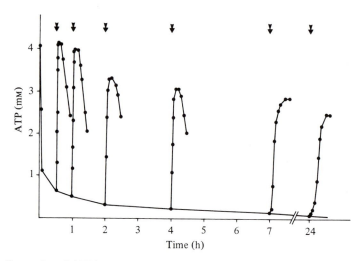

Fig. 5. Generation of ATP in energy-starved *Streptococcus cremoris* upon addition of lactose (indicated by the arrows; 0.9 mM per pulse) at different times after the start of starvation ($t = 0$). Culture conditions are as in Fig. 4, except that the dilution rate was $0.14\,h^{-1}$. (Unpublished data of R. Otto & J. Vije.)

(Kashket, 1981*b*). Any decay of the internal ATP pool (and the phosphate potential) following lactose starvation will therefore have an effect on the magnitude of $\Delta\bar{\mu}_{H^+}$. The fate of the membrane potential in *S. cremoris* cells following lactose starvation has recently been investigated (Otto *et al.*, 1983; R. Otto, unpublished). When the lactose supply to a lactose-limited chemostat culture of the organism growing at pH 7.6 was stopped, the membrane potential ($=\Delta\psi$ component of the $\Delta\bar{\mu}_{H^+}$) collapsed within 30 min (Fig. 6*a*). Since the transmembrane pH gradient (ΔpH) was zero under these conditions, these results show that at pH 7.6 the $\Delta\bar{\mu}_{H^+}$ is completely dissipated in approximately half an hour. When the cells were grown at pH 5.7, the $\Delta\psi$ also disappeared within 30 min; however, the ΔpH component of the protonmotive force, which was about 70 mV under these conditions, remained constant for at least an hour (Fig. 6*b*). The rapid decrease in membrane potential in the absence of lactose is probably due to the low electrical capacity ($10^{-2}\,F\,m^{-2}$) of the cytoplasmic membrane. In contrast, the internal buffering capacity of the cytoplasm is sufficiently high to ensure a constant ΔpH.

In many bacteria (Konings & Michels, 1980), including *S. cremoris* (Otto *et al.*, 1983), a protonmotive force is required to energize

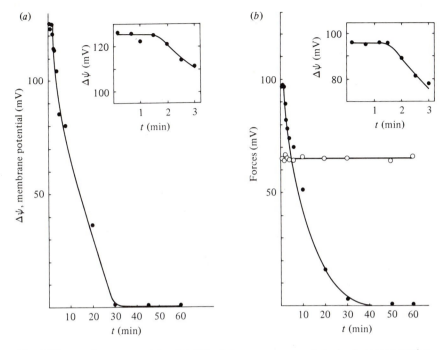

Fig. 6. The membrane potential and ΔpH in *Streptococcus cremoris* during lactose starvation. The organism was grown in a lactose-limited chemostat at pH 7.6 (*a*) or pH 5.7 (*b*) at a dilution rate of 0.15 h^{-1}. The $\Delta\psi$ (●) and ΔpH (○) were measured as described by Otto *et al.* (1983). The insets show the time course of the decay of $\Delta\psi$ on an extended time scale. (Reproduced from Otto *et al.* (1983), with permission.)

uptake of a number of amino acids and possibly also to maintain intracellular pools of these compounds. Consequently one would expect that this pool will rapidly disappear during lactose starvation as has been observed for energy-starved cells of *Staphylococcus epidermidis* (Horan *et al.*, 1981). However, this was not the case for the leucine and glutamic acid pools in *S. cremoris* (R. Otto, unpublished). Apparently in this organism, as in other organisms, dissipation of energy reservoirs following starvation does not immediately lead to drainage of metabolite pools; the proportion of their metabolite and ion pools that remain may help in sustaining viability despite the absence of a membrane potential. This ability of the cells to maintain concentration gradients of essential metabolites may be due to gating phenomena by which transport systems are blocked when the protonmotive force decreases below a certain level (see Konings & Veldkamp, 1983). As a result of this gating the (low) protonmotive force is no longer able to drive carrier-mediated solute uptake and accumulated solutes are unable to leave the cells via transport carriers. It is entirely possible that through this type of gating phenomenon many bacteria manage to survive very long periods of starvation.

Reappearance of an energy source in the environment of starved gram-negative bacteria may lead to substrate-accelerated death (Postgate & Hunter, 1964). This applies to a situation in which addition of more of a substrate whose exhaustion had limited growth to a starving suspension of cells in non-nutrient buffer leads to greatly accelerated death rates. Postgate & Hunter (1964) reported that the phenomenon can be prevented by magnesium ions. Since magnesium ions are quite abundant in many natural environments this indicates that substrate-accelerated death may be of little ecological importance. Survivors of substrate-accelerated death generally exhibit long division lags in recovery media. As this is reminiscent of the recovery of organisms which have been subjected to repression of enzyme synthesis, a phenomenon that in *E. coli* can be alleviated with cyclic AMP, it was of interest to study the effect of cyclic AMP on glycerol- and lactose-accelerated death. Substrate-accelerated death in *Klebsiella aerogenes* is associated with a rapid decrease of the internal cyclic AMP pool and in this organism addition of cyclic AMP protected in both cases the population of organisms (Calcott & Postgate, 1972; Calcott, Montague & Postgate, 1972).

In the case of lactose-accelerated death the underlying mechanism is thought to function as follows. Uptake of lactose by starved cells leads to a (partial) depolarization of the cytoplasmic membrane (Dijkhuizen & Hartl, 1978) and this in turn deactivates adenylate cyclase (Peterkofsky & Gazdar, 1979) which converts ATP into cyclic AMP. Cyclic phosphodiesterase on the other hand remains active and this leads to a decrease in the cyclic AMP level (Calcott & Calvert, 1981). This latter enzyme is inhibited by magnesium ions, which explains the protective effect of these ions against substrate-accelerated death. Although there is currently a wealth of information concerning the regulation of adenylate cyclase, phosphodiesterase and release of cyclic AMP from bacterial cells, it is as yet unclear in which way a decreased level of cyclic AMP results in lactose-accelerated death. Glycerol-accelerated death is thought to be due to the accumulation of methylglyoxal (Freedberg, Kistler & Lin, 1971), a metabolite which is toxic for glycolytic enzymes such as fructose 1,6-bisphosphate aldolase in particular (Leoncini, Maresca & Bonsignore, 1980). However, in this case also, cyclic AMP exerts a protective effect (Calcott et al., 1972), the mechanistic details of which are equally unclear.

CONCLUDING REMARKS

Studies of the various ways in which micro-organisms adapt themselves to changes in their environment have formed a focal point in microbiology for many decades. After it became clear that it was necessary to distinguish between an adaptation in a population of cells of a single species which was due to a phenotypic change affecting all the cells, and an adaptation that resulted from mutation and selection, much of the ambiguity present in the early literature was resolved and since then both aspects of adaptation have been studied in great detail. Our present discussion has centred around the ways in which microbes respond in a phenotypic fashion to environmental changes that may be expected to occur frequently in nature, and much to our regret we have had to limit the scope of our contribution to a few selected topics which, not surprisingly, reflect our own interests. Thus we have ignored adaptation of microbes to more extreme environments (Kushner, 1980), important processes such as the tactic response of organisms to chemical or physical stimuli (Carlile, 1980; Koshland, 1981) and the role of membrane-associated processes in metabolic regulation in microbes (Konings & Veldkamp, 1980, 1983). This last aspect in particular appears of prime importance because the cytoplasmic membrane is the major barrier between the contents of an organism and its (potentially) hostile environment. Many environmental changes initially affect the electrochemical gradient across the membrane and there is now increasing evidence to suggest that changes in this gradient are often at the root of the phenotypic responses of a microbial cell. Experimental tools are now available to probe how the cell monitors environmental changes and these no doubt will help to clarify how microbes control the fluxes of the various nutrients required for balanced growth and survival.

We are grateful to Dr R. Otto for valuable suggestions and critical comments and for allowing us to use some of his unpublished observations. The help of Marry Pras in the preparation of the manuscript is greatly appreciated.

REFERENCES

ABDELAL, A. T. (1979). Arginine catabolism by microorganisms. *Annual Review of Microbiology*, **33**, 139–68.

ARBIGE, M. & CHESBRO, W. (1982). *relA* and related loci are growth rate determinants for *Escherichia coli* in a recycling fermenter. *Journal of General Microbiology*, **128**, 693–703.

BACKMAN, K. C., CHEN, Y.-M., UENO-NISHIO, S. & MAGASANIK, B. (1983). The product of *glnL* is not essential for regulation of bacterial nitrogen assimilation. *Journal of Bacteriology*, **154**, 516–19.

BARKER, S. L. (1977). Effects of sodium ions on the electrical and pH gradient across the membrane of *Streptococcus lactis* cells. *Journal of Supramolecular Structure*, **6**, 383–8.

BAUMBERG, S. (1981). The evolution of metabolic regulation. In *Molecular and Cellular Aspects of Microbial Evolution*, ed. M. J. Carlile, J. F. Collins & B. E. B. Moseley, *Symposium of the Society for General Microbiology*, **32**, pp. 229–72. Cambridge University Press.

BOTSFORD, J. L. (1981). Cyclic nucleotides in procaryotes. *Microbiological Reviews*, **45**, 620–42.

BOYLEN, C. W. & ENSIGN, J. C. (1970). Long-term starvation survival of rod and spherical cells of *Arthrobacter crystallopoietes*. *Journal of Bacteriology*, **103**, 569–77.

BROWN, A. D. (1976). Microbial water stress. *Bacteriological Reviews*, **40**, 803–46.

BROWN, C. M. (1976). Nitrogen metabolism in bacteria and fungi. In *Continuous Culture*, vol 6, *Applications and New Fields*, ed. A. C. R. Dean, D. C. Ellwood, C. G. T. Evans & J. Melling, pp. 170–83. Chichester: Ellis Horwood.

BROWN, C. M., MACDONALD-BROWN, D. S. & MEERS, J. L. (1974). Physiological aspects of microbial inorganic nitrogen metabolism. *Advances in Microbiological Physiology*, **11**, 1–52.

BRUINENBERG, P. G., VEENHUIS, M., DIJKEN, J. P. VAN, DUINE, J. A. & HARDER, W. (1982). A quantitative analysis of selective inactivation of peroxisomal enzymes in the yeast *Hansenula polymorpha* by high-performance liquid chromatography. *FEMS Microbiology Letters*, **15**, 45–50.

BURNS, R. G. (1983). Extracellular enzyme–substrate interactions in soil. In *Microbes in their Natural Environments*, ed. J. H. Slater, R. Whittenbury & J. W. T. Wimpenny, *Symposium of the Society for General Microbiology*, **34**, pp. 249–98. Cambridge University Press.

CALCOTT, P. H. (1982). Cyclic AMP and cyclic GMP control of synthesis of constitutive enzymes in *Escherichia coli*. *Journal of General Microbiology*, **128**, 705–12.

CALCOTT, P. H. & CALVERT, T. J. (1981). Characterization of 3':5'-cyclic AMP phosphodiesterase in *Klebsiella aerogenes* and its role in substrate-accelerated death. *Journal of General Microbiology*, **122**, 313–21.

CALCOTT, P. H., MONTAGUE, W. & POSTGATE, J. R. (1972). The levels of cyclic AMP during substrate-accelerated death. *Journal of General Microbiology*, **73**, 197–200.

CALCOTT, P. H. & POSTGATE, J. R. (1972). On substrate-accelerated death in *Klebsiella aerogenes*. *Journal of General Microbiology*, **70**, 115–22.

CALHOUN, D. H. & HATFIELD, G. W. (1975). Autoregulation of gene expression. *Annual Review of Microbiology*, **29**, 275–99.

CARLILE, M. J. (1980). Positioning mechanisms – the role of motility, taxis and tropism in the life of microorganisms. In *Contemporary Microbial Ecology*, ed. D. C. Ellwood, J. N. Hedger, M. J. Latham, J. M. Lynch & J. H. Slater, pp. 55–74. New York & London: Academic Press.

CHAPMAN, A. G., FALL, L. & ATKINSON, D. E. (1971). Adenylate energy charge in *Escherichia coli* during growth and starvation. *Journal of Bacteriology*, **108**, 1072–86.

CLARK, J. B. (1979). Sphere–rod transitions in *Arthrobacter*. In *Developmental Biology of Prokaryotes*, ed. J. H. Parish, pp. 73–92. Oxford: Blackwell Scientific.

CLARKE, P. H., HOULDSWORTH, M. A. & LILLY, M. D. (1968). Catabolic repression and the induction of amidase synthesis by *Pseudomonas aeruginosa* 8602 in continuous culture. *Journal of General Microbiology*, **51**, 225–34.

CLARKE, P. H. & ORNSTON, L. N. (1975). Metabolic pathways and regulation. In *Genetics and Biochemistry of Pseudomonas*, ed. P. H. Clarke & M. H. Richmond, pp. 191–340. New York: Wiley.

CRAWFORD, I. P. & STAUFFER, G. V. (1980). Regulation of tryptophan biosynthesis. *Annual Review of Biochemistry*, **49**, 163–95.

DAWES, E. A. (1976). Endogenous metabolism and the survival of starved prokaryotes. In *The Survival of Vegetative Microbes*, ed. T. R. G. Gray & J. R. Postgate, *Symposium of the Society for General Microbiology 26*, pp. 19–53. Cambridge University Press.

DAWES, E. A., MIDGLEY, M. & WHITING, P. H. (1976). Control of transport systems for glucose, gluconate and 2-oxo-gluconate and of glucose metabolism in *Pseudomonas aeruginosa*. In *Continuous Culture*, vol. 6, *Applications and New Fields*, ed. A. C. R. Dean, D. C. Ellwood, C. G. T. Evans & J. Melling, pp. 195–207. Chichester: Ellis Horwood.

DAWES, E. A. & RIBBONS, D. W. (1964). Some aspects of the endogenous metabolism of bacteria. *Bacteriological Reviews*, **28**, 126–49.

DAWES, E. A. & RIBBONS, D. W. (1965). Studies on the endogenous metabolism of *Escherichia coli*. *Biochemical Journal*, **95**, 332–43.

DAWES, E. A. & SENIOR, P. J. (1973). The role and regulation of energy reserve polymers in micro-organisms. *Advances in Microbial Physiology*, **10**, 135–266.

DEAN, A. C. R. (1972). Influence of environment on the control of enzyme synthesis. *Journal of Applied Chemistry and Biotechnology*, **22**, 245–59.

DIJKEN, J. P. VAN, OTTO, R. & HARDER, W. (1976). Growth of *Hansenula polymorpha* in a methanol-limited chemostat. *Archives of Microbiology*, **111**, 137–44.

DIJKHUIZEN, D. & HARTL, D. (1978). Transport by the lactose permease of *Escherichia coli* as the basis of lactose killing. *Journal of Bacteriology*, **135**, 876–82.

DIJKHUIZEN, L., WERF, B. VAN DER & HARDER, W. (1980). Metabolic regulation in *Pseudomonas oxalaticus* OX1. Diauxic growth on mixtures of oxalate and formate or acetate. *Archives of Microbiology*, **124**, 261–8.

DOW, C. S., WHITTENBURY, R. & CARR, N. G. (1983). The 'shut down' or 'growth precursor' cell. An adaptation for survival in a potentially hostile environment. In *Microbes in their Natural Environments*, ed. J. H. Slater, R. Whittenbury & J. W. T. Wimpenny, *Symposium of the Society for General Microbiology 34*, pp. 187–247. Cambridge University Press.

DUNDAS, I. E. D. (1977). Physiology of Halobacteriaceae. *Advances in Microbial Physiology*, **15**, 85–120.

ELLWOOD, D. C. & TEMPEST, D. W. (1972). Effects of environment on bacterial wall content and composition. *Advances in Microbial Physiology*, **7**, 83–117.

FREEDBERG, W. B., KISTLER, W. S. & LIN, E. C. C. (1971). Lethal synthesis of methylglyoxal by *Escherichia coli* during unregulated glycerol metabolism. *Journal of Bacteriology*, **108**, 137–44.

GALINSKI, E. A. & TRÜPER, H. G. (1982). Betaine, a compatible solute in the extremely halophilic phototrophic bacterium *Ectothiorhodospira halochloris*. *FEMS Microbiology Letters*, **13**, 357–60.

GALLANT, J. (1979). Stringent control in *E. coli*. *Annual Review of Genetics*, **13**, 393–415.

GINSBURG, A. & STADTMAN, E. R. (1973). Regulation of glutamine synthetase in *Escherichia coli*. In *The Enzymes of Glutamine Metabolism*, ed. S. Prusiner & E. R. Stadtman, pp. 9–44. New York & London: Academic Press.

GLENN, A. R. (1976). Production of extracellular proteins by bacteria. *Annual Review of Microbiology*, **30**, 41–62.

GOTTSCHAL, J. C. & KUENEN, J. G. (1980). Selective enrichment of facultatively chemolithotrophic thiobacilli and related organisms in continuous culture. *FEMS Microbiology Letters*, **7**, 241–7.

HANSON, R. S. (1979). The physiology and diversity of bacterial endospores. In *Developmental Biology of Prokaryotes*, ed. J. H. Parish, pp. 37–56. Oxford: Blackwell Scientific.

HARDER, W. & DIJKHUIZEN, L. (1982). Strategies of mixed substrate utilization in microorganisms. *Philosophical Transactions of the Royal Society of London Series B*, **297**, 459–80.

HARDER, W. & DIJKHUIZEN. L. (1983). Physiological responses to nutrient limitation. *Annual Review of Microbiology*, **37**, 1–23.

HARDER, W. & VELDKAMP, H. (1967). A continuous culture study of an obligately psychrophilic *Pseudomonas* species. *Archiv für Mikrobiologie*, **59**, 123–30.

HARDER, W. & VELDKAMP, H. (1968). Physiology of an obligately psychrophilic marine *Pseudomonas* species. *Journal of Applied Bacteriology*, **31**, 12–23.

HARDER, W. & VELDKAMP, H. (1969). Impairment of protein synthesis in an obligately psychrophilic *Pseudomonas* species grown at superoptimal temperatures. In *Continuous Cultivation of Microorganisms*, ed. I. Malek, pp. 59–69. Prague: Academia.

HERBERT, D. (1961). The chemical composition of microorganisms as a function of their environment. In *Microbial Reaction to Environment*, ed. G. G. Meynell & H. Gooder, *Symposium of the Society for General Microbiology 11*, pp. 391–418. Cambridge University Press.

HOLZER, H. (1976). Catabolite inactivation in yeast. *Trends in Biochemical Sciences*, **1**, 178–81.

HORAN, N. J., MIDGLEY, M. & DAWES, E. A. (1981). Effect of starvation on transport, membrane potential and survival of *Staphylococcus epidermidis* under anaerobic conditions. *Journal of General Microbiology*, **127**, 223–30.

KASHKET, E. R. (1981*a*). Effects of aerobiosis and nitrogen source on the proton motive force in growing *Escherichia coli* and *Klebsiella pneumoniae* cells. *Journal of Bacteriology*, **146**, 377–84.

KASHKET, E. R. (1981*b*). Proton motive force in growing *Streptococcus lactis* and *Staphylococcus aureus* cells under aerobic and anaerobic conditions. *Journal of Bacteriology*, **146**, 369–76.

KLUYVER, A. J. (1956). Life's flexibility: microbial adaptation. In *The Microbe's Contribution to Biology*, ed. A. J. Kluyver & C. B. van Niel, p. 93. Cambridge, Mass.: Harvard University Press.

KOCH, A. L. (1976). How bacteria face depression, recession and derepression. *Perspectives in Biology and Medicine*, **20**, 44–63.

KOCH, A. L. (1979). Microbial growth in low concentrations of nutrients. In *Strategies of Microbial Life in Extreme Environments*, ed. M. Shilo, *Dahlem Conference Life Sciences Research Report 13*, pp. 261–79. Weinheim & New York: Verlag Chemie.

KONINGS, W. N., HELLINGWERF, K. J. & ROBILLARD, G. T. (1981). Transport across bacterial membranes. In *Membrane Transport*, ed. S. L. Bonting & J. J. H. H. M. de Pont, pp. 257–83. Amsterdam: Elsevier/North-Holland.

KONINGS, W. N. & MICHELS, P. A. M. (1980). Electron-transfer-driven solute translocation across bacterial membranes. In *Diversity of Bacterial Respiratory Systems*, vol. 1, ed. C. J. Knowles, pp. 33–86. Boca Raton: CRC Press.

KONINGS, W. N. & ROBILLARD, G. T. (1982). Physical mechanism for regulation of proton solute symport in *Escherichia coli. Proceedings of the National Academy of Sciences, USA*, **79**, 5480–4.

KONINGS, W. N. & VELDKAMP, H. (1980). Phenotypic responses to environmental change. In *Contemporary Microbial Ecology*, ed. D. C. Ellwood, J. N. Hedger, M. J. Latham, J. M. Lynch & J. H. Slater, pp. 161–91. New York & London: Academic Press.

KONINGS, W. N. & VELDKAMP, H. (1983). Energy transduction and solute transport mechanisms in relation to environments occupied by micro-organisms. In *Microbes in their Natural Environments*, ed. J. H. Slater, R. Whittenbury & J. W. T. Wimpenny, *Symposium of the Society for General Microbiology 34*, pp. 153–86. Cambridge University Press.

KORNBERG, H. L. (1966). The role and control of the glyoxylate cycle in *Escherichia coli. Biochemical Journal*, **99**, 1–11.

KOSHLAND, D. E. JR (1981). Biochemistry of sensing and adaptation in a single bacterial system. *Annual Review of Biochemistry*, **50**, 765–82.

KUSHNER, D. J. (1980). Extreme environments. In *Contemporary Microbial Ecology*, ed. D. C. Ellwood, J. N. Hedger, M. J. Latham, J. M. Lynch & J. H. Slater, pp. 29–54. New York & London: Academic Press.

LARSEN, H. (1967). Biochemical aspects of extreme halophilism. *Advances in Microbial Physiology*, **1**, 97–132.

LEONCINI, G., MARESCA, M. & BONSIGNORE, A. (1980). The effects of methylglyoxal on the glycolytic enzymes. *FEBS Letters*, **117**, 17–18.

LUSCOMBE, B. M. & GRAY, T. R. G. (1974). Characteristics of *Arthrobacter* grown in continuous culture. *Journal of General Microbiology*, **82**, 213–22.

MAALØE, O. & KJELDGAARD, N. O. (1966). *Control of Macromolecular Synthesis*. New York & Amsterdam: W. A. Benjamin.

MCFARLAND, N., MCCARTER, L., ARTZ, S. & KUSTU, S. (1981). Nitrogen regulatory locus '*glnR*' of enteric bacteria is composed of cistrons *ntrB* and *ntrC*: identification of their protein products. *Proceedings of the National Academy of Sciences, USA*, **78**, 2135–9.

MAGASANIK, B. (1961). Catabolite repression. *Cold Spring Harbor Symposia on Quantitative Biology*, **26**, 245–56.

MAGASANIK, B. (1976). Classical and postclassical modes of regulation of the synthesis of degradative bacterial enzymes. *Progress in Nucleic Acids Research and Molecular Biology*, **17**, 99–115.

MAGASANIK, B. (1980). Regulation in the *hut* system. In *The Operon*, 2nd edn, ed. J. H. Miller & W. S. Reznikoff, pp. 373–87. Cold Spring Harbor, NY: Cold Spring Harbor Laboratory.

MAGASANIK, B. (1982). Genetic control of nitrogen assimilation in bacteria. *Annual Review of Genetics*, **16**, 135–68.

MALCOLM, N. L. (1968). A temperature-induced lesion in amino acid-transfer ribonucleic acid attachment in a psychrophile. *Biochimica et Biophysica Acta*, **157**, 493–500.

MATIN, A. (1979). Microbial regulatory mechanisms at low nutrient concentrations as studied in chemostat. In *Strategies of Microbial Life in Extreme Environments*, ed. M. Shilo, *Dahlem Conference Life Sciences Research Report 13*, pp. 323–39. Weinheim & New York: Verlag Chemie.

MEASURES, J. C. (1975). Role of amino acids in osmoregulation of non-halophilic bacteria. *Nature, London*, **257**, 398.

MEIBERG, J. B. M., BRUINENBERG, P. M. & HARDER, W. (1980). Effect of dissolved oxygen tension on the metabolism of methylated amines in *Hyphomicrobium* X in the absence and presence of nitrate: evidence for 'aerobic' denitrification. *Journal of General Microbiology*, **120**, 453–63.

MEULEN, H. J. VAN DER & HARDER, W. (1975). Production and characterization of the agarase of *Cytophaga flevensis*. *Antonie van Leeuwenhoek*, **41**, 431–47.

MEULEN, H. J. VAN DER & HARDER, W. (1976*a*). Characterization of the neoagaro-tetra-ase and neoagarobiase of *Cytophaga flevensis*. *Antonie van Leeuwenhoek*, **42**, 81–94.

MEULEN, H. J. VAN DER & HARDER, W. (1976*b*). The regulation of agarase production by resting cells of *Cytophaga flevensis*. *Antonie van Leeuwenhoek*, **42**, 277–86.

MINK, R. W., PATTERSON, J. A. & HESPELL, R. B. (1982). Changes in viability, cell composition, and enzyme levels during starvation of continuously cultured (ammonia-limited) *Selenomonas ruminantium*. *Applied and Environmental Microbiology*, **44**, 912–22.

MINNIKIN, D. E. & ABDOLRAHIMZADEH, H. (1974). The replacement of phospha-tidylethanolamine and acidic phospholipids by an ornithine-amide lipid and a minor phosphorus-free lipid in *Pseudomonas fluorescens* NCMB129. *FEBS Letters*, **43**, 257–60.

MONOD, J. (1942). *Recherches sur la croissance des cultures bactériennes*. Paris: Hermann.

MORITA, R. Y. (1975). Psychrophilic bacteria. *Bacteriological Reviews*, **39**, 144–67.

NEIJSSEL, O. M., HUETING, S., CRABBENDAM, K. J. & TEMPEST, D. W. (1975). Dual pathways of glycerol assimilation in *Klebsiella aerogenes* NCIB418. Their role and possible functional significance. *Archives of Microbiology*, **104**, 83–7.

NG, F. M.-W. & DAWES, E. A. (1973). Chemostat studies on the regulation of glucose metabolism in *Pseudomonas aeruginosa* by citrate. *Biochemical Journal*, **132**, 129–40.

NIERLICH, D. P. (1978). Regulation of bacterial growth, RNA and protein synthesis. *Annual Review of Microbiology*, **32**, 393–432.

NOVITSKY, J. A. & MORITA, R. Y. (1978). Possible strategy for the survival of marine bacteria under starvation conditions. *Marine Biology*, **48**, 289–95.

O'BRIEN, R. W., NEIJSSEL, O. M. & TEMPEST, D. W. (1980). Glucose phosphoenol-pyruvate phosphotransferase activity and glucose uptake rate of *Klebsiella aerogenes* growing in chemostat culture. *Journal of General Microbiology*, **116**, 305–14.

OTTO, R., BRINK, B. TEN, VELDKAMP, H. & KONINGS, W. N. (1983). The relation between growth rate and electrochemical proton gradient of *Streptococcus cremoris*. *FEMS Microbiology Letters*, **16**, 69–74.

PADAN, E., ZILBERSTEIN, D. & SCHULDINER, S. (1981). pH homeostasis in bacteria. *Biochimica et Biophysica Acta*, **650**, 151–66.

PAHEL, G., ROTHSTEIN, D. M. & MAGASANIK, B. (1982). Complex *glnA–glnL–glnG* operon of *Escherichia coli*. *Journal of Bacteriology*, **150**, 202–13.

PARDEE, A. B. (1961). Response of enzyme synthesis and activity to environment. In *Microbial Reaction to Environment*, ed. G. G. Meynell & H. Gooder, *Symposium of the Society for General Microbiology 11*, pp. 19–40. Cambridge University Press.

PARRA, F., JONES-MORTIMER, M. C. & KORNBERG, H. L. (1983). Phosphotrans-ferase-mediated regulation of carbohydrate utilization in *Escherichia coli* K12: the nature of the *iex* (*crr*) and *gsr* (*tgs*) mutations. *Journal of General Microbiology*, **129**, 337–48.

PETERKOFSKY, A. & GAZDAR, C. (1979). *Escherichia coli* adenylated cyclase complex: regulation by the proton electrochemical gradient. *Proceedings of the National Academy of Sciences, USA*, **76**, 1099–102.

POSTGATE, J. R. & HUNTER, J. R. (1964). Accelerated death of *Aerobacter aerogenes* starved in the presence of growth-limiting substrates. *Journal of General Microbiology*, **34**, 459–73.

POSTMA, P. W. (1982). Regulation of sugar transport in *Salmonella typhimurium*. *Annales de microbiologie (Institut Pasteur)*, **133A**, 261–7.

POSTMA, P. W. & ROSEMAN, S. (1976). The bacterial phosphoenolpyruvate: sugar phosphotransferase system. *Biochimica et Biophysica Acta*, **57**, 213–57.

REID, D. S. (1980). Water activity as the criterion of water availability. In *Contemporary Microbial Ecology*, ed. D. C. Ellwood, J. N. Hedger, M. J. Latham, J. M. Lynch & J. H. Slater, pp. 15–27. New York & London: Academic Press.

ROBERTSON, B. R. & BUTTON, D. K. (1979). Phosphate-limited continuous culture of *Rhodotorula rubra*: kinetics of transport, leakage, and growth. *Journal of Bacteriology*, **138**, 884–95.

ROBERTSON, L. A. & KUENEN, J. G. (1983). Aerobic denitrification. Poster presented at the 34th Symposium of the Society for General Microbiology, Warwick, UK.

ROBILLARD, G. T. & KONINGS, W. N. (1981). Physical mechanisms for regulation of phosphoenolpyruvate-dependent glucose transport activity in *Escherichia coli*. *Biochemistry*, **20**, 5025–32.

SAIER, M. H. (1977). Bacterial phosphoenolpyruvate: sugar phosphotransferase systems: structural, functional and evolutionary interrelationships. *Bacteriological Reviews*, **41**, 856–71.

SANWAL, B. D. (1970). Allosteric controls of amphibolic pathways in bacteria. *Bacteriological Reviews*, **34**, 20–39.

SAVAGEAU, M. A. (1979). Autogenous and classical regulation of gene expression: a general theory and experimental evidence. In *Biological Regulation and Development*, vol. 1, *Gene Expression*, ed. R. F. Goldberger, pp. 57–108. New York & London: Plenum Press.

SMITH, A. L. & KELLY, D. P. (1979). Competition in the chemostat between an obligately and a facultatively chemolithotrophic *Thiobacillus*. *Journal of General Microbiology*, **115**, 377–84.

SMITH, M. W. & NEIDHARDT, F. C. (1983*a*). Proteins induced by anaerobiosis in *Escherichia coli*. *Journal of Bacteriology*, **154**, 336–43.

SMITH, M. W. & NEIDHARDT, F. C. (1983*b*). Proteins induced by aerobiosis in *Escherichia coli*. *Journal of Bacteriology*, **154**, 344–50.

STANIER, R. Y. (1942). The Cytophaga group: a contribution to the biology of Myxobacteria. *Bacteriological Reviews*, **6**, 143–96.

STANLEY, S. O. & BROWN, C. M. (1974). Influence of temperature and salinity on the amino-acid pools of some marine pseudomonads. In *Effect of the Ocean Environment on Microbial Activities*, ed. R. R. Colwell & R. Y. Morita, pp. 92–103. Baltimore: University Park Press.

STEWART, W. D. P. (1983). Natural environments – challenges to microbial success and survival. In *Microbes in their Natural Environments*, ed. J. H. Slater, R. Whittenbury & J. W. T. Wimpenny, *Symposium of the Society for General Microbiology 34*, pp. 1–35. Cambridge University Press.

STRANGE, R. E. (1976). Microbial response to mild stress. In *Patterns of Progress*, ed. J. G. Cook. Shildon: Meadowfield Press.

SWITZER, R. L. (1977). The inactivation of microbial enzymes *in vivo*. *Annual Review of Microbiology*, **31**, 135–57.

TEMPEST, D. W. & HUNTER, J. R. (1965). The influence of temperature and pH value on the macromolecular composition of magnesium-limited and glycerol-limited *Aerobacter aerogenes* growing in a chemostat. *Journal of General Microbiology*, **41**, 267–73.

TEMPEST, D. W., MEERS, J. L. & BROWN, C. M. (1970). Influence of environment on the content and composition of microbial free amino acid pools. *Journal of General Microbiology*, **64**, 171–85.

TEMPEST, D. W. & NEIJSSEL, O. M. (1978). Eco-physiological aspects of microbial growth in aerobic nutrient-limited environments. *Advances in Microbial Ecology*, **2**, 105–53.

TEMPEST, D. W. & NEIJSSEL, O. M. (1981). Metabolic compromises involved in the growth of micro-organisms in nutrient-limited (chemostat) environments. In *Basic Life Sciences*, vol. 18, ed. A. Hollaender, pp. 335–56. New York: Plenum Press.

TEMPEST, D. W., NEIJSSEL, O. M. & ZEVENBOOM, W. (1983). Properties and performance of microorganisms in laboratory culture: their relevance to growth in natural ecosytems. In *Microbes in their Natural Environments*, ed. J. H. Slater, R. Whittenbury & J. W. T. Wimpenny, *Symposium of the Society for General Microbiology 34*, pp. 119–52. Cambridge University Press.

THOMAS, T. D. (1976). Regulation of lactose fermentation in group N streptococci. *Applied and Environmental Microbiology*, **32**, 474–8.

THOMAS, T. D. & BATT, R. D. (1969a). Synthesis of protein and ribonucleic acid by starved *Streptococcus lactis* in relation to survival. *Journal of General Microbiology*, **58**, 363–9.

THOMAS, T. D. & BATT, R. D. (1969b). Metabolism of exogenous arginine and glucose by starved *Streptococcus lactis* cells in relation to survival. *Journal of General Microbiology*, **58**, 371–80.

THOMPSON, J. (1978). *In vivo* regulation of glycolysis and characterization of sugar: phosphotransferase systems in *Streptococcus lactis*. *Journal of Bacteriology*, **136**, 465–76.

TIEDJE, J. M., SEXSTONE, A. J., MYROLD, D. D. & ROBINSON, J. A. (1982). Denitrification: ecological niches, competition and survival. *Antonie van Leeuwenhoek*, **48**, 569–83.

TRAVERS, A. A., KARI, C. & MACE, H. A. F. (1981). Transcriptional regulation by bacterial RNA polymerase. In *Genetics as a Tool in Microbiology*, ed. S. W. Glover & D. A. Hopwood, *Symposium of the Society for General Microbiology 31*, pp. 111–30. Cambridge University Press.

TRONICK, S. R., CIARDI, J. E. & STADTMAN, E. R. (1973). Comparative biochemical and immunological studies of bacterial glutamine synthetases. *Journal of Bacteriology*, **115**, 858–68.

TYLER, B. (1978). Regulation of the assimilation of nitrogen compounds. *Annual Review of Biochemistry*, **47**, 1127–62.

UMBARGER, H. E. (1978). Amino acid biosynthesis and its regulation. *Annual Review of Biochemistry*, **47**, 533–606.

VEENHUIS, M., DIJKEN, J. P. VAN & HARDER, W. (1976). Cytochemical studies on the localization of methanol oxidase and other oxidases in peroxisomes of methanol-grown *Hansenula polymorpha*. *Archives of Microbiology*, **111**, 123–35.

VEENHUIS, M., DIJKEN, J. P. VAN & HARDER, W. (1983a). The significance of peroxisomes in the metabolism of one-carbon compounds in yeasts. *Advances in Microbial Physiology*, **24**, 1–78.

VEENHUIS, M., DOUMA, A., HARDER, W. & OSUMI, M. (1983b). Degradation and turnover of peroxisomes in the yeast *Hansenula polymorpha* induced by selective inactivation of peroxisomal enzymes. *Archives of Microbiology*, **134**, 193–203.

VEENHUIS, M., HARDER, W., DIJKEN, J. P. VAN & MAYER, F. (1981). Substructure of crystalline peroxisomes in methanol-grown *Hansenula polymorpha*: evidence for an *in vivo* crystal of alcohol oxidase. *Molecular and Cellular Biology*, **1**, 949–57.

VELDKAMP, H. (1977). *Continuous Culture in Microbial Physiology and Ecology.* Durham: Meadowfield Press.

VELDKAMP, H. & JANNASCH, H. W. (1972). Mixed culture studies with the chemostat. *Journal of Applied Chemistry and Biotechnology*, **22**, 105–23.

WHITTENBURY, R. & DOW, C. S. (1977). Morphogenesis and differentiation in *Rhodomicrobium vannielii* and other budding and prosthecate bacteria. *Bacteriological Reviews*, **1**, 754–808.

WHOOLEY, M. A. & McLOUGHLIN, A. J. (1983). The proton motive force in *Pseudomonas aeruginosa* and its relationship to exoprotease production. *Journal of General Microbiology*, **129**, 989–96.

WHOOLEY, M. A., O'CALLAGHAN, J. A. & McLOUGHLIN, A. J. (1983). Effect of substrate on the regulation of exoprotease production by *Pseudomonas aeruginosa* ATCC 10145. *Journal of General Microbiology*, **129**, 981–8.

WIERSMA, M., HANSEN, T. A. & HARDER, W. (1978*b*). Effect of environmental conditions on the production of two extracellular proteolytic enzymes by *Vibrio* SA1. *Antonie van Leeuwenhoek*, **44**, 129–40.

WIERSMA, M. & HARDER, W. (1978). A continuous culture study of the regulation of extracellular protease production in *Vibrio* SA1. *Antonie van Leeuwenhoek*, **44**, 141–55.

WIERSMA, M., VERSTEEGH, G., ASSINK, H. A., WELLING, G. W. & HARDER, W. (1978*a*). Purification and some properties of two extracellular proteolytic enzymes produced by *Vibrio* SA1. *Antonie van Leeuwenhoek*, **44**, 157–69.

WIMPENNY, J. W. T. (1981). Spatial order in microbial ecosystems. *Biological Reviews*, **56**, 295–342.

WIMPENNY, J. W. T., LOVITT, R. W. & COOMBS, J. P. (1983). Laboratory model systems for the investigation of spatially and temporally organized microbial ecosystems. In *Microbes in their Natural Environments*, ed. J. H. Slater, R. Whittenbury & J. W. T. Wimpenny, *Symposium of the Society for General Microbiology 34*, pp. 67–117. Cambridge University Press.

WOOD, A. P. & KELLY, D. P. (1977). Heterotrophic growth of *Thiobacillus* A2 on sugars and organic acids. *Archives of Microbiology*, **113**, 257–64.

WYMAN, J. (1972). On allosteric models. *Current Topics in Cellular Regulation*, **6**, 209–26.

YANOFSKY, C. (1981). Attenuation in the control of expression of bacterial operons. *Nature, London*, **289**, 751–8.

YOUNG, K., HONG, K. C., DUCKWORTH, M. & YAPHE, W. (1971). Enzymic hydrolysis of agar and properties of bacterial agarases. In *Proceedings of the Seventh International Seaweed Symposium*, pp. 469–72. University of Tokyo Press.

MICROBES IN THE OCEANIC ENVIRONMENT

HOLGER W. JANNASCH

Woods Hole Oceanographic Institution, Woods Hole, MA 02543, USA

Environmental microbiology has come of age rather recently. Although solidly rooted in Winogradski's and Beijerinck's early physiological-ecological work, it has grown relatively slowly in the shade of hygienically oriented medical microbiology. Originally under the wings of agricultural research, limnology and marine biology, environmental microbiology started to develop rapidly into a field in its own right about 20 to 25 years ago along with the growing public awareness of environmental problems and perils. Under the label of microbial ecology, it blossomed into workshops, symposia, courses, books and new journals, just short of generating a new discipline. Environmental or ecological microbiology is now generally defined as applying concepts of general microbiology – comprising microbial physiology, biochemistry, biophysics, taxonomy, genetics, etc. – to studies concerning the natural occurrence of prokaryotes and their interaction with the environment, with each other, and with eukaryotes.

Interest in the offshore and pelagic ocean is based not so much on environmental concerns but on the fact that it constitutes the major portion of the biosphere. If deliberately defined as the region overlying depths of 1000 m or more, the oceanic environment, the pelagic and deep sea, covers 62% of the earth's surface and comprises more than 90% of the biosphere. If considered in terms of the area and volume of the oceans only, and if the shelf is extended to the 1000 m depth line, the oceanic environment contains 97.6% of the total amount of sea water. In view of this immense volume and the microbes' primary role in the global biogeochemical cycling of matter, their metabolic activities, however small, and their adaptations to overcome the characteristic environmental constraints of the deep ocean become research topics of utmost ecological importance.

The *Challenger* expedition (1873–6), commonly seen as the beginning of modern oceanic biology, occurred just before the concepts of microbial ubiquity, physiological versatility and metabolic efficiency developed through the works of Pasteur, Winogradski and Beijerinck. This extremely influential expedition did not, therefore, include microbiological investigations in its programme, except for revealing the non-existence of T. H. Huxley's 'Urschleim' organism, *Bathybius haecklei*: an affair amiably recounted by Rehbock (1975). During the *Travailleur* and *Talisman* expeditions (1882–3), Certes (1884*a*, *b*) looked for bacteria in sediment samples from depths of up to 5000 m. He noted their survival at great pressures and hypothesized their *in situ* state of suspended animation.

While deep-sea studies were not further pursued until the 1940s, Fischer's (1886, 1894) and Russell's (1892, 1893) work on the distribution of micro-organisms in surface waters resulted in the general conclusion that bacteria are not indigenous to the open ocean but are imported from land and may be able to grow in the richer inshore waters only. Observations on luminescent bacteria and psychrophilic growth behaviour were first recorded in these early studies. These beginnings were reviewed in two monographs (Benecke, 1933; ZoBell, 1946) which are now credited with having introduced marine microbiology as an independent field of research.

During the last decade, studies on the pelagic and deep sea have focused on the physiological strategies of microbial growth at the characteristically low nutrient levels of sea water (Jannasch, 1967, 1979; Poindexter, 1981) and on microbial responses to hydrostatic pressure (Marquis & Matsumura, 1978; Landau & Pope, 1980). Recent advances profited from technical progress in the design of pressure vessels for the sampling and culture of micro-organisms in the absence of decompression. In addition, a new area of deep-sea microbiology opened up with the unexpected discovery of active submarine vulcanism which results in the release of inorganic sources of energy for bacterial growth. The following discussion focuses on these two recent research topics.

HYDROSTATIC PRESSURE AS AN ENVIRONMENTAL FACTOR

The input of organic carbon as the source of energy for microbial life in the pelagic and deep sea is determined by the photosynthetic production of planktonic algae in the upper 100–300 m of the water column. It is dependent on the availability of light and nutrients at the various geographic locations. The dissolved portion of this organic material and part of the particulate portion, at least 80–90% of the total, are turned over in the surface layer. The transport of the remaining, primarily particulate, organic material to the bottom has recently been the subject of an extensive research effort using sediment traps (Turner, 1979; Honjo, Manganini & Cole, 1982). The results indicate that the consumption of this energy extends over the entire water column, diminishing with depth and exhibiting another peak of decomposition in the uppermost layer of the bottom sediments.

In this scenario, the two major physical parameters affecting life processes are – next to the absence of light in most of the water column – the rather constant and relatively low temperatures (2–3 °C) and the hydrostatic pressure, which increases by approximately 1 atm for every 10 m of depth (the deviation from this relation, amounting to 27 atm at a depth of 10 000 m, is discussed by Saunders & Fofonoff, 1976).

The existence of psychrophilic (i.e. low-temperature-adapted) bacteria (Ingraham & Stokes, 1959; Schmidt-Lorenz, 1967) and their occurrence in the deep sea have been well established (Morita, 1975). Their sensitivity to temperatures above 20 °C requires particular precautions during the recovery of samples through warm surface waters and during handling aboard ship. In contrast, the existence of decompression-sensitive bacteria has rarely been considered, although growth-limiting effects of pressure have been well documented (reviewed by Marquis & Matsumura, 1978). The fact that high numbers of bacteria viable at 1 atm can be retrieved from the surface of deep-sea sediments, and that these organisms can survive several compression–decompression cycles, led to the general feeling that decompression has no lethal or irreversibly damaging effect. ZoBell (1968) cautioned:

Although many bacteria from the deep sea survived, this observation fails to prove that some bacteria, possibly the most sensitive ones, were not destroyed by decompression. Answering this question may require the examination of deep sea bacteria at *in situ* pressure without subjecting them to decompression.

Bacteria are principally barotolerant to various degrees. In other words the maximum pressures at which growth ceases cover a wide range. Growth of some isolates is arrested at 200 atm, while in extremely barotolerant strains the ultimate cut-off pressure is not known. In these organisms optimal growth occurs at 1 atm (Fig. 1A

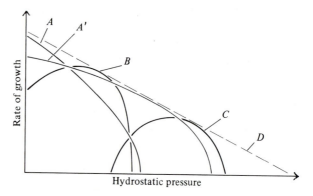

Fig. 1. Scheme of microbial growth responses to hydrostatic pressure, based on available data. *A* and *A'* show barotolerance varying by degree and range; *B* and *C* show barophilic characteristics as defined by increased rates of growth at pressures higher than 1 atm. The obligate barophilic response, *C*, is defined by the inability to grow at 1 atm. *D* indicates the general decline in metabolic activity with increasing pressure. The quantitative relationships between the responses *A* to *D* as well as the shapes of the curves are arbitrary since the actual data available have not been obtained under comparable conditions.

and *A'*). In contrast, barophilic bacteria are adapted to a specific pressure higher than 1 atm, and growth is inhibited at ranges of pressure both above and below the optimal value (Fig. 1*B*). When no growth occurs at 1 atm, the organism is defined as obligately barophilic (Fig. 1*C*). In the scheme of Fig. 1 the quantitative growth relationships between these four types, especially their approximation to a joint and linear slope, are hypothetical.

Most of the data available in the literature cannot yet be arranged in the above fashion since they have not been obtained under comparable conditions. This refers especially to the different types and concentrations of carbon sources used. Most isolates of barophilic bacteria (Yayanos, Dietz & Van Boxtel, 1979, 1981; Deming, Tabor & Colwell, 1981) were obtained from deep-sea animals and grown on relatively rich peptone yeast extract media. If data from a recent paper by Yayanos *et al.* (1982) are plotted as in Fig. 1, the decline of total activity with the increasing degree of barophilism does agree well with the hypothetical scheme, although the data exclude barotolerant organisms. Even if determined under comparable growth conditions, the potential growth advantages of the different organisms in such a scheme do not necessarily determine the actual outcome of competition under natural conditions where other factors, especially the nature and the concentration of nutrients, will be decisive.

In their early work on barophilic bacteria, ZoBell & Morita (1957) observed that sulphate-reducing bacteria in deep-sea cores lost their barophilic characteristics after several transfers involving decompression. In 1968 ZoBell summarized:

Considerable difficulty has been experienced in trying to maintain barophilic bacteria in the laboratory. Most of the enrichment cultures from the deep sea lost their viability after two or three transplants to new media, although some barophiles have survived in sediment samples for several months when stored at *in situ* pressures and temperatures. Even at refrigeration temperatures, the deep sea barophiles die off much more rapidly at 1 atm than when compressed to *in situ* pressures.

Growth rates of bacteria in a natural population from the intestinal tract of amphipods collected at a depth of 7000 m were found by Schwarz, Yayanos & Colwell (1976) to be almost equal when measured in a starch medium at 700 atm and at 1 atm. In pure culture isolates, however, barotolerance decreased during subsequent transfers. A study on death of a barophilic isolate during incubation at 1 atm was done by Yayanos & Dietz (1983).

The probable physiological and molecular reasons for the loss of barophilism and barotolerance during decompression are not the topic of this paper. They were discussed on the molecular level by Pope *et al.* (1975*a*, *b*, 1976) and Landau & Pope (1980), on the enzymatic level by Jaenicke (1981) and in a general treatise by Marquis (1982). In principle, the nature of barophilism, especially that of 'obligate' barophilism, implies growth inhibition at pressures below a certain optimum. Hypothetically, this inhibition may result in a loss of barophilic bacteria from decompressed natural populations during sampling, as well as in a loss of barophilic characteristics in culture during purification and transfers. Consequently, the potential inactivation of decompression-sensitive bacteria has to be considered in the study of deep-sea bacteria.

In situ *incubation experiments*

One simple way to avoid decompression is to conduct growth experiments directly on the deep-sea floor. Pressure changes will then take place only after certain transformation processes are completed within given periods of time. The *in situ* inoculation of bottles precharged with growth media, or the injection of sediment cores with labelled substrates attached to free-falling tripods, can be fully automated at practically any depth. Timed or acoustically triggered release of weights after periods of hours, days, weeks or months for ascent and pick-up at the surface are common oceanographic techniques. Many such experiments have been done in depths of 1000–6000 m with the general result that, within a given time period, growth or transformations of labelled substrates into cell material and carbon dioxide are one to two orders of magnitude lower than in the controls (i.e. parallel samples incubated at 1 atm and *in situ* temperature) (Jannasch & Wirsen, 1973). The same results were found with the decomposition of polymers such as starch, agar and chitin (Wirsen & Jannasch, 1976). These observations imply that, whatever the relative proportion of barotolerant to barophilic types of bacteria may be, the slope depicted in Fig. 1 does exist and is considerable.

In situ incubation experiments suffer from the major limitation that the data represent mere end-point measurements which do not facilitate rate determinations. To accomplish the latter, true time course experiments with one and the same sample and inoculum are necessary. Beyond that, pure culture experiments with undecompressed isolates would be necessary for the ultimate proof of decompression-sensitivity in barophilic bacteria.

Studies with pressure-retaining equipment

A 1 litre pressure-retaining sampler was built and used extensively for studying natural microbial populations from depths of up to 6000 m (Jannasch, Wirsen & Winget, 1973; Jannasch, Wirsen & Taylor, 1976). When triggered on the cable by a messenger, the 1 litre sample enters slowly and without a pressure differential. There is sufficient insulation in the heavy steel housing to prevent harmful temperature changes during retrieval. A nitrogen gas cushion compensates for the small volume increases which occur during retrieval due to vessel expansion and during the attachment of a pressure gauge when determining the exact depth of sampling. A simple, hand-operated transfer unit allows for the introduction and removal of 13 ml subsamples without affecting the pressure within the chamber. Thus the sampler, equipped with a magnetic stirring bar, is turned into a culture chamber for time course experiments measuring the uptake and respiration of radiolabelled substrates.

Subsequently, in order to increase the number of samples that could be taken and studied aboard ship and ashore, an *in situ*

EXPLANATION OF PLATES

PLATE 1

(*a*) Turbid and milky-bluish water at *c.* 15 °C (as measured about 0.5 m above the bottom) emitted from a vent surrounded by various invertebrates. ('Mussel Bed' vent site, Galapagos Rift; approximately 0° 48′ N, 86° W; depth 2550 m; dive 887; photograph by J. F. Grassle, reproduced with permission.) (*b*) Bed of the 'giant' white clam (*Calyptogena magnifica*, Boss & Turner, 1980; average length 20–25 cm) in the immediate vicinity of hot 'black smokers'. The pink colour of the extended clam feet is caused by (cellular) haemoglobin. (21° N, 109° W vent site; depth 2610 m; dive 1214; photograph by C. O. Wirsen, reproduced with permission.)

PLATE 2

Vestimentiferan tube worms (*Riftia pachyptila*, Jones, 1980; length in this picture approximately 0.5–1.0 m) at the base of a hot 'black smoker'. Gills containing (non-cellular) haemoglobin are withdrawn into some of the tubes. An array of six 200 ml syringes, precharged with $^{14}CO_2$, is in the process of being filled (intake orifice on the opposite side) by the submersible *Alvin's* mechanical arm for *in situ* incubation. (21° N, 109° W vent site; depth 2610 m; dive 1223; photograph by D. M. Karl, reproduced with permission.)

PLATE 1

(a)

(b)

PLATE 2

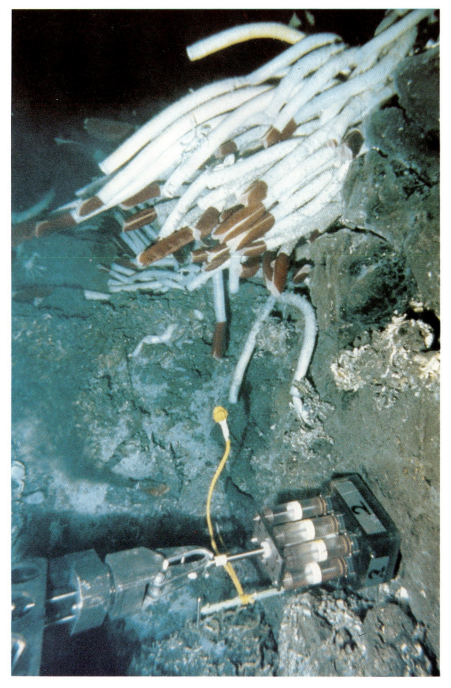

filtering and pressure-retaining device was developed for a 1:200 concentration of a 3 litre sample over a Nucleopore filter (Jannasch & Wirsen, 1977). This instrument is resterilized at sea and allows collection of as many undecompressed samples on a single cruise as there are transfer/storage units available. These units are comparatively inexpensive and are used to store the sample-concentrates at about $-2\,°C$ for later use in the laboratory. They are then injected into the prepressurized 1 litre sampler, now used as incubation chamber, containing seawater media with the specific substrates to be studied.

Time course data on the uptake and respiration of ^{14}C-labelled glutamate, casamino acids, glucose and acetate in samples from different depths indicated, in brief, that the generally barotolerant activity of natural microbial populations of sea water decreased with depth and varied greatly with the type of substrate used (Jannasch & Wirsen, 1982). Glutamate conversion was the most affected by pressure while acetate utilization was 90–100% barotolerant, i.e. unaffected by pressure (Fig. 2). Metabolism of casamino acids and glucose showed intermediate effects of pressure. Tabor *et al.* (1981) developed a somewhat different sampling system for similar studies and also found that the rates and degree of substrate utilization were generally stimulated by decompression. It should be noted that experiments conducted with decompressed samples of gut contents of deep-sea animals resulted in barophilic uptake responses (Deming & Colwell, 1982). It is conceivable that in niches of high substrate concentration, such as the digestive tracts of bottom-foraging deep-sea animals, microbial growth and genetic transformations occur at relatively high rates, resulting in a population of predominantly pressure-adapted bacteria.

Marquis & Matsumura (1978) reviewed the literature on growth at various pressures of bacterial pure culture isolates collected from the deep sea and isolated at 1 atm. The notion that psychrophilic bacteria might exhibit generally higher degrees of barotolerance and vice versa was only partly confirmed in a study by Wirsen & Jannasch (1975). The complexity of pressure effects is illustrated by the observations that the degree of barotolerance appears to be substrate dependent, that highly barotolerant bacteria could be isolated from garden soil (Kriss, 1962), and that small pressure increases above 1 atm may stimulate bacterial growth (Marquis & Matsumura, 1978). Truly barophilic bacteria were isolated by Yayanos *et al.* (1979) from deep-sea amphipods using pressurized

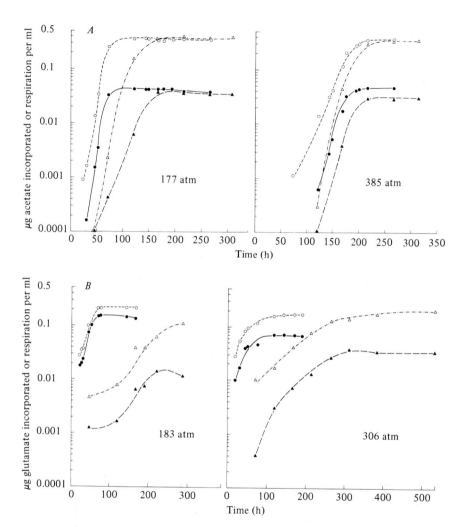

Fig. 2. Incorporation (filled symbols) and respiration (open symbols) of radiolabelled carbon of sodium acetate (A) and sodium glutamate (B) added to seawater (initial concentration $0.5 \, \mu g \, ml^{-1}$) collected from two depths and incubated at *in situ* pressure without prior decompression (triangles) and at 1 atm (circles). (From Jannasch & Wirsen, 1982.)

shake tubes (Dietz & Yayanos, 1978). With this technique it was found that barophilism and psychrophilism did generally occur together in the same organism. Except for the above-mentioned recent work on a decompression-sensitive barophilic isolate (Yayanos & Dietz, 1983), no systematic study on the effect of exposure of barotolerant and barophilic bacteria to 1 atm pressure for various periods of time has yet been done.

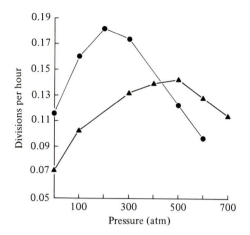

Fig. 3. Growth of two barophilic bacteria versus pressure in sea water supplemented with 0.1% peptone yeast extract. Strain 72 (circles) was isolated in the absence of decompression from a water sample collected at a depth of 4900 m; strain 51A-1 (triangles) was isolated from a decaying amphipod collected at a depth of 5100 m, decompressed during retrieval, and recompressed for a 3 week incubation at 500 atm.

One prerequisite for such a study is preventing possible preselection for decompression-insensitive organisms during isolation at 1 atm. The development of a purification procedure in the absence of decompression using pressure-retaining filter sampler described above was not as difficult as it may sound. Taylor (1978, 1979a, b) had found that colony growth of marine isolates on agar plates in hyperbaric chambers under a helium/oxygen gas phase was not inhibited compared with growth of cell suspensions at hydrostatic pressure, i.e. in the absence of a gas phase. This observation permitted streaking of the undecompressed filter concentrate of a deep-water sample on suitable agar dishes in a prepressurized chamber. Nine such dishes fastened to a running belt can be moved individually under a viewing port. After the introduction of a sample from one of the transfer/storage units, it can be streaked and re-streaked on agar media by a sterilizable (chargeable resistance wire) loop. Colonies can be transferred within the chamber into one of four small (8 ml) vials containing liquid medium and withdrawn and transferred to one of the above-mentioned 1 litre (prepressurized) culture chambers. The instrument and procedure have been described (Jannasch, Wirsen & Taylor, 1982) and a number of barotolerant and barophilic isolates have now been obtained.

As an example, Fig. 3 shows one of the barophilic organisms

(strain 72) which was sampled at a depth of 4900 m and isolated in the absence of decomposition at 490 atm and 3 °C. Strain 51A–1 on the other hand, was isolated after two decompressions from a decaying deep-sea amphipod collected at 5100 m. Before this isolation the amphipod was incubated at 500 atm and 3 °C for three weeks. The resulting cell suspension was then mixed into a slush agar (0.3%) tube supplemented with peptone yeast extract and incubated again at 500 atm and 3 °C.

At present, it appears that many barophilic bacteria can indeed sustain brief and repeated exposures to atmospheric pressure. Their occurrence in the deep sea seems to be limited to habitats of relatively high substrate concentrations. Few of them have so far been grown on defined media that would facilitate studies with radiolabelled substrates. The quantitative contribution of barophilic bacteria to the turnover of organic matter in the deep sea may be limited compared with that of the more or less barotolerant bacteria carried in by sinking particulate nutrients (Turner, 1979). At present the problem rests with establishing the actual growth advantage of barophilic organisms over less pressure-adapted bacteria under natural conditions of food supply and by studying the dependence of pressure adaptation upon substrate specificity and affinity.

MICROBES OF DEEP-SEA THERMAL VENTS

Chemosynthetic primary production

The general notion that the deep sea represents the ultimate low-nutrient environment was suddenly challenged when, about 7 years ago, copious populations of animals were discovered at depths where the amounts of living biomass are normally orders of magnitude lower. These discoveries were the result of a systematic search for active deep-sea vulcanism at submarine spreading centres as predicted by the theory of plate tectonics. There is good evidence now that the earth's crust, rather than being one continuous shell, is broken into continent-sized plates which are moving in relation to each other. The concept of plate tectonics was developed 30 or 40 years ago, at about the same time as that of molecular biology, and its impact on every aspect of the earth sciences has been similar to that of molecular biology on the biological sciences.

Spreading centres are zones where two plates move apart, at rates of 2–15 cm per year. They are mostly submarine and located beyond the shelf. Due to thermal contraction, the freshly formed crust at these deep-sea spreading centres is highly permeable. Sea water therefore penetrates several kilometres down into the crust where reactions with the hot basalts take place at temperatures in excess of 360 °C. The resulting altered sea water, a highly reduced hydrothermal fluid, ascends to the sea floor where it emerges as warm (8–25 °C) or hot (160–350 °C) vents.

When first discovered with the aid of the research submersible *Alvin* at the Galapagos Rift spreading centre (Ballard, 1977; Lonsdale, 1977; Corliss *et al.*, 1979), the warm vents were observed to emit irridescent, milky-bluish water (Plate 1*a*) with a sulphide content of up to 160 μmol l^{-1}. Since no mechanism for a photosynthetic sustenance of the unusually dense populations of invertebrates was conceivable, a sulphur-based, chemosynthetically supported food chain was proposed (Jannasch & Wirsen, 1979; Karl, Wirsen & Jannasch, 1980). Mussels of unusual size were clustered around the vent openings in a density equal to that found in highly productive mussel beds of some inshore waters. The occurrence of vestimentiferan tube worms (a new family Riftidae, *Riftia pachyptila*: Jones, 1980) up to 2.6 m long and 5 cm thick (Plate 2) in similar concentration indicated the extent of vent plumes. 'Giant' white clams, 20–26 cm long (*Calyptogena magnifica*: Boss & Turner, 1980), were observed also in small lava fissures at greater distances (20–30 m) from the vents (Plate 1*b*). In such cases, however, it is never quite clear whether or not weak or intermittent emissions do occur in these crevices. Size and growth rate of the bivalves, the latter determined by shell marking experiments (Rhoads *et al.*, 1981) and by analysis of natural radionuclides in the shell (Turekian & Cochran, 1981), indicate an abundant albeit narrowly localized food source. As a minor part of the overall biomass, there is also a specific planktonic and nektonic (i.e. actively swimming fish, invertebrate larvae, siphonophores, etc.) fauna associated with the vent ecosystems.

There is now sufficient evidence to show that the existence of this rich fauna is based on an aerobic and anaerobic microbial chemosynthetic reduction of carbon dioxide to organic carbon using geothermally reduced inorganic compounds as the source of energy. This chemosynthetic food supply, in a permanently dark environment, occurs in the form of (1) bacterial suspensions contained in the emitted vent waters, (2) microbial mats growing in the immediate surroundings of the vents exposed to the plume of emitted hydrothermal fluid, and (3) symbiotic associations of chemosynthetic prokaryotes with the most abundant vent invertebrates.

While chemosynthesis is known to occur widely in many interface situations where suitable reduced and oxidized compounds come into contact, it always depends ultimately on photosynthetic processes. Therefore, chemosynthetic production has been termed 'secondary' as compared with the primary photosynthetic production. Basing this terminology on the energy source, rather than on the energy sink, the animal communities found at the deep-sea vents can be described as living principally on chemosynthetic primary production, and on terrestrial rather than solar energy. The fact, however, that these circumstances led to the development of exceptionally rich populations of unusually large organisms, unrivalled in the photosynthetic realm, was entirely unexpected.

In situ *observations, enrichments and pure culture studies*

Carbon dioxide fixation was measured *in situ* in 'syringe arrays' (Plate 2). These consist of six 200 ml chambers which are precharged with [^{14}C]bicarbonate and can be filled through a single orifice by the submersible's mechanical arm. They can be left for incubation *in situ* or brought to the surface for studies in the ship's laboratory under controlled conditions. The particular array shown in Plate 2 was filled with 23 °C vent water but, because of the small size of the vent plume, incubated in 3 °C ambient water. The rates of carbon dioxide incorporation (Fig. 4) in shipboard control experiments at 3 °C showed a slight decrease, possibly indicating barophilic behaviour. A representative isolate from these cell suspensions, a member of the genus *Thiomicrospira*, was, however, moderately barotolerant (Fig. 5).

The data of Fig. 4 also show that supplementation with thiosulphate led to a considerable increase in carbon dioxide fixation, implying a rate limitation by reduced sulphur in this particular water sample. A third point to emerge from this set of data is the mesophilic growth response. Carbon dioxide fixation by the natural microbial population of vent water increased by almost two orders of magnitude when the temperature was raised from 3 °C to 23 °C. The temperature optimum for growth of the *Thiomicrospira* sp. was 25 °C (Ruby & Jannasch, 1982). When vent water was collected with

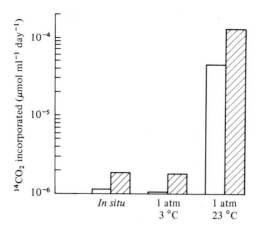

Fig. 4. Rate of carbon dioxide fixation by turbid vent water incubated *in situ* (260 atm, 3 °C; see Plate 2) and in the ship's laboratory at two temperatures. Hatched columns indicate a 1 mM thiosulphate supplement.

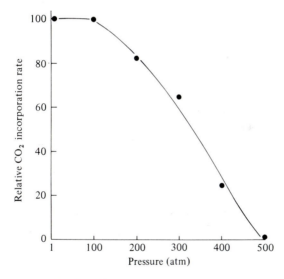

Fig. 5. Relative rate of carbon dioxide fixation versus pressure in *Thiomicrospira* (strain L-12) in a sea water medium supplemented with 10 mM thiosulphate. (From Ruby & Jannasch, 1982.)

the pressure-retaining sampler as described earlier and incubated with labelled bicarbonate in time course experiments at 25 °C, similar incorporation rates to those measured *in situ* (Fig. 4) were found and the degree of barotolerance was confirmed. Since most of the warm (8–25 °C) vent plumes were small and quickly cooled after

Table 1. *Microbial nucleotides at a hydrothermal vent site (Galapagos Rift)*

Sample	ATP[a]	A_T	GTP:ATP
Surface sea water[b] (50 m)	$130 \pm 32\,\mathrm{ng\,l^{-1}}$	$340 \pm 112\,\mathrm{ng\,l^{-1}}$	0.16 ± 0.08
Deep sea water[b] (2400 m)	$1.7 \pm 0.3\,\mathrm{ng\,l^{-1}}$	$3.4 \pm 0.3\,\mathrm{ng\,l^{-1}}$	0.075
'Garden of Eden' vent[b] (2500 m)	$491 \pm 151\,\mathrm{ng\,l^{-1}}$	$1494 \pm 553\,\mathrm{ng\,l^{-1}}$	0.86 ± 0.17
'Garden of Eden' vent[c] (2500 m)	$1943 \pm 1143\,\mathrm{ng\,g^{-1}}$	$4248 \pm 2031\,\mathrm{ng\,g^{-1}}$	0.89 ± 0.35

Data from Karl *et al.* (1980).
[a]ATP, adenosine 5'-triphosphate; A_T, total adenylates; GTP:ATP, ratio of guanosine 5'-triphosphate to adenosine 5'-triphosphate.
[b]Filtered.
[c]Settled particles.

emission, the bulk of the bacterial suspension must be assumed to have grown at elevated temperatures within the subsurface vent system (Tuttle, Wirsen & Jannasch, 1983).

Cell counts in vent water, obtained by using acridine orange staining and epifluorescence microscopy, varied greatly from $10^5\,\mathrm{ml^{-1}}$ (Jannasch & Wirsen, 1979) to $10^9\,\mathrm{ml^{-1}}$ (Corliss *et al.*, 1979). As a general indicator of biomass in these samples, ATP concentrations were found to be two to three times higher than those obtained in photosynthetically productive surface waters (Karl *et al.*, 1980). The GTP:ATP ratio in Table 1 has been measured as an indicator of growth rate (Karl, 1980). Activity of ribulose bisphosphate carboxylase was found in cell suspensions as well as in microbial mats covering all types of surfaces (lava rocks, clam shells, crab carapaces, etc.) exposed to vent plumes (J. H. Tuttle, unpublished).

These microbial mats were composed of a few predominant cell types (Figs 6a–c, 7a and b). The most prominent, relatively large, septated and trichome-like structures (Fig. 6a–c) presented a puzzle. If sectioned they resemble sheathed cyanobacteria-like organ-

Fig. 6. (a) Example of the microbial mats that cover surfaces which are exposed to plumes of hydrothermal fluid in the immediate vicinity of warm vents. The three predominant cell types are: coccoid, stalked cells and sheathed trichomes. (Scale bar, 10 μm.) (b) Section of (a). Manganese–iron oxides are deposited around empty cell envelopes and within the polysaccharide matrix; the smallest round structures represent sections through the stalks of *Hyphomicrobium*-type cells (Fig. 7a). (Scale bar, 1 μm.) (c) Section of (a). Sheathed cyanobacterium-like trichome encrusted with manganese–iron oxides. (Scale bar, 1 μm.)

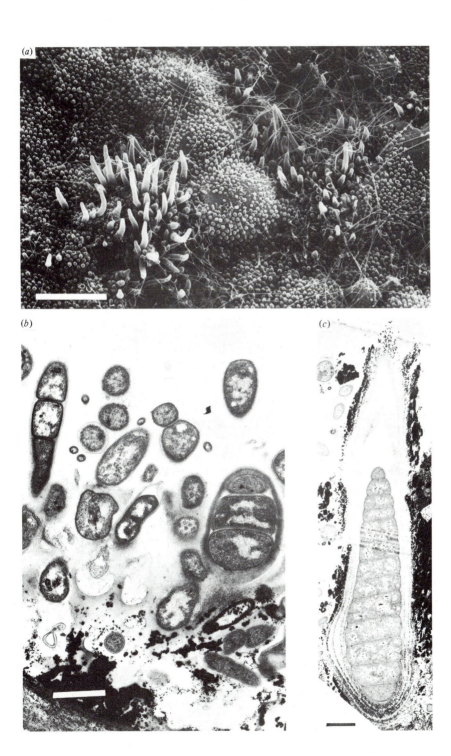

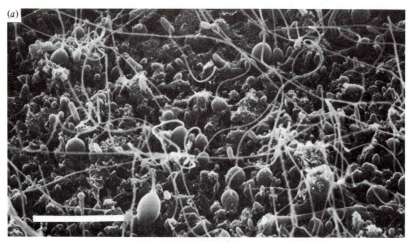

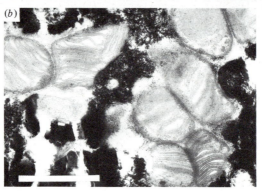

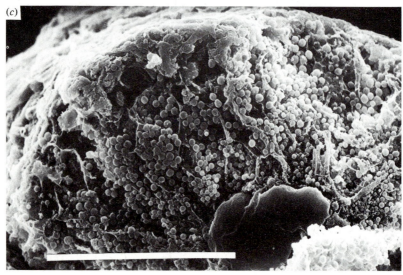

isms such as *Calothrix* (Jannasch & Wirsen, 1981), and their type of metabolism in permanent darkness, possibly including iron oxidation and nitrogen fixation, is open to much speculation. The omnipresent thin filaments (*c.* $0.1\,\mu m$ in diameter: Figs. 6*a*, *b* and 7*a*) represent stalks of *Hyphomicrobium* and *Hyphomonas* species which have indeed been isolated from mat material (S. R. Poindexter, personal communication). They are, along with various coccoid cells, often deeply embedded in layers of manganese–iron oxides and hydroxides. The likely participation of these organisms in the formation of these deposits from the metal-supersaturated hydrothermal fluid is an open question, although an actively manganese-oxidizing bacterium was isolated (Ehrlich, 1983). From the independence of the binding of manganese ions to specific absorbants not found in manganese nodule isolates, Ehrlich concludes that this vent isolate does indeed contribute to the primary production of organic carbon by means of manganese oxidation.

Up to 15% of the cells within the mats, as observed by transmission electron microscopy, exhibited extensive intracellular membrane systems (Fig. 7*b*) which resemble those of Type I methylotrophic bacteria (R. Whittenbury, personal communication). Methane does indeed occur in vent water in considerable, although variable, quantities (Welhan & Craig, 1979) and pure cultures of a *Methylococcus*-like organism have recently been isolated from a variety of samples (R. S. Hanson, personal communication). The origin of methane within the vent system is believed to be geothermal. Recently, however, Jones *et al.* (1983) isolated a thermophilic, methanogenic bacterium of the genus *Methanococcus* from metal sulphide deposits collected in the immediate vicinity of hot vents. This archaebacterium exhibits a 30 min doubling time at 86 °C.

Enrichments with various reduced sulphur compounds yielded over 200 isolates of obligate and facultative chemoautotrophic as well as mixotrophic strains generally falling in the known categories of sulphur bacteria (Ruby, Wirsen & Jannasch, 1981). The prominence of the genus *Thiomicrospira* among the obligate and acid-producing chemoautotrophs was surprising (Ruby & Jannasch,

Fig. 7. (*a*) Area of microbial mat as in Fig. 6. Note the predominance of *Hyphomicrobium*-type cells on and within heavy manganese–iron deposits. (Scale bar, 5 μm.) (*b*) Section of (*a*). Cells exhibiting extensive intraplasmatic membrane systems are common within the most dense metal oxide deposits. (Scale bar, 1 μm.) (*c*) 'Lobe' of *Riftia pachyptila* trophosome. (Scale bar, 100 μm.)

1982). A recent isolate from the '21°N' vent area exhibited a doubling time of 70 min at 36 °C (D. C. Nelson, unpublished). The typically 'oligonitrophilic' behaviour of most isolates raised the suspicion of a nitrogen-fixing ability in these organisms. This was substantiated, however, only in one case. Many of the isolates were facultatively chemoautotrophic and oxidized sulphide and thiosulphate to polythionates (Ruby, Wirsen & Jannasch, 1981). As observed in earlier studies on oxic/anoxic interfaces in the Black Sea and other marine environments (Tuttle & Jannasch, 1973) these non-acidophilic organisms appear to have a competitive growth advantage in the well-buffered sea water over the acidophilic and obligate chemoautotrophic sulphur bacteria. Within the subsurface vent systems growth may take place under very different competitive conditions. The pH values of water emitted from the warm vents were mildly acidic (Edmond et al., 1982).

There is ample visual evidence of the presence of filamentous sulphur bacteria such as Beggiatoa and Thiothrix, but isolation attempts have so far been unsuccessful. As a valuable by-product of our preparation for these attempts, pure cultures of marine Beggiatoa from estuaries were obtained and their ability to fix dinitrogen (Nelson, Waterbury & Jannasch, 1982) and to grow chemoautotrophically (Nelson & Jannasch, 1983) was demonstrated.

The apparent preoccupation with sulphur-based chemosynthesis stems from the preponderance of sulphur compounds among the reduced inorganic materials in the hydrothermal fluid emitted from the warm vents studied so far (Edmond et al., 1982). An exception has been found recently at a new vent site at 2000 m depth in the Gulf of California where ammonia concentrations of up to 15 mM have been found (J. Edmond, personal communication). Future studies will have to deal here with the possibility of a food chain based on chemosynthesis by nitrifying bacteria.

Although no thermophilic sulphur bacteria have so far been isolated from warm vent emissions, there are indications that other types of highly thermophilic bacteria are associated with the 'hot' vents. These discharges of 350 °C hydrothermal fluid at rates of up to $2 \, m/s^{-1}$ produce 'chimneys' of deposited polymetal sulphides. The sudden precipitation of these materials from the hot and metal-saturated water in contact with the 2 °C ambient sea water generates black clouds of suspended metal sulphides which led to the term 'black smokers'. Boiling would not occur below 460 °C at this depth (2600 m). The above-mentioned isolation of a thermophilic methane-producing bacterium (Jones *et al.*, 1983) was achieved from a sample of metal sulphide mud collected from the immediate vicinity (2 °C) of a black smoker.

In comparison with recent isolations of highly thermophilic bacteria (capable of growth beyond 100 °C) from terrestrial and shallow marine springs (Zillig, Schnabel & Tu, 1982; Stetter, 1982), the attempts made so far to obtain clear evidence for the existence of such bacteria at the deep-sea hot vents have been less successful. The production of methane, carbon monoxide and hydrogen in certain media inoculated with hot vent water and incubated at 100 °C, accompanied by increasing cell counts (acridine staining and epifluorescence microscopy) was observed by Baross, Lilley & Gordon (1982). More recently microbial growth at 250 °C was claimed (J. A. Baross, personal communication) in a study that still awaits confirmation.

As new deep-sea vents are discovered, their widely differing chemical and biological characteristics become apparent. They have one phenomenon in common which seems to be the actual basis for their enormous productivity: namely various types of symbioses between the most abundant vent invertebrates and chemoauto-trophic prokaryotes. Although research in this area is in its infancy, its many novel and unique aspects justify a brief appraisal of its present standing.

Chemoautotrophic symbionts

The general notion that microbial chemosynthesis may replace photosynthesis for supporting the existence of a thriving ecosystem is still disturbingly novel to many scientists as well as to laymen. Indeed, the small size of the visibly turbid plumes of the warm and heavily populated vents was not convincing. This was especially true for the massive productivity of the large white clams which often occur 10–20 m away from a noticeable vent where no plume or turbidity was perceptible (Plate 1*b*). A model was constructed proposing gathering of particulate food by thermoconvective sys-tems (Enright *et al.*, 1981). This failed to offer a solution to the problem mainly because of the physical-oceanographic impossibility of transporting to this depth the necessary amounts of food material produced photosynthetically at the surface, but also because man-ganese and helium data indicated the limited extent of the vent water plumes (Lupton *et al.*, 1980; Jannasch, 1983).

The solution finally came from histological (Cavanaugh *et al.*, 1981) and enzymological (Felbeck, 1981) studies on the vestimentiferan tube worms and clams. It appears obvious now that the filtration of a chemosynthetically produced bacterial suspension is far less efficient than transporting dissolved oxygen and hydrogen sulphide (and possibly other reduced compounds such as ammonia and methane) to certain sites or specialized organelles where chemosynthesis may take place *within* the animals. Thus the wasteful filtration process, involving an enormous loss of suspended bacterial cells into the surrounding sea water, is replaced by much more effective symbiotic associations.

Vestimentiferan tube worms are known to lack any ingestive and digestive apparatus, i.e. they have neither mouths nor guts. The species known hitherto, which are only a few millimetres in length, were assumed to take up dissolved organic nutrients through the epidermis. This appeared an insufficient mechanism for growth of specimens the size of *Riftia pachyptila* for various reasons. Anatomical studies finally revealed that the so-called trophosome tissue in the body cavity of the worms consists of cells that appear to be prokaryotic (Cavanaugh *et al.*, 1981) and may, in fact, amount to 60% of the animal's body weight (Fig. 7c). Felbeck (1981) described the 'chemoautotrophic potential' of the trophosome tissue by giving an account of the enzymes catalysing the synthesis of ATP using energy from the oxidation of reduced sulphur compounds (rhodanese, adenosine phosphosulphate reductase and ATP sulphurylase) as well as the Calvin–Benson cycle enzymes ribulose bisphosphate carboxylase and ribulose-5-phosphate kinase (later ADP sulphurylase and phosphoenolpyruvate carboxylase were also found). None of these enzymes was detected in the muscle tissue of the worms. The necessary and simultaneous transport of oxygen and hydrogen sulphide from the retractable gill plume to the trophosome is carried out by the extracellular haemoglobin of the annelid-type blood system and has been the topic of several studies, the most recent ones being by Arp & Childress (1983) and Powell & Somero (1983).

First attempts to isolate and study the relatively large prokaryotic cells of the trophosome led to the general result that, unlike the example of root nodule symbiosis, no axenic or even highly predominant population existed. Many different facultatively chemoautotrophic sulphur-oxidizing and heterotrophic bacteria were isolated from trophosome material by us, and a methylotrophic organism by R. Hanson (personal communication). There is no indication that any of the present isolates is identical with the organism producing the predominant large cells pictured in Fig. 7c. Eggs of the worms did not contain symbiotic prokaryotes. Speculations on the ontogeny of this symbiosis are boundless at this time. Microscopic studies on the 'giant' white clam (Plate 1b) revealed extra- as well as intracellular prokaryotes on and in the gill tissue. The clams do also possess red blood but, in contrast to the pogonophores' blood system, the mollusc haemoglobin is contained in erythrocytes. Its high oxygen affinity has been studied by Terwilliger, Terwilliger & Arp (1983). Comparative analytical gene cloning and nucleotide sequencing completed so far using trophosome material as well as clam gill tissue demonstrated closest similarities to the 5S ribosomal RNA of the above-mentioned *Thiomicrospira* isolate (N. Pace, personal communication).

Because of the present impossibility of studying living and healthy deep-sea animals experimentally under their normal environmental conditions, the actual work confirming the symbiotic relationship between these vent invertebrates and chemoautotrophic prokaryotes is minimal. In fact, it would be insufficient for the rather definite statements made above if the same type of symbiosis had not been intuitively suspected and then discovered in various genera of marine bivalves and pogonophores in locations unrelated to geothermal emissions (Cavanaugh et al., 1981; Felbeck, Childress & Somero, 1981; Southward et al., 1981; Southward, 1982; Cavanaugh, 1983). In these cases, of course, the source of sulphide is not geothermal, but is dependent on the photosynthetic primary production of organic matter. Isolation of the symbionts from these animals, often found in shallow waters, has a good chance of being successful soon and will be helpful in the study of deep-sea vent symbiosis.

The principal notion that animals and whole ecosystems can exist on this globe (and possibly elsewhere) on the basis of chemosynthesis and in the permanent absence of light, is certainly new. One can argue that radiation may still be needed to generate free oxygen as the required electron acceptor by photohydrolysis. The reduced inorganic materials as terrestrial sources of energy are very different, however, from the partly photosynthetically, partly geothermally produced materials (coal, oil, gas) or from geothermal heat sources. Efforts are now being made to use hydrogen sulphide, one of the most common and hazardous ('acid rain') by-products of mining operations, for the production of bacterial biomass in aquaculture (work inspired by the observations on deep-sea vent mussels) or for fermentation to synfuels (Jannasch, 1979; C. D. Taylor & H. W. Jannasch, unpublished).

OUTLOOK

Environmental microbiology seems to have recaptured some of the last century's exploratory spirit. Its close alliance with geological and geochemical research, including expedition-style field work, confers a new face on microbiology, a subject generally better known for white-coat laboratory work than for blue-water oceanography or operations using deep-sea submersibles. Furthermore, the last two decades have seen many new discoveries of physiological as well as morphological types of micro-organisms, ranging from *Bdellovibrio* to *Desulfonema*, and are rivalling Beijerinck's era of the first overwhelmingly productive exploitation of the enrichment culture technique. As a rich source of new finds, the channels of Delft have been challenged by the Solar Lake near Elat in Israel, where the competition between the chemotrophic and the phototrophic way of life among prokaryotes provided evolutionary insights, especially through the discovery of anoxygenic cyanobacteria.

There is no reason to assume that this recent rapid expansion of basic and novel microbiological information is going to stop at this point. The interdisciplinary exploration of biogeochemical features such as fossil and recent stromatolites, polar ice pockets, terrestrial and marine hot springs, as well as the intricate relationships in oligospecies consortia and multi-species microbial mats and sediments, initiated new physiological-ecological studies. The intriguing work on microbial growth at temperatures beyond 100 °C and increased pressures as well as on the symbioses between chemosynthetic prokaryotes and marine invertebrates, has only just begun.

For invaluable help in the preparation of this paper and its illustrations I thank my colleagues C. O. Wirsen, D. C. Nelson, J. B. Waterbury, S. J. Molyneaux and Jane M. Peterson. The work was supported by the National Science Foundation (Grants OCE80-24253 and OCE81-17560). Contribution No. 5387 from the Woods Hole Oceanographic Institution.

REFERENCES

ARP, A. G. & CHILDRESS, J. J. (1983). Sulfide binding by the blood of the hydrothermal vent tube worm *Riftia pachyptila. Science*, **219**, 295–7.

BALLARD, R. D. (1977). Notes on a major oceanographic find. *Oceanus*, **20**, 35–44.

BAROSS, J. A., LILLEY, M. D. & GORDON, L. I. (1982). Is the Ch_4, H_2 and CO venting from submarine hydrothermal systems produced by thermophilic bacteria? *Nature, London*, **298**, 366–8.

BENECKE, W. (1933). Bakteriologie des Meeres. *Abderhaldens Handbuch der biologischen Arbeitsmethoden*, **9** (5), 717–854.

BOSS, K. J. & TURNER, R. D. (1980). The giant white clam from the Galapagos Rift; *Calyptogena magnifica* species novum. *Malacologia*, **20**, 161–94.

CAVANAUGH, C. M. (1983). Symbiotic chemotrophic bacteria in marine invertebrates from sulphide-rich habitats. *Nature, London*, **302**, 58–61.

CAVANAUGH, C. M., GARDINER, S. L., JONES, M. L., JANNASCH, H. W. & WATERBURY, J. B. (1981). Procaryotic cells in the hydrothermal vent tube worm *Riftia pachyptila* Jones: possible chemoautotrophic symbionts. *Science*, **213**, 340–1.

CERTES, A. (1884*a*). Sur la culture, à l'abri des germes atmosphériques, des eaux et des sédimentes rapportes par les expéditions due *Travailleur* et du *Talisman. Comptes Rendus de l'Académie des Sciences, Paris*, 690–3.

CERTES, A. (1884*b*). Note relative à l'action des hautes pressions sur la vitalité des microorganismes d'eau douce et l'eau de mer. *Comptes Rendus de l'Académie des Sciences, Paris*, **36**, 220–2.

CORLISS, J. B., DYMOND, J., GORDON, L. I., EDMOND, J. M., VON HERZEN, R. P., BALLARD, R. D., GREEN, K., WILLIAMS, D., BAINBRIDGE, A., CRANE, K. & VAN ANDEL, T. H. (1979). Submarine thermal springs on the Galapagos Rift. *Science*, **203**, 1073–83.

DEMING, J. W. & COLWELL, R. R. (1982). Barophilic bacteria associated with deep sea animals. *BioScience*, **31**, 507–11.

DEMING, J. W., TABOR, P. S. & COLWELL, R. R. (1981). Barophilic growth of bacteria from intestinal tracts of deep-sea invertebrates. *Microbial Ecology*, **7**, 85–94.

DIETZ, A. S. & YAYANOS, A. A. (1978). Silica gel media for isolating and studying bacteria under hydrostatic pressure. *Applied and Environmental Microbiology*, **36**, 966–8.

EDMOND, J. M., VON DAMM, K. L., McDUFF, R. E. & MEASURES, C. I. (1982). Chemistry of hot springs on the East Pacific rise and their effluent dispersal. *Nature, London*, **297**, 187–91.

EHRLICH, H. (1983). Manganese oxidizing bacteria from a hydrothermally active area on the Galapagos Rift. *Ecological Bulletin*, **35**, 357–66.

ENRIGHT, J. T., NEWMAN, W. A., HESSLER, R. R. & McGOWAN, J. A. (1981). Deep-ocean hydrothermal vent communities. *Nature, London*, **289**, 219–21.

FELBECK, H. (1981). Chemoautotrophic potentials of the hydrothermal vent tube worm, *Riftia pachyptila* (Ventimentifera). *Science*, **213**, 336–8.

FELBECK, H., CHILDRESS, J. J. & SOMERO, J. N. (1981). Calvin–Benson cycle and sulphide oxidation enzymes in animals from sulphide-rich habitats. *Nature, London*, **293**, 291–3.

FISCHER, B. (1886). Bakteriologische Untersuchungen auf einer Reise nach Westindien. *Zeitschrift für Hygiene*, **1**, 421–64.

FISCHER, B. (1894). Die Bakterien des Meeres nach den Untersuchungen der Planktonexpedition unter gleichzeitiger Berückichtigung einiger älterer und neuerer Untersuchungen. *Centralblatt für Bakteriologie*, **15**, 657–66.

HONJO, S., MANGANINI, S. J. & COLE, J. J. (1982). Sedimentation of biogenic matter in the deep ocean. *Deep-Sea Research*, **29**, 609–25.

INGRAHAM, J. L. & STOKES, J. L. (1959). Psychrophilic bacteria. *Bacteriological Reviews*, **23**, 97–108.

JAENICKE, R. (1981). Enzymes under extremes of physical conditions. *Annual Reviews of Biophysics and Bioengineering*, **10**, 1–67.

JANNASCH, H. W. (1967). Growth of marine bacteria at limiting concentrations of organic carbon in seawater. *Limnology and Oceanography*, **12**, 264–71.

JANNASCH, H. W. (1979). Microbial ecology of aquatic low-nutrient habitats. In *Strategy of Life in Extreme Environments*, ed. M. Shilo, pp. 243–60. Berlin: Dahlem Konferenzen.

JANNASCH, H. W. (1983). Microbial processes at deep sea hydrothermal vents. In *Hydrothermal Processes at Sea Floor Spreading Centers*, ed. R. A. Rona. New York: Plenum Press (in press).

JANNASCH, H. W. & WIRSEN, C. O. (1973). Deep sea microorganisms: *in situ* response to nutrient enrichment. *Science*, **180**, 641–3.

JANNASCH, H. W. & WIRSEN, C. O. (1977). Retrieval of concentrated and undecompressed microbial populations from the deep sea. *Applied and Environmental Microbiology*, **33**, 642–6.

JANNASCH, H. W. & WIRSEN, C. O. (1979). Chemosynthetic primary production at East Pacific Ocean floor spreading centers. *BioScience*, **29**, 492–8.

JANNASCH, H. W. & WIRSEN, C. O. (1981). Morphological survey of microbial mats near deep sea thermal vents. *Applied and Environmental Microbiology*, **42**, 528–38.

JANNASCH, H. W. & WIRSEN, C. O. (1982). Microbial activities in undecompressed and decompressed deep sea water samples. *Applied and Environmental Microbiology*, **43**, 1116–24.

JANNASCH, H. W., WIRSEN, C. O. & TAYLOR, C. D. (1976). Undecompressed microbial populations from the deep sea. *Applied and Environmental Microbiology*, **32**, 360–7.

JANNASCH, H. W., WIRSEN, C. O. & TAYLOR, C. D. (1982). Deep sea bacteria: isolation in the absence of decompression. *Science*, **216**, 1315–17.

JANNASCH, H. W., WIRSEN, C. O. & WINGET, C. L. (1973). A bacteriological, pressure-retaining, deep-sea sampler and culture vessel. *Deep-Sea Research*, **20**, 661–4.

JONES, M. L. (1980). *Riftia pachyptila*, n. gen., n. sp., the vestimentiferan worm from the Galapagos Rift geothermal vents (Pogonophora). *Proceedings of the Biological Society of Washington*, **93**, 1205–313.

JONES, W. G., LEIGH, J. A., MAYER, F., WOESE, C. R. & WOLFE, R. S. (1983). *Methanococcus jannaschii* sp. nov., an extremely thermophilic methanogen from a submarine hydrothermal vent. *Archives of Microbiology* (in press).

KARL, D. M. (1980). Cellular nucleotide measurements and applications in microbial ecology. *Microbiology Reviews*, **44**, 739–96.

KARL, D. M., WIRSEN, C. O. & JANNASCH, H. W. (1980). Deep sea primary production at the Galapagos hydrothermal vents. *Science*, **207**, 1345–7.

KRISS, A. E. (1962). *Marine Microbiology (Deep Sea)*. London: Wiley-Interscience.

LANDAU, J. V. & POPE, D. H. (1980). Recent advances in the area of barotolerant protein synthesis in bacteria and implications concerning barotolerant and barophilic growth. *Advances in Aquatic Microbiology*, **2**, 49–76.

LONSDALE, P. F. (1977). Clustering of suspension-feeding macrobenthos near abyssal hydrothermal vents at oceanic spreading centers. *Deep-Sea Research*, **24**, 857–63.

LUPTON, J. E., KLINKHAMMER, G., NORMARK, W., HAYMON, R., MACDONALD, K., WEISS, R. & CRAIG, H. (1980). Helium-3 and manganese at the 21° N East Pacific Rise hydrothermal site. *Earth and Planetary Science Letters*, **50**, 115–27.

MARQUIS, R. E. (1976). High-pressure microbial physiology. *Advances in Microbial Physiology*, **14**, 158–241.

MARQUIS, R. E. (1982). Microbial barobiology. *BioScience*, **32**, 267–71.

MARQUIS, R. E. & MATSUMURA, P. (1978). Microbial life under pressure. In *Microbial Life in Extreme Environments*, ed. D. J. Kushner. New York: Academic Press.

MORITA, R. Y. (1975). Psychrophilic bacteria. *Bacteriological Reviews*, **39**, 144–67.

NELSON, D. C. & JANNASCH, H. W. (1983). Chemoautotrophic growth of a marine *Beggiatoa* in sulfide-gradient cultures. *Archives of Microbiology* (in press).

NELSON, D. C., WATERBURY, J. B. & JANNASCH, H. W. (1982). Nitrogen fixation and nitrate utilization by marine and freshwater *Beggiatoa*. *Archives of Microbiology*, **133**, 172–7.

POINDEXTER, J. S. (1981). Oligotrophy: feast and famine existence. *Advances in Microbial Ecology*, **5**, 67–93.

POPE, D. H., SMITH, W. P., SCHWARZ, R. W. & LANDAU, J. V. (1975a). Role of bacterial ribosomes in barotolerance. *Journal of Bacteriology*, **121**, 664–9.

POPE, D. H., CONNORS, N. T. & LANDAU, J. V. (1975b). Stability of *Escherichia coli* polysomes at high hydrostatic pressure. *Journal of Bacteriology*, **121**, 753–68.

POPE, D. H., SMITH, W. P., ORGRINC, M. A. & LANDAU, J. V. (1976). Protein synthesis at 680 atm: is it related to environmental origin, physiological type, or taxonomic group? *Applied and Environmental Microbiology*, **31**, 1001–2.

POWELL, M. A. & SOMERO, G. N. (1983). Blood component prevents sulfide poisoning of respiration of the hydrothermal vent tube worm *Riftia pachyptila*. *Science*, **219**, 297–9.

REHBOCK, P. F. (1975). Huxley, Haeckel, and the oceanographers: the case of *Bathybius haeckeli*. *Isis*, **66**, 504–33.

RHOADS, D. C., LUTZ, R. A., REVELAS, E. C. & CERRATO, R. M. (1981). Growth of bivalves at deep sea hydrothermal vents along the Galapagos Rift. *Science*, **214**, 911–12.

RUBY, E. G. & JANNASCH, H. W. (1982). Physiological characteristics of *Thiomicrospira* sp. L-12 isolated from deep sea hydrothermal vents. *Journal of Bacteriology*, **149**, 161–5.

RUBY, E. G., WIRSEN, C. O. & JANNASCH, H. W. (1981). Chemolithotrophic sulfur-oxidizing bacteria from the Galapagos Rift hydrothermal vents. *Applied and Environmental Microbiology*, **42**, 317–24.

RUSSELL, H. L. (1892). Untersuchungen über im Golf von Neapel lebende Bacterien. *Zeitschrift für Hygiene*, **11**, 165–206.

RUSSELL, H. L. (1893). The bacterial flora of the Atlantic Ocean in the vicinity of Woods Hole. *Botanical Gazette*, **18**, 383–95.

SAUNDERS, P. M. & FOFONOFF, N. P. (1976). Conversion of pressure to depth in the ocean. *Deep-Sea Research*, **23**, 109–11.

SCHMIDT-LORENZ, W. (1967). Behaviour of microorganisms at low temperatures. *Bulletin of the International Institute of Refrigeration*, **2** and **4**, 1–59.

SCHWARZ, J. R., YAYANOS, A. A. & COLWELL, R. R. (1976). Metabolic activities of the intestinal microflora of a deep sea invertebrate. *Applied and Environmental Microbiology*, **31**, 46–8.

SOUTHWARD, A. J., SOUTHWARD, E. C., DANDO, P. R., RAU, G. H., FELBECK, H. & FLUGEL, H. (1981). Bacterial Symbionts and low $^{13}C/^{12}C$ ratios in tissues of Pogonophora indicate unusual nutrition and metabolism. *Nature, London*, **293**, 616–20.

SOUTHWARD, E. C. (1982). Bacterial symbionts in Pogonophora. *Journal of the Marine Biological Association of the United Kingdom*, **62**, 889–906.

STETTER, K. O. (1982). Ultrathin mycelia-forming organisms from submarine volcanic areas having an optimum growth temperature of 105 °C. *Nature, London*, **300**, 258–60.

TABOR, P. S., DEMING, J. W., OHWADA, K., DAVIS, H., WAXMAN, M. & COLWELL, R. R. (1981). A pressure-retaining deep ocean sampler and transfer system for measurements of microbial activity in the deep sea. *Microbial Ecology*, **7**, 51–65.

TAYLOR, C. D. (1978). The effect of pressure upon the solubility of oxygen in water. *Archives of Biochemistry and Biophysiology*, **191**, 375–84.

TAYLOR, C. D. (1979a). Growth of a bacterium under a high pressure oxy-helium atmosphere. *Applied and Environmental Microbiology*, **37**, 42–9.

TAYLOR, C. D. (1979b). Solubility of oxygen in a seawater medium in equilibrium with a high pressure oxy-helium atmosphere. *Undersea Biomedical Research*, **6**, 147–51.

TERWILLIGER, R. C., TERWILLIGER, N. B. & ARP, A. (1983). Thermal vent clam (*Calyptogena magnifica*) hemoglobin. *Science*, **219**, 981–3.

TUREKIAN, K. K. & COCHRAN, J. K. (1981). Growth rate of a vesocomyid clam from the Galapagos spreading center. *Science*, **214**, 909–10.

TURNER, J. T. (1979). Microbial attachment to copepod fecal pellets and its possible ecological significance. *Transactions of the American Microscopical Society*, **98**, 131–5.

TUTTLE, J. H. & JANNASCH, H. W. (1973). Dissimilatory reduction of inorganic sulfur by facultatively anaerobic marine bacteria. *Journal of Bacteriology*, **115**, 732–7.

TUTTLE, J. H., WIRSEN, C. O. & JANNASCH, H. W. (1983). Microbial activities in the emitted hydrothermal vent waters of the Galapagos Rift vents. *Marine Biology*, **73**, 293–9.

WELHAN, J. A. & CRAIG, H. (1979). Methane and hydrogen in East Pacific Rise hydrothermal fluid. *Geophysical Research Letters*, **6**, 829.

WIRSEN, C. O. & JANNASCH, H. W. (1975). Activity of marine psychrophilic bacteria at elevated hydrostatic pressures and low temperatures. *Marine Biology*, **31**, 201–8.

WIRSEN, C. O. & JANNASCH, H. W. (1976). Decomposition of solid organic materials in the deep sea. *Environmental Science and Technology*, **10**, 880–7.

YAYANOS, A. A. & DIETZ, A. S. (1983). Death of a hadal deep sea bacterium after decompression. *Science*, **220**, 497–8.

YAYANOS, A. A., DIETZ, A. S. & VAN BOXTEL, R. (1979). Isolation of a deep-sea barophilic bacterium and some of its growth characteristics. *Science*, **205**, 808–10.

YAYANOS, A. A., DIETZ, A. S. & VAN BOXTEL, R. (1981). Obligately barophilic bacterium from the Mariana Trench. *Proceedings of the National Academy of Sciences, USA*, **78**, 5212–15.

YAYANOS, A. A., DIETZ, A. S. & VAN BOXTEL, R. (1982). Dependence of reproduction rate on pressure as a hallmark of deep sea bacteria. *Applied and Environmental Microbiology*, **44**, 1356–61.

ZILLIG, W., SCHNABEL, R. & TU, J. (1982). The phylogeny of archaebacteria, including novel anaerobic thermoacidophiles in the light of RNA polymerase structure. *Naturwissenschaften*, **69**, 197–204.

ZOBELL, C. E. (1946). *Marine Microbiology*. Waltham, MA: Chronica Botanica Company.

ZOBELL, C. E. (1968). Bacterial life in the deep sea. *Bulletin of the Misaki Marine Biological Institute of Kyoto University*, **12**, 77–96.

ZOBELL, C. E. & MORITA, R. Y. (1957). Barophilic bacteria in some deep sea sediments. *Journal of Bacteriology*, **73**, 563–8.

METABOLISM OF CHEMOTROPHIC ANAEROBES: OLD VIEWS AND NEW ASPECTS

RUDOLF K. THAUER* AND J. GARETH MORRIS†

*Fachbereich Biologie, Mikrobiologie, Philipps-Universität, D-3550 Marburg (Lahn), Federal Republic of Germany, and †Dept of Botany and Microbiology, University College of Wales, Aberystwyth SY23 3DA, UK

INTRODUCTION

During the period when interest in the pathways of intermediary metabolism was at its height, anaerobic bacteria were star performers in the repertoire of the microbial physiologist. Their metabolic ingenuity was impressive and instructive and they proved to be the source of many novel reactions, enzymes and cofactors. More importantly, several of the studies of anaerobes that were initially provoked by curiosity at their ability to utilize apparently unlikely substrates ended up by making major contributions to our knowledge of fundamental biochemical processes common to all life forms (Morris, 1977). Thereafter, when the focus of research activity shifted to investigations of gene expression and its control, and to aspects of microbial cell behaviour and development that were best studied using mutant strains carrying defined genetic lesions, it was manifestly sensible to select as experimental subjects microbes that were more easily manipulated and which were amenable to current techniques of genetic analysis. For these and other reasons the anaerobes were for a time relegated to playing a largely supporting role, continuing to contribute examples of the diversity of life forms and occasionally supplying the most apt system with which to study some timely problem in bioenergetics or biodegradation. However, during the past decade or so there has been a marked resurgence of interest in the anaerobes; so much so, that they now seem to have reclaimed their former prominence in the esteem of microbial physiologists (Thauer, Jungermann & Decker, 1977; Gottschalk & Andreesen, 1979).

In large measure this must be attributed to the fact that, in the

Dedicated to Professor Dr P. Karlson on the occasion of his sixty-fifth birthday.

interim, techniques of isolating and handling strict anaerobes had been improved and simplified to the point that they had become routine procedures in all good microbiological laboratories. Quite suddenly, or so it appeared, anaerobes achieved enhanced status in the eyes of the clinical microbiologists who had formerly tended to overlook their existence in any but the most obvious disease states associated with anaerobic bacterial sepsis or intoxication (Miraglia, 1974). The availability of metronidazole as the first comprehensive anti-anaerobic bactericidal agent greatly accelerated this reassessment of their ubiquitous distribution and important contribution to human health as well as to disease (Edwards *et al.*, 1982). In the context of general microbiology the improved facilities for handling anaerobes arrived most propitiously at the same time as the revival of interest in anaerobic microbial ecology (especially in the relations between aerobes and anaerobes in mixed microbial communities) and in the biotechnological exploitation of anaerobic bacteria both as a means of undertaking potentially profitable biotransformations and as a source of useful enzymes. One of the more obvious manifestations of this burgeoning interest in the anaerobes is apparent in the microbiological literature of the past few years, wherein the isolation of new families and genera of anaerobic bacteria has been reported with a frequency that was formerly more characteristic of the description of new species and strains.

Faced with this explosive increase in research effort devoted to anaerobic bacteria, we recognized from the outset that our contribution to this volume could not hope to be comprehensive in its coverage of either the totality of anaerobic metabolism or the attendant literature. We have therefore chosen, in an avowedly arbitrary manner, to consider merely a selection of topics in which there is a current lively research interest. These should be viewed as being exemplary and not exhaustive, whilst the supporting bibliography is necessarily incomplete.

The following topics are discussed: (i) size of the biological energy quantum; (ii) novel mechanisms of free energy conservation; (iii) the stoichiometric ATP gain; (iv) fermentation products and the intracellular pH; (v) catabolism of acetate; (vi) degradation of aromatic compounds; (vii) autotrophic fixation of carbon dioxide; (viii) mechanistically anomalous reactions; (ix) unusual trace elements; and (x) novel coenzymes. Oxidation reactions with nitrate or nitrite have been disregarded since these electron acceptors are of minor ecological importance in strictly anaerobic environments.

SIZE OF THE BIOLOGICAL ENERGY QUANTUM

Free energy transduction in the living cell proceeds via the ATP system. The reactions that supply the free energy have traditionally been considered to be stoichiometrically coupled to the formation of ATP, so leading to a quantization of the free energy transfer. This concept has, in turn, given rise to the view that in biological systems free energy is conserved and utilized in these quantal 'packets'. The 'biological energy quantum' has therefore been considered to be equivalent to the minimum amount of free energy required to drive the synthesis of 1 mol ATP *in vivo*. This is approximately $-44\,kJ\,mol^{-1}$ (i.e. $-10.5\,kcal\,mol^{-1}$) under quasi-reversible conditions (Decker, Jungermann & Thauer, 1970; Thauer *et al.*, 1977).

Many chemotrophic anaerobes can obtain energy for growth from reactions whose associated free energy change is very small. Indeed, the free energy made available by the consumption of 1 mol of substrate is sometimes considerably less than that required for the synthesis of 1 mol of ATP. It follows that, in such cases, some mechanism must exist whereby free energy is conserved and accumulated for transduction into a specific yield of ATP which is less than 1 mol ATP per mole of substrate utilized. An 'energized' membrane could provide such a mechanism whose operation could result in a fractional specific yield of ATP.

ATP synthesis via electron-transport-linked phosphorylation

Chemotrophic anaerobes synthesize ATP by substrate-level phosphorylation or by electron-transport-linked phosphorylation, or by both of these mechanisms (Thauer *et al.*, 1977). Several recent findings indicate that the mechanism of electron-transport-linked phosphorylation may be such as to allow fractional stoichiometries of ATP formation:

1. It is now recognized that transmembrane vectorial translocation of protons can be accomplished via proteinaceous proton pumps and is not entirely reliant on the 'direct' mechanisms of impelled proton movement supplied by redox loops or the protonmotive Q cycle. Indeed, it has been suggested that proton pumps are the predominant means of effecting the energy-linked transfer of protons from one side of a membrane to the other (Kell *et al.*, 1981*a*). Now, whereas the $\rightarrow H^+/e^-$ stoichiometry of a redox

loop is necessarily 1, a proton pump may, theoretically, effect the translocation of one proton per two electrons (or, more generally, of n protons per electron).

2. More than two, very possibly three, vectorial protons may be required to drive the synthesis of one ATP molecule via the membrane-associated H^+-ATP synthase (proton-translocating ATPase).

3. Evidence of variable degrees of 'uncoupling' at the energized membrane, of which 'molecular slip in the protonmotive pump' is one instance, introduces additional inexactitudes into calculations of $\rightarrow H^+/ATP$ stoichiometries.

Evidence for the existence of proton pumps and of a $\rightarrow H^+/ATP$ stoichiometry of 3, has come from measurements of $\rightarrow H^+/e^-$ and $\rightarrow H^+/ATP$ ratios in mitochondria and chloroplasts (Hauska & Trebst, 1977; von Jagow & Engel, 1980; Ferguson & Sorgato, 1982). Since these 'endosymbionts' are phylogenetically related to Eubacteria (Margulis, 1981; Seewaldt & Stackebrandt, 1982) these findings can probably be extended to prokaryotes. Maloney (1983) recently provided evidence that the $\rightarrow H^+/ATP$ stoichiometry of the bacterial H^+-ATP synthase is 3 rather than 2 (Maloney, 1982). However, virtually all of the calculations so far made of $\rightarrow H^+/ATP$ in bacteria have been based on measurements of $\Delta\mu H^+$ (protonmotive force) and ΔG_p (phosphorylation potential) and not of actual stoichiometries of ATP-driven proton translocation under level flow conditions. Irrespective of the correctness, or otherwise, of variously deduced $\rightarrow H^+/ATP$ stoichiometries, it is now evident that electron transport can, in principle, effect the net formation of fractional numbers of moles of ATP per $2e^-$, thus explaining how it is that chemotrophic anaerobes reliant on electron-transport-linked phosphorylation can couple redox processes with the net synthesis of less than 1 mol ATP per mole of substrate oxidized. The following example illustrates this point.

Thermoproteus neutrophilus is an anaerobic archaebacterium that can grow on hydrogen and elemental sulphur as sole energy source (Fischer *et al.*, 1983). The reduction of elemental sulphur by hydrogen to give hydrogen sulphide is associated in the standard state with a free energy change of only $-33\,\mathrm{kJ\,mol^{-1}}$, which is insufficient for the formation of 1 mol ATP per mole of hydrogen oxidized.* Organisms utilizing hydrogen as sole electron donor can

---

* $\Delta G^{0\prime}$ for the different reactions was calculated from Δgf^0 values compiled by Thauer *et al.* (1977). Carbon dioxide, hydrogen sulphide and methane were considered to be in the gaseous state.

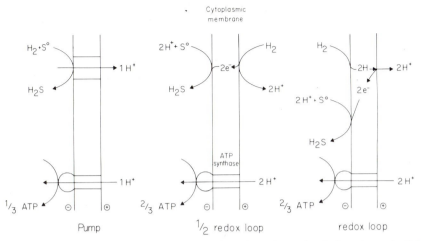

Fig. 1. Three possible mechanisms of generation of an electrochemical proton potential during sulphur reduction by hydrogen, allowing for fractional ATP/2e ratios. The →H$^+$/ATP ratio of the ATP synthase is assumed to be 3.

synthesize ATP only via electron-transport-linked phosphorylation. In Fig. 1 are depicted three possible topological arrangements of the electron transport chain with associated ATP generation, each of which assumes, for the sake of argument, a →H$^+$/ATP ratio of 3. Possibility 1 assumes that, per mole of hydrogen oxidized, one proton is electrogenically pumped across the membrane; the proton in turn then drives the synthesis of $\frac{1}{3}$ mol ATP when it returns via the H$^+$-ATP synthase. Possibilities 2 and 3 assume 'direct' proton translocation via a redox loop with a →H$^+$/2e$^-$ stoichiometry of 1; the ATP/2e$^-$ stoichiometry in both cases is therefore $\frac{2}{3}$.

ATP synthesis via substrate-level phosphorylation

Many chemotrophic anaerobes synthesize ATP solely by substrate-level phosphorylation reactions. Although, by the partitioning of intermediates between different branches of a fermentation pathway, it is possible to obtain a fractional specific yield of ATP (e.g. 3.3 mol ATP per mole of glucose utilized by *Clostridium pasteurianum*: Thauer *et al.*, 1977), it is, at first sight, not easy to explain why certain of these bacteria are able to grow on substrates whose fermentation is not sufficiently exergonic to yield 1 mol ATP per mole of substrate consumed. In such instances (i.e. when ΔG (fermentation) $<\Delta G$ (ATP synthesis)) it is evident that free energy must be 'accumulated' in some biophysical state, such as an energized membrane, in order to effect the necessary ATP production. For example, some part of the ATP generated by substrate-level phosphorylation may be used to drive a thermodynamically unfavourable redox reaction. Such ATP-driven electron transport is mechanistically the reverse of spontaneous electron-transport-linked phosphorylation. We will assume that the H^+-ATPase which is involved has a $\rightarrow H^+ / ATP$ stoichiometry of 3, on the basis that the membrane-associated H^+-ATPase of fermentative bacteria and the H^+-ATP synthase of bacteria undertaking electron-transport-linked phosphorylation are, in many ways, very similar (Clarke, Fuller & Morris, 1979; Clarke & Morris, 1979).

Reversed electron transport was first observed in mitochondria effecting reduction of NAD^+ ($E^{0'} = -320\,mV$) by succinate ($E^{0'} = +30\,mV$) (Klingenberg & Schollmeyer, 1960), and was later found to be similarly involved in reduction of NAD^+ in many aerobic chemolithotrophs and anaerobic phototrophs (Aleem, 1977; Knaff, 1978). Reversed electron transport has not, as yet, been experimentally demonstrated in chemotrophic anaerobes but is very likely to occur in this group of bacteria also. It must, for example, occur in methanogens growing on hydrogen plus carbon dioxide (Kell *et al.*, 1981*b*).

Another fermentation in which, we believe, reversed electron transport should be implicated, is that whereby *Syntrophomonas wolfei*, when syntrophically associated with hydrogen-consuming anaerobes which maintain a very low hydrogen concentration in its environment, grows on butyrate and converts this fatty acid to two acetate plus two hydrogen (McInerney *et al.*, 1981):

$$\text{Butyrate}^- + 2H_2O \rightarrow 2\ \text{acetate}^- + H^+ + 2H_2 \quad \Delta G^{0'} = +48\,kJ\,mol^{-1}$$

Even at a hydrogen concentration of $0.1\ \mu M$ and an acetate concentration of $10\ \mu M$, the fermentation is exergonic only to the extent of $-15\,kJ\,mol^{-1}$ and can thus effect synthesis of, at most, $\frac{1}{3}$ mol ATP per mole of butyrate utilized. Lower concentrations of hydrogen and of acetate cannot easily be maintained by hydrogen- and acetate-consuming organisms since their K_s values for hydrogen and acetate are of the order of $1\ \mu M$ and $0.1\ mM$, respectively (Kristjansson, Schönheit & Thauer, 1982; Schönheit, Kristjansson & Thauer, 1982). Acetate formation from butyrate in *S. wolfei* most probably proceeds via butyryl-CoA, crotonyl-CoA, β-hydroxybutyryl-CoA, acetoacetyl-CoA and two acetyl-CoA, and hence via substrate-level phosphorylation with the net synthesis of 1 mol ATP per mole of butyrate consumed. The oxidation of butyryl-CoA to crotonyl-CoA

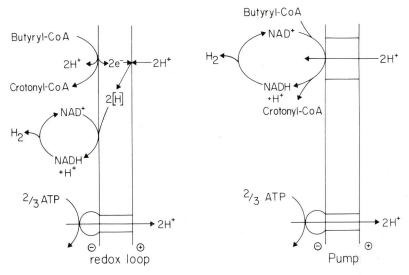

Fig. 2. Two hypothetical mechanisms of reversed electron transport by which the endergonic oxidation of butyryl-CoA to crotonyl-CoA and hydrogen could be driven by the hydrolysis of $\frac{2}{3}$ mol ATP.

$(E^{0\prime} = -15\,\text{mV})$ with protons as the electron acceptor $(E' \approx -200\,\text{mV}$ at a hydrogen concentration of $0.1\ \mu\text{M})$ is thermodynamically only feasible if driven by the hydrolysis of $\frac{2}{3}$ mol ATP. Possible mechanisms are shown in Fig. 2, which explains the origin of the net $\frac{1}{3}$ mol ATP per mole butyrate which was the maximum specific yield anticipated from the thermodynamics of the fermentation of butyrate to acetate plus hydrogen.

Variable $\rightarrow H^+/e^-$ and $\rightarrow H^+/ATP$ stoichiometries

Theoretical arguments have been advanced (Hill, 1977; Stucki, 1982) in support of the view that the $\rightarrow H^+/e^-$ stoichiometry of proton pumps and the $\rightarrow H^+/ATP$ stoichiometry of the H^+-ATP synthase/ATPase should be variable and subject to regulation by the prevailing electrochemical proton potential. (For a discussion of the proton-variable stoichiometry of various transport systems see Konings & Booth, 1981.) It follows that (exergonic) free energy increments per mole of substrate utilized that are less than the equivalent required to synthesize even $\frac{1}{3}$ mol ATP could be harnessed by organisms to drive the synthesis of ATP; almost any exergonic redox reaction could thus sustain growth, provided that electron transport (normal or reversed) at an energy-transducing membrane was involved. This would render meaningless the term 'biological energy quantum' as formerly conceived.

NOVEL MECHANISMS OF FREE ENERGY CONSERVATION

Until recently it was assumed that chemotrophic anaerobes conserve free energy by two mechanisms only: viz. substrate-level phosphorylation and electron-transport-linked phosphorylation (Thauer et al., 1977). In the latter, free energy is first conserved as a transmembrane electrochemical proton potential which then drives ATP synthesis via a H^+-ATP synthase. In the last few years two additional mechanisms have come to light whereby free energy can be conserved in anaerobes by the generation of a transmembrane electrochemical gradient of ions. The first of these, which we refer to as being 'transport-coupled', directly generates a protonmotive force (Michels et al., 1979); the second, 'sodium-coupled' mechanism initially conserves free energy as a transmembrane sodium ion potential difference (Dimroth, 1980). In principle, both mechanisms can drive ATP synthesis via a membrane H^+-ATP synthase, though when they operate in fermentative organisms it could be considered that they merely spare such ATP (produced by substrate-level phosphorylation) as would otherwise have to be consumed to sustain the necessary cell membrane energization. In either case the outcome would be the same: more ATP per mole of substrate utilized would be available for biomass synthesis.

Transport-coupled membrane energization

Certain of the fermentation products formed by chemotrophic anaerobes leave the cell by carrier-mediated excretion (e.g. lactate or succinate). In a theoretical study of lactate efflux, Michels et al. (1979) proposed that membrane-located carriers effect the electrogenic excretion of such metabolic end-products in symport with protons. These substances thus accumulate in the cell until the outwardly directed driving force supplied by the chemical product gradient exceeds the inwardly directed driving force supplied by the electrochemical gradient of protons. The excretion of end-products will then lead to the generation of an electrochemical proton gradient which could conceivably be used to drive the synthesis of ATP (Fig. 3). Experimental evidence consistent with this hypothesis

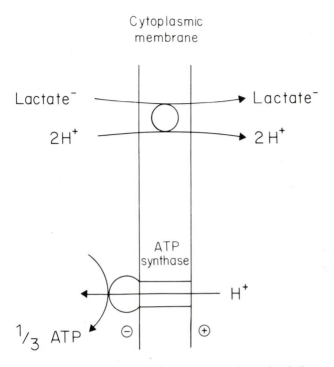

Fig. 3. Coupling of lactate efflux with the generation of an electrochemical proton gradient across the cytoplasmic membrane which could be used to drive the synthesis of ATP.

was obtained by Konings' group when they demonstrated that lactate efflux from de-energized cells of *Streptococcus cremoris* could drive the uptake of leucine (Otto *et al.*, 1980*b*; ten Brink & Konings, 1982). A similar observation was made with membrane vesicles of *Escherichia coli* (ten Brink & Konings, 1980). It was also found that the molar growth yield of *S. cremoris* on lactose was substantially increased when the external concentration of lactate was kept low (Otto *et al.*, 1980*a*; see also Otto *et al.*, 1983). This observation has ecological consequences in that a lactate-forming bacterium when grown in the presence of a lactate-consuming organism conserves more free energy from its fermentation process than when it is grown in pure culture.

Sodium-coupled membrane energization

Oxaloacetate decarboxylase from *Klebsiella aerogenes* (Dimroth, 1980, 1982*a*, *b*), methylmalonyl-CoA decarboxylase from *Veil-*

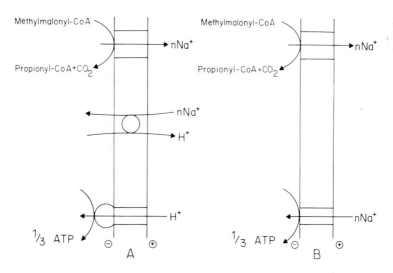

Fig. 4. Coupling of methylmalonyl-CoA decarboxylation with the generation of an electrochemical sodium ion potential across the cytoplasmic membrane which could be used to drive the synthesis of ATP via two possible mechanisms (A and B).

lonella alcalescens (Hilpert & Dimroth, 1982), and glutaconyl-CoA decarboxylase from *Acidaminococcus fermentans*, *Clostridium symbiosum* and *Peptococcus aerogenes* (Buckel & Semmler, 1982) are all membrane-associated, biotin-containing enzymes which are activated by sodium ions. Recent work in the laboratories of Dimroth and Buckel has demonstrated that these decarboxylases function as electrogenic sodium ion pumps whereby the free energy of the decarboxylation reaction ($\Delta G^{0\prime} = -30\,\mathrm{kJ\,mol^{-1}}$) is conserved in an electrochemical gradient of sodium ions. Via a sodium/hydrogen ion antiporter (Lanyi, 1979) the 'sodium ion motive force' can, in principle, be converted into a protonmotive force and thus used to drive the synthesis of ATP (Fig. 4A) (Dimroth, 1982c). Or, as another possibility, the 'sodium ion motive force' could directly poise a sodium-dependent ATP synthase (Heefner & Harold, 1982) (Fig. 4B).

Schink & Pfennig (1982a) have isolated a strictly anaerobic sodium-requiring bacterium, *Propionigenium modestum*, that can grow on succinate as sole energy source by converting this dicarboxylic acid to propionate plus carbon dioxide:

$$\text{Succinate}^{2-} + \text{H}^+ \rightarrow \text{propionate}^- + \text{CO}_2 \quad \Delta G^{0\prime} = -25\,\mathrm{kJ\,mol^{-1}}$$

Methylmalonyl-CoA is an intermediate; by serving as a sodium ion pump the methylmalonyl-CoA decarboxylase provides a mechanism for free energy conservation in this organism (B. Schink, personal communication).

Prompted by such findings, it might be prudent to consider whether other anaerobes which show a growth requirement for sodium ions could not also be deriving benefit from the operation of similar sodium ion pumps. Methane formation from carbon dioxide, methanol and acetate has recently been found to be strictly dependent on provision of sodium ions (Perski, Moll & Thauer, 1981; Perski, Schönheit & Thauer, 1982). The role of sodium ions in free energy conservation in methanogenic bacteria is, however, not yet understood.

Inorganic pyrophosphate as an energy source for growth

Recently, evidence has been forthcoming that species of *Desulfotomaculum* can utilize inorganic pyrophosphate (PP_i) as a source of energy for growth (Liu, Hart & Peck, 1982):

$$PP_i + H_2O \rightarrow 2P_i \quad \Delta G^{0\prime} = -22\,\text{kJ}\,\text{mol}^{-1}$$

It further appears that this unique type of energy metabolism may not be restricted to these sulphate-reducing bacteria (Peck *et al.*, 1982; Varma & Peck, 1983). Though it has been proposed that ATP synthesis at the expense of pyrophosphate proceeds via acetyl phosphate as an intermediate (Peck & LeGall, 1982), this mechanism is thermodynamically not very appealing since the synthesis of acetyl phosphate from acetate and pyrophosphate is highly endergonic ($\Delta G^{0\prime} = +23\,\text{kJ}\,\text{mol}^{-1}$). Inorganic-pyrophosphate-driven ATP synthesis in liposomes containing membrane-bound inorganic pyrophosphatase and ATP synthase has recently been described (Nyrén & Baltscheffsky, 1983). For a review of 'How useful is the energy in inorganic pyrophosphate?' see Reeves (1976).

THE STOICHIOMETRIC ATP GAIN

Derivation of the stoichiometric ATP gain from measured growth yields

Bauchop & Elsden (1960) observed that approximately $10\,\text{g}$ dry weight of bacterial cells would be produced per mole of ATP generated during growth of a batch culture (i.e. $Y_{\text{ATP}}^{\text{obs}} = 10$). This ratio was subsequently used with many chemotrophic anaerobes as the basis for calculating their stoichiometric ATP gains (per mole of substrate utilized) from their observed molar growth yields. Yet, more recently, estimations of the ATP requirement for cell synthesis (Stouthamer, 1979) have revealed that when growth takes place in a glucose minimal medium, nearly 30 g dry weight of cells should theoretically be formed per mole of ATP available ($Y_{ATP}^{theor} = 30$). Correction for the maintenance energy requirement of the growing cells (Pirt, 1975, 1982) made it possible to calculate a maximum value for the growth yield ($Y_{substrate}^{max}$) representing the yield of cells which would be produced per mole of substrate utilized at an infinitely high specific growth rate. However, a substantial gap still remained between the experimental (Y_{ATP}^{max}) and theoretical (Y_{ATP}^{theor}) growth yields, which led to the conclusion being drawn that the apparent ATP gain ($k \cdot n$) in growing bacterial cells is always smaller than the stoichiometric ATP gain (n) (see Harder, van Dijken & Roels, 1981).

$$Y_{substrate}^{max} = k \cdot n \cdot Y_{ATP}^{theor}$$

The yield coefficient (k), though always less than 1, is not a constant and its value varies with growth conditions. When the rate of bacterial growth is limited by the rate of free energy provision (e.g. by the supplied concentration of that substrate which serves as the energy source) then k has values between 0.4 and 0.8 (Stouthamer, 1979). These values have been derived from the maximal growth yields of bacteria undertaking fermentations whose stoichiometric ATP gains (n) were thought both to be precisely known and to be integral numbers. However, since n for a given fermentation may have a fractional value (see the section on the size of the biological energy quantum above) and since processes such as solute export may contribute unpredictably but significantly to the ATP gain (see the section on novel mechanisms of free energy conservation), the values of k can only be empirical, for there is, at present, no means of determining k independently of n. Therefore even when obtained under the correct experimental conditions (growth rate limited by the energy supply), growth yield data can only give a rough estimate of the stoichiometric ATP gain. The accuracy of this determination does not, in fact, exceed that of the theoretical estimation of the value of n from the free energy change of the energy-supplying reaction. This can be illustrated by reference to findings made with *Vibrio succinogenes* (*Wolinella succinogenes*: Tanner *et al.*, 1981), whose relatively straightforward metabolism has probably been better investigated than that of any other chemotrophic anaerobe.

This rumen bacterium grows on formate and fumarate as sole source of energy and synthesizes ATP only by electron-transport phosphorylation:

$$\text{Formate}^- + H^+ + \text{fumarate} \rightarrow \text{succinate} + CO_2$$
$$\Delta G^{0\prime} = -80 \, \text{kJ mol}^{-1}$$

The stoichiometric ATP gain was measured *in vitro* at low phosphorylation potentials by two methods and was found to be 1 ± 0.3 mol $\text{ATP}/2e^-$ (Kröger & Winkler, 1981). A theoretical $Y^{\text{max}}_{\text{ATP}}$ value of 18 ± 2 g dry weight of cells per mole of ATP has been calculated for *V. succinogenes* (Bronder *et al.*, 1982) following elucidation of the major biosynthetic pathways in this organism and their ATP requirements. To obtain the value $Y^{\text{max}}_{\text{substrate}}$, *V. succinogenes* was grown at various dilution rates in formate-limited chemostat culture. Since formate was used as a source of neither carbon nor hydrogen in cellular synthesis, it was deemed likely that when the formate concentration was growth-rate-limiting the organism was growing under conditions of energy limitation. The values of $Y^{\text{obs}}_{\text{formate}}$ so obtained were found to be related to the imposed growth rates as predicted by the Pirt equation, and $Y^{\text{max}}_{\text{formate}}$, obtained by extrapolation, was equal to 14 g dry weight of cells per mole of formate (Mell, Bronder & Kröger, 1982). Calculation of the stoichiometric ATP gain (n) from the above values of $Y^{\text{theor}}_{\text{ATP}}$ and $Y^{\text{max}}_{\text{formate}}$ (on the assumption that k equalled 0.4–0.8), gave values for n of between 1 and 2. Thus the growth yield data predicted that, for *V. succinogenes*, between 1 and 2 mol ATP are formed per mole of fumarate reduced to succinate when formate is the electron donor.

Derivation of the stoichiometric ATP gain from thermodynamic considerations

The free energy change associated with the conversion of substrate to product, $\Delta G(S \rightarrow P)$, and the free energy required to drive the synthesis of ATP from ADP and inorganic phosphate (P_i) under quasi-reversible conditions in the cell, $\Delta G(\text{ADP} \rightarrow \text{ATP}) = +44 \, \text{kJ mol}^{-1}$, are the two values which determine the maximum number of moles of ATP that can be formed per mole of substrate consumed. It is clear, however, that the free energy conservation associated with microbial growth cannot be computed as though it occurs under reversible conditions, for living cells are open systems, life proceeds irreversibly, and a portion of the transformed energy is always dissipated as heat. Even so, a theoretical basis for the derivation of the efficiency of free energy conservation during bacterial growth is beginning to emerge (Westerhoff *et al.*, 1982; Hellingwerf *et al.*, 1982; Westerhoff, Hellingwerf & van Dam, 1983). It is generally observed than η, the thermodynamic efficiency of ATP synthesis, is, in chemotrophic anaerobes, between 40 and 80% (Thauer *et al.*, 1977):

$$\eta = \frac{\text{rate of ADP phosphorylation} \times \Delta G(\text{ADP} \to \text{ATP})}{\text{rate of substrate catabolism} \times \Delta G(\text{S} \to \text{P})} \times 100$$

or

$$\eta = \frac{n \, \Delta G(\text{ADP} \to \text{ATP})}{\Delta G(\text{S} \to \text{P})} \times 100$$

where n is the stoichiometry of ATP gain, i.e. moles of ATP formed per mole of substrate utilized.

Using the values of η so calculated, it can be estimated that in *V. succinogenes* between 0.7 and 1.5 mol ATP are formed per mole of fumarate reduced to succinate.

Conclusion

It is evident, therefore, that despite considerable experimental efforts to obtain the most reliable growth yield data, the stoichiometric ATP gain determined by this means is no more precise than the theoretical estimate of its value that may be made on thermodynamic grounds alone. The limits of accuracy of the value of n derived from cell yield measurements may indeed be even broader in the case of other bacteria, especially when their growth rates are not definitely known to be limited by the rate of energy provision. However, it should not be concluded that measurements of molar growth yields are of no value. The apparent ATP gain $(k \cdot n)$ derived from such measurements mirrors the degree of coupling between catabolism and biosynthesis under the prevailing growth conditions (Hellingwerf *et al.*, 1982; Westerhoff *et al.*, 1982, 1983). This pragmatic information is of great importance, not least in the contribution it makes to our understanding of the energetics of bacterial growth (Tempest, 1978; Hobson & Wallace, 1982).

FERMENTATION PRODUCTS AND
THE INTRACELLULAR pH

It is generally accepted that most living cells strive to maintain a nearly constant intracellular (cytoplasmic) pH in the face of all environmental stresses that would tend to cause it to change. The most obvious of these would be a substantial alteration in the external pH, and several studies have shown that aerobically respiring bacteria are able to regulate their cytoplasmic pH within narrow limits over a wide range of external pH values (Konings & Veldkamp, 1983). Thus respiring cells of *Escherichia coli* maintain an internal pH (pH_i) of 7.4–7.8 when the external pH (pH_{ext}) moves from 5.5 to 9 (Slonczewski *et al.*, 1981). Such homeostasis of the cytoplasmic pH, though about somewhat different mean values, is even sustained by acidophiles and alkalophiles (Padan, Zilberstein & Schuldiner, 1981). Indeed, pH_i homeostasis is such a general phenomenon that, especially in cells of higher organisms, it is suggested that perturbations in pH_i may serve as regulatory signals which trigger certain metabolic and/or developmental responses (Nuccitelli & Heiple, 1982).

The mechanism(s) that regulate pH_i in bacteria are not clearly understood, though it is generally assumed that transmembrane movements of cations such as potassium and sodium could play a major role (Harold, 1982; Booth & Kroll, 1983). Ultimately, pH_i homeostasis is accomplished only at the expense of expenditure of free energy, and it is therefore easy to understand how, in an organism such as *E. coli*, it is better sustained under aerobic than under anaerobic conditions (Slonczewski *et al.*, 1981). Even so, maintenance of a constant pH_i has been reported in several fermentative anaerobes including various species of *Streptococcus*; for example, in *S. cremoris* growing in lactose-limited chemostat culture, pH_i remained at 6.7–7.0 over a pH_{ext} range of 5.7–7.0 (Otto *et al.*, 1983). On the other hand, in glycolysing cells of *Mycoplasma mycoides*, although a pH difference across the membrane (ΔpH) was generated, it amounted only to 0.8 pH unit at a pH_{ext} of 5.7 and then decreased with increasing pH_{ext} until it vanished at a pH_{ext} of 7.9 (Benyoucef, Rigaud & Leblanc, 1981).

As was noted previously (see section on novel mechanisms of free energy conservation), the lactate produced by *S. cremoris* leaves the cells in symport with protons in a process which actually aids the maintenance of a ΔpH (interior alkaline) whilst the growth medium is progressively acidified. But what of the organism whose fermentation produces a weak acid which in its undissociated form is freely diffusible across the cell membrane? An example is provided by *Clostridium pasteurianum* whose fermentation of glucose produces both acetate and butyrate. The ability of acetic acid to penetrate bacterial cell membranes has led to its being used, in low concentrations, to measure the ΔpH across such membranes (Padan *et al.*, 1981). The same property of acetic acid, which it shares with other weak acids such as propionic and butyric acid, means that in higher concentrations it acts to diminish the ΔpH. During batch growth of *C. pasteurianum* on excess glucose, the pH of the medium can quickly fall to pH 5 due to the accumulation of acetic and butyric acids. Under such growth conditions, though efflux of acetic acid could be construed as being helpful to the sustenance of a ΔpH, the reverse influx of acetic acid would serve to dissipate any such differential pH between the cytoplasm and the external medium. Thus, so long as the cell membrane was sufficiently freely permeable to undissociated acetic acid, as has actually been demonstrated in the case of *C. pasteurianum* (Kell *et al.*, 1981c), any ΔpH would become more difficult to maintain as the pH_{ext} fell and the concentration of the fermentation product mounted in the medium. Measurement of the ΔpH in cells of *C. pasteurianum* showed that its value was relatively small (0.4–0.8 pH units) and that pH_i changed from 7.5 to 5.9 as pH_{ext} fell from 7.1 to 5.1 (Riebeling, Thauer & Jungermann, 1975). Actual evidence of the ability of acetate at an acidic pH_{ext} to dissipate the ΔpH has been obtained with several other bacteria, both aerobes and anaerobes. For example, in studies of the induction of butane-2,3-diol formation in *Klebsiella aerogenes*, Thomson & Booth (1982) found that, when added at concentration that induced the pathway, acetate but not pyruvate collapsed the ΔpH when the pH_{ext} was 6.0.

A parallel decrease in the pH_i and pH_{ext} of an anaerobic bacterium is of especial interest when the fall in pH_{ext} apparently provokes some profound metabolic change. Such is the case in glucose-utilizing batch cultures of *Clostridium acetobutylicum*, whose switch from an acidogenic fermentation (producing acetic and butyric acids) to a solventogenic fermentation (yielding acetone and butanol) is traditionally signalled by a fall in pH_{ext} to a critically low 'break-point' value (Spivey, 1978; Andersch, Bahl & Gottschalk, 1982; Bahl, Andersch & Gottschalk, 1982b). In fact, in this instance, the switch to solventogenesis appears to be triggered not only by the fall in pH_{ext} but also by the accumulation of the acidic products of the fermentation (especially butyric acid) to a threshold concentration (Gottschal & Morris, 1981; Bahl et al., 1982a). When an alternative weak acid known to permeate cell membranes (i.e. dimethyloxazolidine-2,4-dione) was added to a culture of C. aceto-butylicum at pH 5 solventogenesis was not triggered in the same manner (Gottschal & Morris, 1981). It was interesting, however, that the culture continued to grow at pH 5 in the presence of this permeant weak acid at a concentration of 15 mM, thus confirming that the pH_i of C. acetobutylicum could apparently be substantially lowered with no adverse consequences for its growth.

The conclusion may be drawn that maintenance of a constant pH_i close to pH 7 is in many fermentative anaerobes not as likely, nor as necessary to their well-being, as it generally is in aerobes. Possibly the production via fermentation of permeant, low molecular weight fatty acids has caused such organisms to dispense with the benefits of strict pH_i homeostasis.

In anaerobic consortia which contain such acidogens, should the medium become acidified due to the accumulation of such weak-acid fermentation products the consequences for companion anaerobes which are more dependent on maintenance of a pH_i near neutrality could prove severe. The inhibition of methanogenesis caused by 'souring' of an anaerobic digester could be explained on this basis.

CATABOLISM OF ACETATE

Acetic acid is one of the major end-products of the fermentations undertaken by many chemotrophic anaerobes. Anaerobic biotopes would become acid and life within them would cease (see previous section) if anaerobic bacteria were not also present which could metabolize this fatty acid to yield neutral end-products. Such organisms have long been sought, but only in the last few years have they been isolated and characterized. Three metabolic groups of 'acetotrophic' anaerobes are currently known: (i) methanogenic bacteria, (ii) sulphate-reducing bacteria, and (iii) sulphur-reducing bacteria.

Formation of methane and carbon dioxide from acetate

Methanogenic bacteria of the genus Methanosarcina (Balch et al., 1979), Methanothrix soehngenii (Huser, Wuhrmann & Zehnder, 1982) and Methanococcus mazei (Mah, 1980; Touzel & Albagnac,

1983) are capable of fermenting acetate to yield methane and carbon dioxide:

$$\text{Acetate}^- + H^+ \rightarrow CH_4 + CO_2 \quad \Delta G^{0\prime} = -36 \, \text{kJ} \, \text{mol}^{-1}$$

The mechanism of methane formation from acetic acid is still the subject of intense investigation. One certain fact is that trideutero-acetate gives rise to trideuteromethane, which thus excludes the possibility that acetate is oxidized to give two carbon dioxide one of which is then reduced to methane (Pine & Barker, 1956; Blaut & Gottschalk, 1982). It is interesting that in acetate-grown *Methanosarcina barkeri* the specific activity of carbon monoxide dehydrogenase is five times greater than in cells of this organism grown on hydrogen plus carbon dioxide (Krzycki, Wolkin & Zeikus, 1982; Kenealy & Zeikus, 1982), and that the formation of methane from acetate but not from carbon dioxide or from methanol is specifically prevented by cyanide which is an inhibitor of the carbon monoxide dehydrogenase (Eikmanns & Thauer, 1983). The observation that *M. barkeri* contains high activities of acetate kinase and phosphotransacetylase is probably also important (Krzycki *et al.*, 1982; Kenealy & Zeikus, 1982), and on the basis of these findings the mechanism for conversion of acetate to methane and carbon dioxide that is shown in Fig. 5 may be proposed. The key step is the formation from acetyl-CoA of methyl-X (an 'activated methanol') plus carbon monoxide. The reverse of this reaction, whereby acetyl-CoA is synthesized from methyl-X and carbon monoxide, plays an important role in the energy metabolism of acetogenic bacteria (Diekert & Ritter, 1983*a*; Wood, Drake & Hu, 1982; Hu, Drake & Wood, 1982) and in autotrophic fixation of carbon dioxide in methanogenic bacteria (Stupperich *et al.*, 1983). The carbon monoxide formed from the carboxyl group of acetate is oxidized to carbon dioxide in a reaction catalysed by carbon monoxide dehydrogenase with coenzyme F_{420} serving as the electron acceptor (Daniels *et al.*, 1977). Reduced F_{420} is then used to reduce methyl-X (probably methyl-CoM) to methane. The latter redox process must be coupled to the biosynthesis of more than 1 mol ATP, for otherwise the organism could not grow, since 1 mol ATP is consumed in the initial activation of 1 mol acetate to give acetyl-CoA. Zehnder & Brock (1979) have proposed an alternative mechanism for the formation of methane and carbon dioxide from acetate by a route which does not involve a redox process. However, this mechanism is not supported by current experimental evidence.

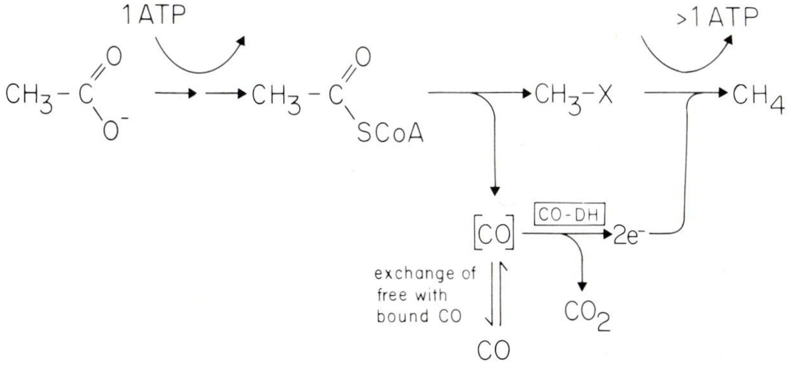

Fig. 5. Proposed mechanism of acetate fermentation to methane and carbon dioxide with bound carbon monoxide, [CO], as an intermediate in acetotrophic methanogenic bacteria. CH_3-X, 'activated methanol'; CO–DH, carbon monoxide dehydrogenase.

Oxidation of acetate by sulphate to give carbon dioxide

Many sulphate-reducing bacteria can grow on acetate and sulphate as the sole source of energy (Pfennig & Widdel, 1981; Widdel, 1980; Widdel & Pfennig, 1981):

$$\text{Acetate}^- + SO_4^{2-} + 3H^+ \rightarrow 2CO_2 + H_2S + 2H_2O$$
$$\Delta G^{0\prime} = -63\,\text{kJ}\,\text{mol}^{-1}$$

Evidence from labelling studies (Gebhardt, Linder & Thauer, 1983) and from enzyme studies (Brandis *et al.*, 1983) indicates that in *Desulfobacter postgatei* acetate is oxidized to carbon dioxide via the tricarboxylic acid cycle (Fig. 6). A thermodynamically problematic step in this cycle is the oxidation of succinate to fumarate ($E^{0\prime} = +30\,\text{mV}$) with sulphite as the electron acceptor (SO_3^{2-}/H_2S: $E^{0\prime} = -170\,\text{mV}$). It is proposed that the oxidation of acetate in sulphate-reducing bacteria involves energy-driven reversed electron transport, consuming part of the free energy conserved in dissimilatory sulphate reduction (Thauer, 1982). In this respect it is of interest that probably more than 1 mol ATP per mole of sulphate reduced is formed in sulphate-reducing bacteria which can grow on hydrogen and sulphate as the sole source of energy (Nethe-Jaenchen & Thauer, 1983).

Oxidation of acetate by sulphur to give carbon dioxide

Desulfuromonas acetoxidans and certain other strictly anaerobic bacteria can grow on acetate and elemental sulphur (S^0) as the sole

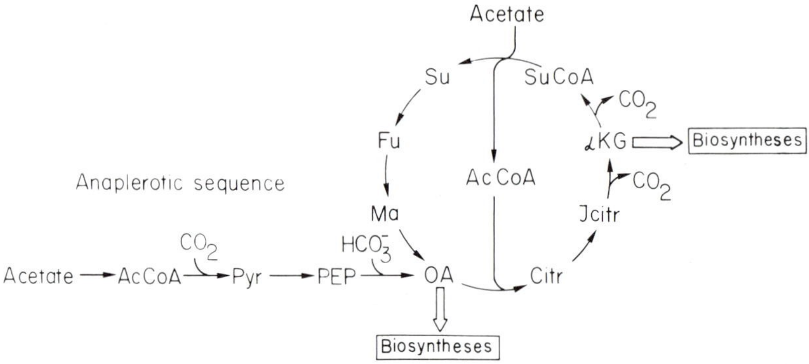

Fig. 6. Acetate oxidation to carbon dioxide with sulphate via the tricarboxylic acid cycle in *Desulfobacter postgatei*. Succinate dehydrogenase and malate dehydrogenase are membrane-associated and probably use menaquinone as coenzyme. Isocitrate dehydrogenase and α-ketoglutarate dehydrogenase are found in the soluble cell fraction and are specific for NADP and ferredoxin, respectively. The synthesis of pyruvate (Pyr) from acetyl-CoA and carbon dioxide is catalysed by a ferredoxin-dependent pyruvate synthase (Brandis *et al.*, 1983). AcCoA, acetyl-CoA; Citr, citrate; F, fumarate; Jcitr, isocitrate; α KG, α-ketoglutarate; Ma, malate; OA, oxaloacetate; PEP, phosphoenol pyruvate; Su, succinate; SuCoA, succinyl-CoA.

source of energy (Pfennig & Biebl, 1976; Pfennig & Widdel, 1981):

$$\text{Acetate}^- + 4S^0 + 2H_2O + H^+ \rightarrow 2CO_2 + 4H_2S \quad \Delta G^{0\prime} = -39\,\text{kJ mol}^{-1}$$

The mechanism of this acetate oxidation has not yet been elucidated, though enzyme studies suggest that the fatty acid is oxidized via the tricarboxylic acid cycle (Kaulfers, 1976).

DEGRADATION OF AROMATIC COMPOUNDS

Aerobic degradation of aromatic compounds is commonplace, with the crucial cleavage of the aromatic ring being accomplished by mono- and di-oxygenase reactions. In contrast, we know relatively little about the mechanism of anaerobic degradation of such compounds. Under anoxic conditions the device of oxygenative ring cleavage manifestly cannot be exploited. However, evidence that some means of anaerobic catabolism of aromatic compounds exists in nature and can accomplish ring cleavage without the participation of molecular oxygen was supplied by Tarvin & Buswell (1934) who reported that, under strictly anaerobic conditions, benzoate, phenylacetate, phenylpropionate and cinnamate were completely utilized, with the production of carbon dioxide and methane, by a mixed microbial population that originated in a sewage sludge inoculum.

Recently, several advances have been made in our understanding of the biochemistry of this phenomenon, and it has been established that in all the cases of aromatic substrates being degraded anaerobically that have been investigated, the micro-organism responsible utilized the device of initial ring reduction. However, the precise mechanism employed has not as yet been elucidated (Evans, 1977; Fina et al., 1978; Keith et al., 1978).

Three metabolic types of chemotrophic anaerobes have been isolated that can degrade aromatic compounds: (i) syntrophic bacteria that can ferment benzoic acid, for example to yield acetate plus carbon dioxide and hydrogen, but only when grown in association with hydrogen-consuming organisms that maintain a low partial pressure of hydrogen in mixed culture (Ferry & Wolfe, 1976; Mountfort & Bryant, 1982); (ii) bacteria that in monoculture can metabolize trihydroxybenzenes to give acetate (Schink & Pfennig, 1982b); (iii) sulphate-reducing bacteria that can grow on benzoate and sulphate as the sole source of carbon and energy (Widdel, 1980).

Syntrophic fermentation of benzoic acid

The fermentation of benzoic acid to yield acetate, carbon dioxide and hydrogen is an endergonic process under standard state conditions (at pH 7):

$$Benzoate^- + 6H_2O \rightarrow 3 \text{ acetate}^- + 2H^+ + CO_2 + 3H_2$$
$$\Delta G^{0\prime} = +53 \text{ kJ mol}^{-1} \text{ benzoate*}$$

This explains why anaerobes can ferment benzoic acid and grow on this compound only when they are syntrophically associated with methanogens or sulphate-reducing bacteria that sustain low culture concentrations of hydrogen and acetate. On the other hand, the degradation of benzoate to give methane and carbon dioxide is an exergonic process even under standard state conditions:

$$Benzoate^- + H^+ + 4.5 H_2O \rightarrow 3.25 CO_2 + 3.75 CH_4$$
$$\Delta G^{0\prime} = -152 \text{ kJ mol}^{-1} \text{ benzoate}$$

The anaerobic degradation of 11 different aromatic compounds by mixed methanogenic cultures has been reported (Ferry & Wolfe,

---

* ΔGf^0 for benzoate$^-$ was calculated from ΔGf^0 for benzoic acid (solid state) (-245 kJ mol^{-1}), from the solubility of benzoic acid at $25\,°C$ (27.8 mM), and from the pK of benzoic acid (4.2) to be $-212.3 \text{ kJ mol}^{-1}$ (see Thauer et al., 1977).

1976; Fina *et al.*, 1978; Healy & Young, 1978; Keith *et al.*, 1978; Healy & Young, 1979; Healy, Young & Reinhard, 1980; Sleat & Robinson, 1983; Mountfort & Bryant, 1982).

Fermentation of trihydroxybenzenes to give acetate

Schink & Pfennig (1982*b*) isolated anaerobic bacteria from limnic and marine mud samples that were capable of fermenting trihydroxybenzenes. The type strain, *Pelobacter acidigallici*, for example, fermented phloroglucinol to yield acetate:

$$\text{Phloroglucinol} + 3H_2O \rightarrow 3 \text{ acetate}^- + 3H^+$$

In the case of such benzene derivatives, the greater the degree of hydroxylation of the aromatic ring the more exergonic is the reaction whereby they give rise to acetate; thus from benzene $\Delta G^{0'} = +70 \text{ kJ mol}^{-1}$, from phenol $+6 \text{ kJ mol}^{-1}$, from dihydroxybenzenes -72 kJ mol^{-1}, and from trihydroxybenzenes -150 kJ mol^{-1} (estimated value). It has been concluded that only the fermentations of dihydroxy- and trihydroxybenzenes are sufficiently exergonic to be able to sustain growth of the appropriate anaerobic bacteria in monoculture.

Oxidation of benzoic acid by sulphate to yield carbon dioxide

Several sulphate-reducing bacteria can utilize benzoate and sulphate as the sole source of carbon and energy for growth (Widdel, 1980).

$$\text{Benzoate}^- + 3.75SO_4^{2-} + 8.5H^+ \rightarrow 7CO_2 + 3.75H_2S + 3H_2O$$
$$\Delta G^{0'} = -254 \text{ kJ mol}^{-1} \text{ benzoate}$$

Degradation of lignin and of benzene

Lignin is a complex aromatic polymer consisting of phenylpropane building blocks held together by irregular carbon- and diaryl-ether linkages. Fermentation of lignin to yield carbon dioxide and methane is probably thermodynamically feasible, but lignin's large molecular size, poor solubility and complex cross-linked structure makes it rather inaccessible to both micro-organisms and secreted enzymes. The polymer is refractory to anaerobic degradation (Zeikus, 1980; Higuchi, 1982; Janshekar & Fiechter, 1982) and must be chemically modified or depolymerized before significant decomposition can occur in anoxic conditions (Zeikus, Wellstein & Kirk, 1982). There is no report of benzene itself being metabolized in anaerobic biotopes, but this is probably for mechanistic rather than thermodynamic reasons, for production of methane and carbon dioxide from benzene would be a highly exergonic process:

$$Benzene + 4.5H_2O \rightarrow 3.75CH_4 + 2.25CO_2$$
$$\Delta G^{0'} = -135\,kJ\,mol^{-1}$$

AUTOTROPHIC FIXATION OF CARBON DIOXIDE

Autotrophy has long been considered to be the prerogative of phototrophs and of a few specialist, aerobic chemolithotrophic bacteria. With the exception of the anaerobic green sulphur bacteria, such as *Chlorobium thiosulfatophilum*, which use a reductive tricarboxylic acid cycle (Buchanan, 1979; Fuchs, Stupperich & Eden, 1980*a*; Fuchs, Stupperich & Jaenchen, 1980*b*; Ivanovsky, Sintsov & Kondratieva, 1980), these organisms assimilate carbon dioxide via the Calvin cycle. The first chemotrophic anaerobe to be recognized as an autotroph was *Methanobacterium thermoautotrophicum* (Zeikus & Wolfe, 1972), but during the past decade a number of other chemotrophic anaerobes have been isolated that can grow on carbon dioxide as the sole source of carbon: viz., most methanogenic bacteria (Balch *et al.*, 1979; Fuchs & Stupperich, 1982); some sulphate-reducing bacteria including *Desulfonema limicola*, *Desulfosarcina variabilis* and *Desulfovibrio baarsii* (Pfennig & Widdel, 1981); sulphur-reducing bacteria such as *Thermoproteus neutrophilus* (Fischer *et al.*, 1983); and most acetogenic bacteria, for example, *Acetobacterium woodii* (Balch *et al.*, 1977) and *Clostridium thermoaceticum* (Wiegel, Braun & Gottschalk, 1981).

Radiotracer and enzyme studies have shown that the autotrophic fixation of carbon dioxide in such chemotrophic anaerobes proceeds neither via the Calvin cycle nor by the reductive tricarboxylic acid cycle (Taylor, Kelly & Pirt, 1976; Fuchs & Stupperich, 1982; Eden & Fuchs, 1982). Evidence has been forthcoming that in *M. thermoautotrophicum* and *A. woodii* acetyl-CoA is probably formed from two carbon dioxide by the mechanism depicted in Fig. 7 (Wood *et al.*, 1982; Stupperich *et al.*, 1983; Stupperich & Fuchs, 1983; Eden & Fuchs, 1983; Rühlemann, Stupperich & Fuchs, 1983; Diekert & Ritter, 1983*a*). Reductive carboxylation of the acetyl-CoA then

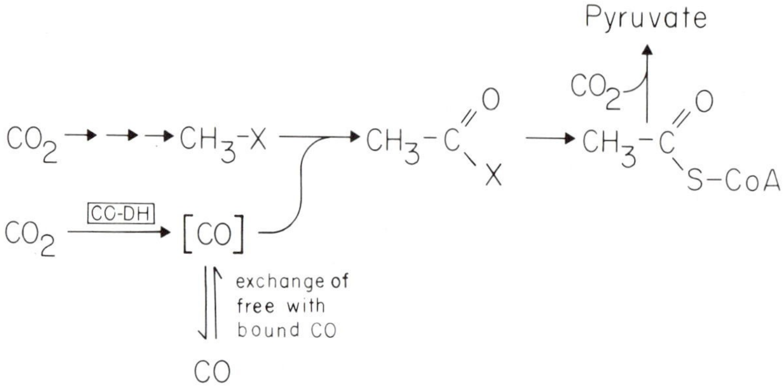

Fig. 7. Proposed mechanism of autotrophic carbon dioxide fixation with bound carbon monoxide, [CO], as an intermediate in *Methanobacterium thermoautotrophicum*. CH_3-X, 'activated methanol'; CO-DH, carbon monoxide dehydrogenase.

leads to the formation of pyruvate from which most of the cell material is synthesized. Intermediates in the route of formation of acetyl-CoA from two carbon dioxide include methanol in an 'activated form' (CH_3-X) and carbon monoxide (probably in a bound form). Unique steps in this pathway are the reduction of carbon dioxide to carbon monoxide, which is catalysed by carbon monoxide dehydrogenase (Stupperich *et al.*, 1983; Diekert & Ritter, 1983*a*), and the carbonylation of methyl-X to give acetyl-X (Hu *et al.*, 1982). Preliminary findings indicate that carbon dioxide is assimilated by this same mechanism in autotrophic sulphate-reducers (K. Jansen & G. Fuchs, unpublished).

The acetogenic bacterium *Clostridium thermoaceticum* was used by Barker & Kamen in 1945 in one of the first studies of microbial fixation of ^{14}C-labelled carbon dioxide. It was thought at that time that resolution of the route of carbon dioxide fixation in this heterotrophic anaerobic bacterium might usefully contribute to knowledge of the autotrophic means of carbon dioxide utilization by green plants. Although this expectation was to prove ill-founded, we see, nearly 40 years later, that Barker & Kamen's selection of *C. thermoaceticum* for such studies was even more apt than they imagined, since it now emerges that a novel pathway of autotrophic carbon dioxide assimilation is indeed operative in some chemotrophic anaerobes (Wood *et al.*, 1982).

A comparison of the ATP requirements for autotrophic carbon dioxide fixation in each of the three presently identified pathways, suggests a possible explanation for the evolution of the acetyl-CoA pathway. Synthesis of 1 mol triose phosphate from three carbon dioxide via the Calvin cycle requires 9 mol ATP, via the reductive tricarboxylic acid cycle requires 5 mol ATP, and via the acetyl-CoA pathway probably only 3 mol ATP (2 mol for the formation of phosphoenolpyruvate from pyruvate (Eyzaguirre, Jansen & Fuchs, 1982) and 1 mol for the formation of bisphosphoglycerate from 3-phosphoglycerate (Jansen, Stupperich & Fuchs, 1982)). Chemotrophic anaerobes have to use ATP sparingly since they are generally energy-limited during growth, which is not usually true of either phototrophs or of chemotrophic aerobes.

MECHANISTICALLY ANOMALOUS REACTIONS

The energy limitation prevalent in anaerobic environments has favoured the evolution in chemotrophic anaerobes of metabolic pathways which enable almost any organic or inorganic compound to be exploited by one or other of such organisms as a source of energy for growth. Even chemically relatively inert substances such as methane (Zehnder & Brock, 1980; Iversen & Blackburn, 1981), ethers (Bache & Pfennig, 1981) and aromatic compounds (see the section on degradation of aromatic compounds) are metabolized in anoxic sediments. The only prerequisite appears to be that the process whereby the compound is utilized is sufficiently exergonic to drive the synthesis of ATP (see the section on the size of the biological energy quantum). With possibly only a few exceptions there appear to be no mechanistic constraints, with the consequence that amongst metabolic reactions undertaken by anaerobes there are to be found several whose mechanisms are puzzling and currently unresolved; their elucidation poses a challenge to both microbiologists and biochemists. We here draw attention to just two such anomalous reactions.

Dehydration of 2-hydroxy acids

Elimination of water from 2-hydroxy acids is an important reaction in the energy metabolism of many chemotrophic anaerobes (Anderson & Wood, 1969; Buckel, 1980; Bühler et al., 1980; Bader, Rauschenbach & Simon, 1982; Pitsch & Simon, 1982; Giesel & Simon, 1983; Akedo, Cooney & Sinskey, 1983):

$$R—CH_2—CHOHCOO^- \rightleftharpoons R—CH=CHCOO^- + H_2O$$

It is almost certain that dehydration proceeds via CoA intermediates (Brockman & Wood, 1975; Tung & Wood, 1975; Akedo *et al.*, 1983) and there is some evidence for catalytic phosphorylation during the reaction (Buckel, 1980). The production of acrylate from lactate by *Clostridium propionicum* (Anderson & Wood, 1969; Akedo *et al.*, 1983), and of glutaconate from 2-hydroxyglutarate by *Acidaminococcus fermentans* (Buckel, 1980), are important and well-studied examples of this process. However, the mechanism whereby the dehydration is accomplished is still not clear. The carboxyl group renders the C-2 of the 2-hydroxy acids nucleophilic and the C-3 electrophilic by induction. Thus a hydride (H^-) or a hydroxyl (OH^-) can more easily be removed from the 3-position than from the 2-position. The converse is true for removal of a proton.

The reaction catalysed by pyruvate–formate lyase

Pyruvate is cleaved to acetyl-CoA and formate in many obligately anaerobic and amphiaerobic bacteria (Knappe, Blaschkowski & Edenharder, 1972; Thauer, Kirchniawy & Jungermann, 1972), the reaction being catalysed by the enzyme pyruvate–formate lyase (Knappe *et al.*, 1974; Pecher *et al.*, 1982):

$$\text{Pyruvate}^- + \text{CoA} \rightarrow \text{acetyl-CoA} + \text{formate}^-$$

Despite considerable efforts during the past years to discover the catalytic mechanism employed, this has not yet been clarified. In contrast, the mechanism of decarboxylation of pyruvate to acetaldehyde and carbon dioxide is well understood. Binding of pyruvate to thiamine pyrophosphate renders the C-2 of pyruvate nucleophilic, allowing carbon dioxide to leave. However, the cleavage of pyruvate to acetate and formate requires that the C-1 of pyruvate is rendered nucleophilic. Textbooks of organic chemistry contain no mechanism which suggests how this might be brought about. The iodoform reaction may proceed by an analogous mechanism (J. Knappe, personal communication):

$$
\begin{array}{ccc}
\text{O} & & \text{O} \\
\parallel & & \parallel \\
\text{R}-\text{C}-\text{CI}_3 + \text{OH}^- & \rightarrow & \text{R}-\text{C} + \text{H}-\text{CI}_3 \\
& & | \\
& & \text{O}^-
\end{array}
$$

Yet it is difficult to envisage how pyruvate–formate lyase could mimic this reaction.

UNUSUAL TRACE ELEMENTS

The growth of many chemotrophic anaerobes is stimulated by trace elements which are not routinely included in growth media; these include selenium, tungsten and nickel. Investigation of these rather unexpected requirements has led to the discovery of selenium-, tungsten- and nickel-enzymes (Ljungdahl, 1976; Stadtman, 1980; Thauer, Diekert & Schönheit, 1980; Andreesen, 1980).

Selenium

The following examples of selenoenzymes have been identified in strictly anaerobic bacteria: (i) formate dehydrogenase from acetogenic bacteria (Wagner & Andreesen, 1977; Ljungdahl & Andreesen, 1978) and from *Methanococcus vannielii* (Jones & Stadtman, 1981); (ii) glycine reductase from several clostridia (Turner & Stadtman, 1973; Stadtman, 1978; Dürre & Andreesen, 1982*a*); (iii) hydrogenase from *Methanococcus vannielii* (Yamazaki, 1982); (iv) nicotinic acid hydroxylase from *Clostridium barkeri* (Imhoff & Andreesen, 1979); (v) xanthine dehydrogenase from certain clostridia (Wagner & Andreesen, 1979; Dürre, Andersch & Andreesen, 1981; Dürre & Andreesen, 1982*b*). In some of these enzymes the selenium has been shown to be present as selenocysteine (Stadtman, 1981). Commercially available sodium sulphide and hydrogen sulphide are contaminated with significant amounts of selenium and, since they are frequently added to culture media for strict anaerobes to establish the requisite reducing conditions, selenium dependence of growth or of enzyme synthesis can easily be overlooked. It is therefore very likely that selenium is of even greater biological importance for chemotrophic anaerobes than has hitherto been suspected.

Tungsten

Tungsten promotes the growth of, and formate dehydrogenase formation in, several acetogenic bacteria (Andreesen & Ljungdahl, 1973; Andreesen, El Ghazzawi & Gottschalk, 1974; Leonhardt &

Andreesen, 1977) and *Methanococcus* species (Jones & Stadtman, 1976). The formate dehydrogenase of *Clostridium thermoaceticum* which in this organism catalyses reduction of carbon dioxide to formate (Thauer, 1972), has been partially purified and has been shown to contain tungsten (Ljungdahl & Andreesen, 1975, 1978; Yamamoto *et al.*, 1983).

Nickel

In the course of nutritional studies it was found that the growth of many anaerobic bacteria was dependent upon their being provided with trace amounts of nickel (Schönheit, Moll & Thauer, 1979; Diekert & Thauer, 1980; Diekert & Ritter, 1982). During the past few years three enzymes have been identified as nickel-containing proteins (for a review see Thauer *et al.*, 1983): (i) carbon monoxide dehydrogenase from *Clostridium pasteurianum* (Diekert, Graf & Thauer, 1979; Drake, 1982), *Clostridium thermoaceticum* (Drake, Hu & Wood, 1980; Diekert & Ritter, 1983*b*; Ragsdale *et al.*, 1983) and *Acetobacterium woodii* (Diekert & Ritter, 1982); (ii) hydrogenase from *Methanobacterium thermoautotrophicum* (Graf & Thauer, 1981; Albracht, Graf & Thauer, 1982; Jacobson *et al.*, 1982), *Vibrio succinogenes* (Unden *et al.*, 1982), *Desulfovibrio gigas* (LeGall *et al.*, 1982; Cammack *et al.*, 1982) and *Desulfovibrio desulfuricans* (Krüger *et al.*, 1982); (iii) methyl-CoM methyl reductase from methanogenic bacteria (Diekert, Klee & Thauer, 1980; Whitman & Wolfe, 1980; Ellefson, Whitman & Wolfe, 1982). In this last instance the nickel is a component of the distinctive prosthetic group, factor F_{430}, of the methyl reductase (see the following section on novel coenzymes).

Nickel is present as a contaminant in iron salts and is also a constituent of stainless steel from which it is released by corrosion. It is therefore difficult to prepare a culture medium which is devoid of nickel. Anaerobic bacteria contain a high-affinity system for the transport of the transition metal (Jarrell & Sprott, 1982), which allows them to take up nickel from the medium at very low concentrations. For these reasons the role of nickel in the metabolism of anaerobes has long been overlooked.

NOVEL COENZYMES

During the past decade a number of novel coenzymes have been discovered in chemotrophic anaerobes. Methanogenic bacteria, in particular, have yielded up a number of intriguing new 'factors' associated with their methane-generating capacity (Wolfe & Higgins, 1979; Balch *et al.*, 1979; Keltjens, 1982).

The first of these to be identified was coenzyme M, which serves as a C-1 carrier in methanogenesis. Taylor & Wolfe (1974) showed it to be a surprisingly simple compound (2-mercaptoethane sulphonic acid) and it was simultaneously proved to be the factor present in rumen fluid that was essential for the growth of *Methanobrevibacter ruminantium* (Taylor *et al.*, 1974). Its structure is:

$$\text{HS}-\overset{\overset{\displaystyle H}{|}}{\underset{\underset{\displaystyle H}{|}}{C}}-\overset{\overset{\displaystyle H}{|}}{\underset{\underset{\displaystyle H}{|}}{C}}-\text{SO}_3^-$$

Other coenzymes of methanogenesis have, however, turned out to be far more complex in their chemical composition. One of these, given the trivial name of methanopterin, is a 2-amino-4-hydroxypterine carrying a complex side chain (Vogels *et al.*, 1982; Keltjens *et al.*, 1983*b*, *d*). A possible role for methanopterin in 'activation' of carbon dioxide was indicated by the identification in methanogenic bacteria of carboxy-5,6,7,8-tetrahydromethanopterin (cTHMP), which suggests that the first step in methanogenesis from carbon dioxide could consist of carboxylation of methanopterin with concomitant reduction of the pterin molecule (Vogels *et al.*, 1982; van Beelen *et al.*, 1983).

The strongly fluorescent coenzyme F_{420} is also present in all methanogenic bacteria, though it is not confined to these anaerobes having also been discovered in *Streptomyces griseus* (Eker *et al.*, 1980; McCormick & Morton, 1982). Its structure was elucidated by Eirich, Vogels & Wolfe (1978) and shown to be that of a deazaflavin derivative (Fig. 8) having the properties of a $1H^+/2e^-$ redox agent with an $E^{0\prime}$ value of $-360\,\text{mV}$. Its relatively low midpoint redox potential enables it to serve as the primary electron acceptor from hydrogen and from formate, but also as a low-potential electron donor in the biosynthesis of pyruvate and 2-oxoglutarate by reductive carboxylation reactions (Balch *et al.*, 1979).

Perhaps the most intriguing of all structures is possessed by factor F_{430} which forms the basis of the prosthetic group of methyl-CoM methyl reductase (Ellefson *et al.*, 1982). It has been shown that F_{430} is a nickel tetrapyrrole (Diekert, Jaenchen & Thauer, 1980; Jaenchen, Diekert & Thauer, 1981; Gilles & Thauer, 1983), the nickel being coordinately bound within a uroporphinoid (type III) ligand

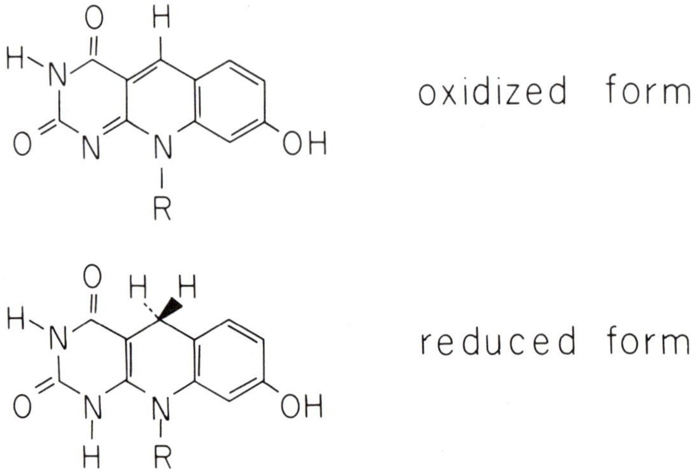

oxidized form

reduced form

Fig. 8. Structure of factor F_{420} in the oxidized and reduced forms. For the complete structure see Eirich *et al.* (1978).

skeleton which contains an additional carbocyclic ring and a chromophoric system that can be considered to be a tetrahydro derivative of the porphin system (Pfaltz *et al.*, 1982). The structure of the chromophore is shown in Fig. 9. It has been suggested that the coenzyme form of the molecule (coenzyme MF_{430}) additionally contains 6,7-dimethyl-8-ribityl-5,6,7,8-tetrahydrolumazine bound to the nickel atom via its N-5, plus coenzyme M attached to the chromophore on the opposite side to the tetrahydrolumazine (Keltjens *et al.*, 1982, 1983a, b).

Still more heat-stable cofactors are present in methanogenic bacteria, but their structure and functions have yet to be determined. They include CDR factor (Romesser & Wolfe, 1982; Leigh & Wolfe, 1983), factor F_{342} (Gunsalus & Wolfe, 1978) and component B (Ellefson & Wolfe, 1980). Although they may not be directly involved in methanogenesis, corrinoids are also found in methanogenic bacteria (Krzycki & Zeikus, 1980; Scherer & Sahm, 1980). The corrinoid in *Methanosarcina barkeri* was found to be aberrant in containing 5-hydroxybenzimidazole in place of the more usual 5,6-dimethylbenzimidazole (Pol, van der Drift & Vogels, 1982; Höllriegl, Scherer & Renz, 1983). Acetotrophic methanogens contain cytochromes (Kühn *et al.*, 1979, 1983; Kühn & Gottschalk, 1983), the chemical structure of which has, however, not yet been analysed.

Other obligate anaerobes have also contributed to the growing list

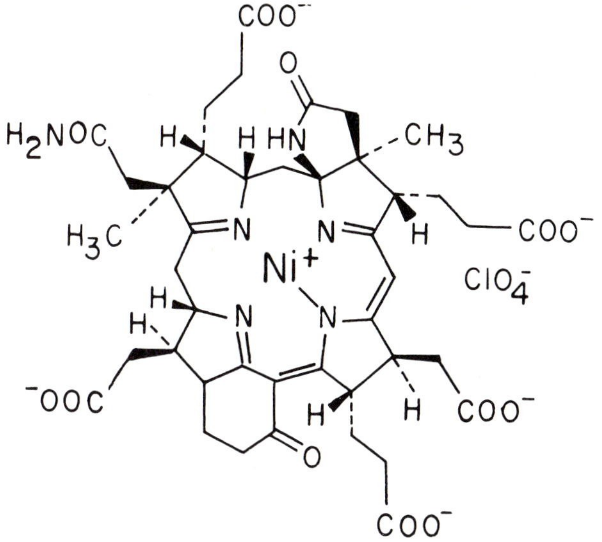

Fig. 9. Structure of factor F_{430} after dissociation from methyl-CoM methyl reductase.

of unusual cofactors. For example, sulphate-reducing bacteria contain a cobalt (III) isobacteriochlorin of unknown function (Moura *et al.*, 1980; Battersby & Sheng, 1982), whilst the carbon monoxide dehydrogenase of acetogenic bacteria possesses a nickel-containing prosthetic group which otherwise displays properties akin to vitamin B_{12} (Diekert & Thauer, 1978; Ragsdale, Ljungdahl & DerVartanian, 1982; Diekert & Ritter, 1983*b*). The discovery of novel coenzymes of anaerobic metabolism followed by elucidation of their structures and functions will undoubtedly prove a fertile field of research for microbiologists and biochemists for many years to come. Their findings will be of particular interest to those whose ambition it is to synthesize 'biomimetic' chemical catalysts so as to accomplish by chemical means some of the important transformations that are currently the prerogative of anaerobic bacteria.

NOTE ADDED IN PROOF

After completion of this manuscript (April 1983) the structures of CDR-factor and of a novel diphospho-*P*,*P'*-diester found in methanogenic bacteria have been elucidated. The CDR-factor has been shown to be a furan derivative (Fig. 10) (J. A. Leigh & R. S. Wolfe,

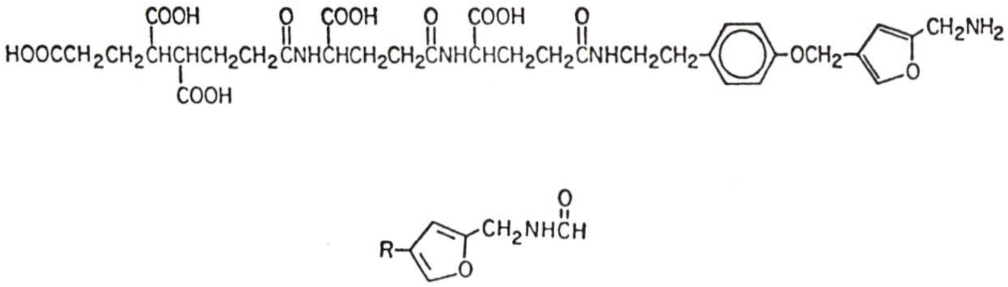

Fig. 10. Structure of CDR-factor (J. A. Leigh & R. S. Wolfe, personal communication; with permission); the formylated form accumulates under certain conditions.

personal communication). The structure of the diphosphodiester has tentatively been established as cyclic 2,3-diphosphoglycerate (Kanodia & Roberts, 1983; Seely & Fahrney, 1983) (Fig. 11).

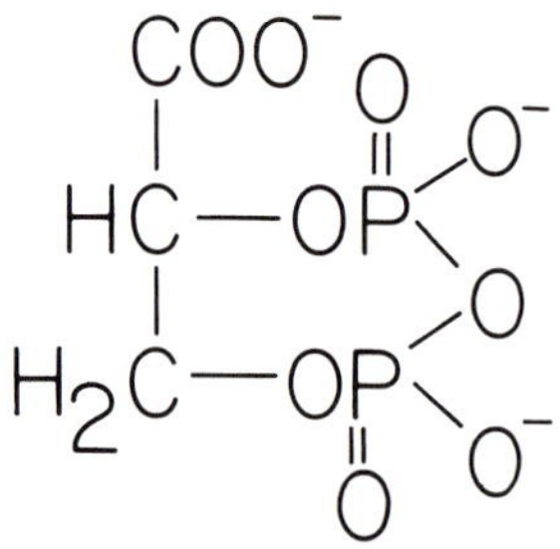

Fig. 11. Structure of a novel diphospho-P,P'-diester found in *Methanobacterium thermoautotrophicum*.

CONCLUDING REMARKS

Our knowledge of anaerobes and their metabolism is still relatively sparse, as is evident from the continual isolation of new genera and species of strictly anaerobic bacteria with quite unexpected properties. However, the necessity that anaerobic processes must be rendered both consistent and predictable before they can be scaled up into satisfactory biotechnological operations (Morris, 1983) is likely to promote extensive investigations of the physiology of at least some selected anaerobes. Further studies of the bioenergetics of anaerobic microbial growth will be undertaken for the same reason. Nor will such studies be confined to monocultures of anaerobes since many of the more desirable anaerobic digestions are best accomplished by heterogeneous mixed cultures. The syntrophic relations that sustain such natural, stable consortia will need to be much more fully comprehended than at present before structured cocultures can sensibly be constructed and most efficiently exploited. But if our knowledge of the microbiology and biochemistry of strict anaerobes is deficient, information regarding their genetics is virtually non-existent. This has become so obvious an obstacle to studies of their physiology that several groups of workers are currently attempting to establish systems for gene transfer and analysis in the methanogens and in species of *Bacteroides* and *Clostridium*.

If there are any predictions that can be made in the certain knowledge that they will be fulfilled, they are that the pace of research into anaerobes will not slacken in the coming decade and that these organisms will continue to surprise us with their wealth of biochemical novelty and instruct us by the variety of solutions that they have adapted to the problems posed by life in anoxic environments.

We wish to thank A. Kröger and D. B. Kell for many stimulating discussions and helpful comments. Work in R. K. Thauer's laboratory was supported by a grant from the Deutsche Forschungsgemeinschaft and by the Fonds der Chemischen Industrie, and J. G. Morris is grateful for grant support from the Biotechnology Directorate of the SERC (UK).

REFERENCES

AKEDO, M., COONEY, C. L. & SINSKEY, A. J. (1983). Direct evidence for lactate–acrylate interconversion in *Clostridium propionicum*. *Bio-Technology*, in press.

ALBRACHT, S. P. J., GRAF, E. G. & THAUER, R. K. (1982). The EPR properties of nickel in hydrogenase from *Methanobacterium thermoautotrophicum*. *FEBS Letters*, **140**, 311–13.

ALEEM, M. I. H. (1977). Coupling of energy with electron transfer reactions in chemolithotrophic bacteria. In *Microbial Energetics*, ed. B. A. Haddock & W. A. Hamilton, *Society for General Microbiology Symposium 27*, pp. 351–81. Cambridge University Press.

ANDERSCH, W., BAHL, H. & GOTTSCHALK, G. (1982). Acetone–butanol production by *Clostridium acetobutylicum* in an ammonium-limited chemostat at low pH values. *Biotechnology Letters*, **4**, 29–32.

ANDERSON, R. L. & WOOD, W. A. (1969). Carbohydrate metabolism in microorganisms. *Annual Review of Microbiology*, **23**, 539–78.

ANDREESEN, J. R. (1980). Role of selenium, molybdenum and tungsten in anaerobes. In *Anaerobes and Anaerobic Infections*, ed. G. Gottschalk, N. Pfennig & H. Werner, pp. 31–40. Stuttgart & New York: Gustav Fischer Verlag.

ANDREESEN, J. R., EL GHAZZAWI, E. & GOTTSCHALK, G. (1974). The effect of ferrous ions, tungstate and selenite on the level of formate dehydrogenase in *Clostridium formicoaceticum* and formate synthesis from CO_2 during pyruvate fermentation. *Archives of Microbiology*, **96**, 103–18.

ANDREESEN, J. R. & LJUNGDAHL, L. G. (1973). Formate dehydrogenase of *Clostridium thermoaceticum:* incorporation of selenium-75, and the effects of selenite, molybdate, and tungstate on the enzyme. *Journal of Bacteriology*, **116**, 867–73.

BACHE, R. & PFENNIG, N. (1981). Selective isolation of *Acetobacterium woodii* on ⁀methoxylated aromatic acids and determination of growth yields. *Archives of Microbiology*, **130**, 255–61.

BADER, J., RAUSCHENBACH, P. & SIMON, H. (1982). On a hitherto unknown fermentation path of several amino acids by proteolytic Clostridia. *FEBS Letters*, **140**, 67–72.

BAHL, H., ANDERSCH, W., BRAUN, K. & GOTTSCHALK, G. (1982a). Effect of pH and butyrate concentration on production of acetone and butanol by *Clostridium acetobutylicum* grown in continuous culture. *European Journal of Applied Microbiology and Biotechnology*, **14**, 17–20.

BAHL, H., ANDERSCH, W. & GOTTSCHALK, G. (1982b). Continuous production of acetone and butanol by *Clostridium acetobutylicum* in a two stage phosphate limited chemostat. *European Journal of Applied Microbiology and Biotechnology*, **15**, 201–5.

BALCH, W. E., FOX, G. E., MAGRUM, L. J., WOESE, C. R. & WOLFE, R. S. (1979). Methanogens: reevaluation of a unique biological group. *Microbiological Reviews*, **43**, 260–96.

BALCH, W. E., SCHOBERTH, S., TANNER, R. S. & WOLFE, R. S. (1977). *Acetobacterium*, a new genus of hydrogen-oxidizing, carbon dioxide-reducing, anaerobic bacteria. *International Journal of Systematic Bacteriology*, **27**, 355–61.

BARKER, H. A. & KAMEN, M. D. (1945). Carbon dioxide utilization in the synthesis of acetic acid by *Clostridium thermoaceticum*. *Proceedings of the National Academy of Sciences, USA*, **31**, 219–25.

BATTERSBY, A. R. & SHENG, Z.-C. (1982). Preparation and spectroscopic properties of Co^{III}-isobacteriochlorins: relationship to the cobalt-containing proteins from *Desulphovibrio gigas* and *D. desulphuricans*. *Journal of the Chemical Society, Chemical Communications*, **24**, 1393–4.

BAUCHOP, T. & ELSDEN, S. R. (1960). The growth of micro-organisms in relation to their energy supply. *Journal of General Microbiology*, **23**, 457–69.

VAN BEELEN, P., THIEMESSEN, H. L., DE COCK, R. M. & VOGELS, G. D. (1983). Methanogenesis and methanopterin conversion by cell-free extracts of *Methanobacterium thermoautotrophicum*. *FEMS Microbiology Letters*, **18**, 135–8.

BENYOUCEF, M., RIGAUD, J.-L. & LEBLANC, G. (1981) The electrochemical proton gradient in *Mycoplasma* cells. *European Journal of Biochemistry*, **113**, 491–8.

BLAUT, M. & GOTTSCHALK, G. (1982). Effect of trimethylamine on acetate utilization by *Methanosarcina barkeri*. *Archives of Microbiology*, **133**, 230–5.

BOOTH, I. R. & KROLL, R. G. (1983). Regulation of cytoplasmic pH (pH$_i$) in bacteria and its relationship to metabolism. *Biochemical Society Transactions*, **11**, 70–2.

BRANDIS-HEEP, A., GEBHARDT, N. A., THAUER, R. K., WIDDEL, F. & PFENNIG, N. (1983). Anaerobic acetate oxidation to CO_2 by *Desulfobacter postgatei* I. Demonstration of all enzymes required for the operation of the citric acid cycle. *Archives of Microbiology*, in press.

TEN BRINK, B. & KONINGS, W. N. (1980). Generation of an electrochemical proton gradient by lactate efflux in membrane vesicles of *Escherichia coli*. *European Journal of Biochemistry*, **111**, 59–66.

TEN BRINK, B. & KONINGS, W. N. (1982). Electrochemical proton gradient and lactate concentration gradient in *Streptococcus cremoris* cells grown in batch culture. *Journal of Bacteriology*, **152**, 682–6.

BROCKMAN, H. L. & WOOD, W. A. (1975). Electron-transferring flavoprotein of *Peptostreptococcus elsdenii* that functions in the reduction of acrylyl-coenzyme A. *Journal of Bacteriology*, **124**, 1447–53.

BRONDER, M., MELL, H., STUPPERICH, E. & KRÖGER, A. (1982). Biosynthetic pathways of *Vibrio succinogenes* growing with fumarate as terminal electron acceptor and sole carbon source. *Archives of Microbiology*, **131**, 216–23.

BUCHANAN, B. B. (1979). Ferredoxin-linked carbon dioxide fixation in photosynthetic bacteria. In *Photosynthesis II, Photosynthetic Carbon Metabolism and Related Processes*, ed. M. Gibbs & E. Latzko, pp. 416–24. Berlin, Heidelberg & New York: Springer Verlag.

BUCKEL, W. (1980). The reversible dehydration of (R)-2-hydroxyglutarate to (E)-glutaconate. *European Journal of Biochemistry*, **106**, 439–47.

BUCKEL, W. & SEMMLER, R. (1982). A biotin-dependent sodium pump: glutaconyl-CoA decarboxylase from *Acidaminococcus fermentans*. *FEBS Letters*, **148**, 35–8.

BÜHLER, M., GIESEL, H., TISCHER, W. & SIMON, H. (1980). Occurrence and the possible physiological role of 2-enoate reductases. *FEBS Letters*, **109**, 244–6.

CAMMACK, R., PATIL, D., AGUIRRE, R. & HATCHIKIAN, E. C. (1982). Redox properties of the ESR-detectable nickel in hydrogenase from *Desulfovibrio gigas*. *FEBS Letters*, **142**, 289–92.

CLARKE, D. J., FULLER, F. M. & MORRIS, J. G. (1979). The proton-translocating adenosine triphosphatase of the obligately anaerobic bacterium *Clostridium pasteurianum*. I. ATP phosphohydrolase activity. *European Journal of Biochemistry*, **98**, 597–612.

CLARKE, D. J. & MORRIS, J. G. (1979). The proton-translocating adenosine triphosphatase of the obligately anaerobic bacterium *Clostridium pasteurianum*. II. ATP synthetase activity. *European Journal of Biochemistry*, **98**, 613–20.

DANIELS, L., FUCHS, G., THAUER, R. K. & ZEIKUS, J. G. (1977). Carbon monoxide oxidation by methanogenic bacteria. *Journal of Bacteriology*, **132**, 118–26.

DECKER, K., JUNGERMANN, K. & THAUER, R. K. (1970). Energy production in anaerobic organisms. *Angewandte Chemie (International edn)*, **9**, 138–58.

DIEKERT, G. B., GRAF, E. G. & THAUER, R. K. (1979). Nickel requirement for carbon monoxide dehydrogenase formation in *Clostridium pasteurianum*. *Archives of Microbiology*, **122**, 117–20.

DIEKERT, G., JAENCHEN, R. & THAUER, R. K. (1980). Biosynthetic evidence for a nickel tetrapyrrole structure of factor F_{430} from *Methanobacterium thermoautotrophicum*. *FEBS Letters*, **119**, 118–20.

DIEKERT, G., KLEE, B. & THAUER, R. K. (1980). Nickel, a component of factor F_{430} from *Methanobacterium thermoautotrophicum*. *Archives of Microbiology*, **124**, 103–6.

DIEKERT, G. & RITTER, M. (1982). Nickel requirement of *Acetobacterium woodii*. *Journal of Bacteriology*, **151**, 1043–5.

DIEKERT, G. & RITTER, M. (1983*a*). Carbon monoxide fixation into the carboxyl group of acetate during growth of *Acetobacterium woodii* on H_2 and CO_2. *FEMS Microbiology Letters*, **17**, 299–302.

DIEKERT, G. & RITTER, M. (1983*b*). Purification of the nickel protein carbon monoxide dehydrogenase of *Clostridium thermoaceticum*. *FEBS Letters*, **151**, 41–4.

DIEKERT, G. & THAUER, R. K. (1978). Carbon monoxide oxidation by *Clostridium thermoaceticum* and *Clostridium formicoaceticum*. *Journal of Bacteriology*, **136**, 597–606.

DIEKERT, G. & THAUER, R. K. (1980). The effect of nickel on carbon monoxide dehydrogenase formation in *Clostridium thermoaceticum* and *Clostridium formicoaceticum*. *FEMS Microbiology Letters*, **7**, 187–9.

DIMROTH, P. (1980). A new sodium-transport system energized by the decarboxylation of oxaloacetate. *FEBS Letters*, **122**, 234–6.

DIMROTH, P. (1982*a*). The generation of an electrochemical gradient of sodium ions upon decarboxylation of oxaloacetate by the membrane-bound and Na$^+$-activated oxaloacetate decarboxylase from *Klebsiella aerogenes*. *European Journal of Biochemistry*, **121**, 443–9.

DIMROTH, P. (1982*b*). The role of biotin and sodium in the decarboxylation of oxaloacetate by the membrane-bound oxaloacetate decarboxylase from *Klebsiella aerogenes*. *European Journal of Biochemistry*, **121**, 435–41.

DIMROTH, P. (1982*c*). Decarboxylation and transport. *Bioscience Reports*, **2**, 849–60.

DRAKE, H. L. (1982). Occurrence of nickel in carbon monoxide dehydrogenase from *Clostridium pasteurianum* and *Clostridium thermoaceticum*. *Journal of Bacteriology*, **149**, 561–6.

DRAKE, H. L., HU, S.-I. & WOOD, H. G. (1980). Purification of carbon monoxide dehydrogenase, a nickel enzyme from *Clostridium thermoaceticum*. *Journal of Biological Chemistry*, **255**, 7174–80.

DÜRRE, P., ANDERSCH, W. & ANDREESEN, J. R. (1981). Isolation and characterization of an adenine-utilizing anaerobic sporeformer, *Clostridium purinolyticum* sp. nov. *International Journal of Systematic Bacteriology*, **31**, 184–94.

DÜRRE, P. & ANDREESEN, J. R. (1982*a*). Selenium-dependent growth and glycine fermentation by *Clostridium purinolyticum*. *Journal of General Microbiology*, **128**, 1457–66.

DÜRRE, P. & ANDREESEN, J. R. (1982*b*). Anaerobic degradation of uric acid via pyrimidine derivatives by selenium-starved cells of *Clostridium purinolyticum*. *Archives of Microbiology*, **131**, 255–60.

EDEN, G. & FUCHS, G. (1982). Total synthesis of acetyl coenzyme A involved in autotrophic CO_2 fixation in *Acetobacterium woodii*. *Archives of Microbiology*, **133**, 66–74.

EDEN, G. & FUCHS, G. (1983). Autotrophic CO_2 fixation in *Acetobacterium woodii*. II. Demonstration of enzymes involved. *Archives of Microbiology*, **135**, 68–73.

EDWARDS, D. I., KNOX, R. J., SKOLIMOWSKI, I. M. & KNIGHT, R. C. (1982). Mode of action of nitroimidazoles. *European Journal of Chemotherapy and Antibiotics*, **2**, 65–72.

EIKMANNS, B. & THAUER, R. K. (1983). Mechanism of methane and CO_2 formation from acetate. Unpublished results.

EIRICH, L. D., VOGELS, G. D. & WOLFE, R. S. (1978). Proposed structure for coenzyme F_{420} from *Methanobacterium*. *Biochemistry*, **17**, 4583–93.

EKER, A. P. M., POL, A., VAN DER MEYDEN, P. & VOGELS, G. D. (1980). Purification and properties of 8-hydroxy-5-deazaflavin derivatives from *Streptomyces griseus*. *FEMS Microbiology Letters*, **8**, 161–5.

ELLEFSON, W. L., WHITMAN, W. B. & WOLFE, R. S. (1982). Nickel-containing factor F_{430}: chromophore of the methylreductase of *Methanobacterium*. *Proceedings of the National Academy of Sciences, USA*, **79**, 3707–10.

ELLEFSON, W. L. & WOLFE, R. S. (1980). Role of component C in the methylreductase system of *Methanobacterium*. *Journal of Biological Chemistry*, **255**, 8388–9.

EVANS, W. C. (1977). Biochemistry of the bacterial catabolism of aromatic compounds in anaerobic environments. *Nature, London*, **270**, 17–21.

EYZAGUIRRE, J., JANSEN, K. & FUCHS, G. (1982). Phosphoenolpyruvate synthetase in *Methanobacterium thermoautotrophicum*. *Archives of Microbiology*, **132**, 67–74.

FERGUSON, S. J. & SORGATO, M. C. (1982). Proton electrochemical gradients and energy-transduction processes. *Annual Review of Biochemistry*, **51**, 185–217.

FERRY, J. G. & WOLFE, R. S. (1976). Anaerobic degradation of benzoate to methane by a microbial consortium. *Archives of Microbiology*, **107**, 33–40.

FINA, L. R., BRIDGES, R. L., COBLENTZ, T. H. & ROBERTS, F. F. (1978). The anaerobic decomposition of benzoic acid during methane fermentation. III. The fate of carbon four and the identification of propanoic acid. *Archives of Microbiology*, **118**, 169–72.

FISCHER, F., ZILLIG, W., STETTER, K. O. & SCHREIBER, G. (1983). Chemolithoautotrophic metabolism of anaerobic extremely thermophilic archaebacteria. *Nature, London*, **301**, 511–13.

FUCHS, G. & STUPPERICH, E. (1982). Autotrophic CO_2 fixation pathway in *Methanobacterium thermoautotrophicum*. In *Archaebacteria*, ed. O. Kandler, pp. 277–88. Stuttgart & New York: G. Fischer Verlag.

FUCHS, G., STUPPERICH, E. & EDEN, G. (1980a). Autotrophic CO_2 fixation in *Chlorobium limicola*. Evidence for the operation of a reductive tricarboxylic acid cycle in growing cells. *Archives of Microbiology*, **128**, 64–71.

FUCHS, G., STUPPERICH, E. & JAENCHEN, R. (1980b). Autotrophic CO_2 fixation in *Chlorobium limicola*. Evidence against the operation of the Calvin cycle in growing cells. *Archives of Microbiology*, **128**, 56–63.

GEBHARDT, N. A., LINDER, D. & THAUER, R. K. (1983). Anaerobic acetate oxidation to CO_2 by *Desulfobacter postgatei*. II. Evidence from ^{14}C-labelling studies for the operation of the citric acid cycle. *Archives of Microbiology*, in press.

GIESEL, H. & SIMON, H. (1983). On the occurrence of enoate reductase and 2-oxo-carboxylate reductase in clostridia and some observations on the amino acid fermentation by *Peptostreptococcus anaerobius*. *Archives of Microbiology*, **135**, 51–7.

GILLES, H. & THAUER, R. K. (1983). Uroporphyrinogen III, an intermediate in the biosynthesis of the nickel containing factor F_{430} in *Methanobacterium thermoautotrophicum*. *European Journal of Biochemistry*, **135**, 109–12.

GOTTSCHAL, J. C. & MORRIS, J. G. (1981). The induction of acetone and butanol production in cultures of *Clostridium acetobutylicum* by elevated concentrations of acetate and butyrate. *FEMS Microbiology Letters*, **12**, 385–9.

GOTTSCHALK, G. & ANDREESEN, J. R. (1979). Energy metabolism in anaerobes. In *Microbial Biochemistry*, vol. 21, ed. J. R. Quayle, pp. 85–115. Baltimore: University Park Press.

GRAF, E. G. & THAUER, R. K. (1981). Hydrogenase from *Methanobacterium thermoautotrophicum*, a nickel-containing enzyme. *FEBS Letters*, **136**, 165–9.

GUNSALUS, R. P. & WOLFE, R. S. (1978). Chromophoric factors F_{342} and F_{430} of *Methanobacterium thermoautotrophicum*. *FEMS Microbiology Letters*, **3**, 191–3.

HARDER, W., VAN DIJKEN, J. P. & ROELS, J. A. (1981). Utilization of energy in methylotrophs. In *Microbial Growth on C_1 Compounds*, ed. H. Dalton, pp. 258–69. London: Heyden.

HAROLD, F. M. (1982). Pumps and currents: a biological perspective. *Current Topics in Membranes and Transport*, **16**, 485–516.

HAUSKA, G. & TREBST, A. (1977). Proton translocation in chloroplasts. In *Current Topics in Bioenergetics*, vol. 6, pp. 151–220. New York & London: Academic Press.

HEALY, J B. JR & YOUNG, L. Y. (1978). Catechol and phenol degradation by a methanogenic population of bacteria. *Applied and Environmental Microbiology*, **35**, 216–18.

HEALY J. B. JR & YOUNG, L. Y. (1979). Anaerobic biodegradation of eleven aromatic compounds to methane. *Applied and Environmental Microbiology*, **38**, 84–9.

HEALY, J. B. JR, YOUNG, L. Y. & REINHARD, M. (1980). Methanogenic decomposition of ferulic acid, a model lignin derivative. *Applied and Environmental Microbiology*, **39**, 436–44.

HEEFNER, D. L. & HAROLD, F. M. (1982). ATP-driven sodium pump in *Streptococcus faecalis*. *Proceedings of the National Academy of Sciences, USA*, **79**, 2798–802.

HELLINGWERF, K. J., LOLKEMA, J. S., OTTO, R., NEIJSSEL, O. M., STOUTHAMER, A. H., HARDER, W., VAN DAM, K. & WESTERHOFF, H. V. (1982). Energetics of microbial growth: an analysis of the relationship between growth and its mechanistic basis by mosaic non-equilibrium thermodynamics. *FEMS Microbiology Letters*, **15**, 7–17.

HIGUCHI, T. (1982). Biodegradation of lignin: biochemistry and potential applications. *Experientia*, **38**, 159–66.

HILL, T. L. (1977). *Free Energy Transduction in Biology*. New York & London: Academic Press.

HILPERT, W. & DIMROTH, P. (1982). Conversion of the chemical energy of methylmalonyl-CoA decarboxylation into a Na^+ gradient. *Nature, London*, **296**, 584–5.

HOBSON, P. N. & WALLACE, R. J. (1982). Microbial ecology and activities in the rumen: II. *CRC Critical Reviews in Microbiology*, **9**, 253–319.

HÖLLRIEGL, V., SCHERER, P. & RENZ, P. (1983). Isolation and characterization of the Co-methyl and Co-aquo derivative of 5-hydroxybenzimidazolylcobamide (factor III) from *Methanosarcina barkeri* grown on methanol. *FEBS Letters*, **151**, 156–8.

HU, S. I., DRAKE, H. L. & WOOD, H. G. (1982). Synthesis of acetyl coenzyme A from carbon monoxide, methyltetrahydrofolate, and coenzyme A by enzymes from *Clostridium thermoaceticum*. *Journal of Bacteriology*, **149**, 440–8.

HUSER, B. A., WUHRMANN, K. & ZEHNDER, A. J. B. (1982). *Methanothrix soehngenii* gen. nov. sp. nov., a new acetotrophic non-hydrogen-oxidizing methane bacterium. *Archives of Microbiology*, **132**, 1–9.

IMHOFF, D. & ANDREESEN, J. R. (1979). Nicotinic acid hydroxylase from *Clostridium barkeri*: selenium-dependent formation of active enzyme. *FEMS Microbiology Letters*, **5**, 155–8.

IVANOVSKY, R. N., SINTSOV, N. V. & KONDRATIEVA, E. N. (1980). ATP-linked citrate lyase activity in the green sulfur bacterium *Chlorobium limicola* forma *thiosulfatophilum*. *Archives of Microbiology*, **128**, 239–41.

IVERSEN, N. & BLACKBURN, T. H. (1981). Seasonal rates of methane oxidation in anoxic marine sediments. *Applied and Environmental Microbiology*, **41**, 1295–300.

JACOBSON, F. S., DANIELS, L., FOX, J. A., WALSH, C. T. & ORME-JOHNSON, W. H. (1982). Purification and properties of an 8-hydroxy-5-deazaflavin-reducing hydrogenase from *Methanobacterium thermoautotrophicum*. *Journal of Biological Chemistry*, **257**, 3385–8.

JAENCHEN, R., DIEKERT, G. & THAUER, R. K. (1981). Incorporation of methionine-derived methyl groups into factor F_{430} by *Methanobacterium thermoautotrophicum*. *FEBS Letters*, **130**, 133–6.

VON JAGOW, G. & ENGEL, W. D. (1980). Struktur und Funktion des energieumwandelnden Systems der Mitochondrien. *Angewandte Chemie*, **92**, 684–700.

JANSEN, K., STUPPERICH, E. & FUCHS, G. (1982). Carbohydrate synthesis from acetyl-CoA in the autotroph *Methanobacterium thermoautotrophicum*. *Archives of Microbiology*, **132**, 355–64.

JANSHEKAR, H. & FIECHTER, A. (1982). On the bacterial degradation of lignin. *European Journal of Applied Microbiology and Biotechnology*, **14**, 47–50.

JARRELL, K. F. & SPROTT, G. D. (1982). Nickel transport in *Methanobacterium bryantii*. *Journal of Bacteriology*, **151**, 1195–203.

JONES, J. B. & STADTMAN, T. C. (1976). *Methanococcus vannielii:* growth and metabolism of formate. In *Symposium on Microbial Production and Utilization of Gases (H₂, CH₄, CO)*, ed. H. G. Schlegel, G. Gottschalk & N. Pfennig, pp. 199–205.

JONES, J. B. & STADTMAN, T. C. (1981). Selenium-dependent and selenium-independent formate dehydrogenases of *Methanococcus vannielii*. Separation of the two forms and characterization of the purified selenium-independent form. *Journal of Biological Chemistry*, **256**, 656–63.

KANODIA, S. & ROBERTS, M. F. (1983). Methanophosphagen: a unique cyclic pyrophosphate isolated from *Methanobacterium thermoautotrophicum. Proceedings of the National Academy of Sciences, USA*, in press.

KAULFERS, P. M. (1976). Untersuchungen über Enzyme des Intermediärstoffwechsels von *Desulfuromonas acetoxidans* Stamm 11070. Thesis, University of Göttingen.

KEITH, C. L., BRIDGES, R. L., FINA, L. R., IVERSON, K. L. & CLORAN, J. A. (1978). The anaerobic decomposition of benzoic acid during methane fermentation. IV. Dearomatization of the ring and volatile fatty acids formed on ring rupture. *Archives of Microbiology*, **118**, 173–6.

KELL, D. B., BURNS, A., CLARKE, D. J. & MORRIS, J. G. (1981*a*). Proteinaceous proton pumps: a minimal model, some properties and their possible universality. *Speculations in Science and Technology*, **4**(2), 109–20.

KELL, D. B., DODDEMA, H. J., MORRIS, J. G. & VOGELS, G. D. (1981*b*). Energy coupling in methanogens. In *Microbial Growth on C₁ Compounds*, ed. H. Dalton, pp. 159–70. London: Heyden.

KELL, D. B., PECK, M. W., RODGER, G. & MORRIS, J. G. (1981*c*). On the permeability to weak acids and bases of the cytoplasmic membrane of *Clostridium pasteurianum. Biochemical and Biophysical Research Communications*, **99**, 81–8.

KELTJENS, J. T. M. (1982). Coenzymes of methanogenesis. Doctoral Thesis, University of Nijmegen, The Netherlands.

KELTJENS, J. T., CAERTELING, C. G., VAN KOOTEN, A. M., VAN DIJK, H. F. & VOGELS, G. D. (1983*a*). 6,7-Dimethyl-8-ribityl-5,6,7,8-tetrahydrolumazine, a proposed constituent of coenzyme MF₄₃₀ from methanogenic bacteria. *Biochimica et Biophysica Acta*, **743**, 351–8.

KELTJENS, J. T., CAERTELING, C. G., VAN KOOTEN, A. M., VAN DIJK, H. F. & VOGELS, G. D. (1983*b*). Chromophoric derivatives of coenzyme MF₄₃₀, a proposed coenzyme of methanogenesis in *Methanobacterium thermoautotrophicum. Archives of Biochemistry and Biophysics*, **223**, 235–53.

KELTJENS, J. T., DANIELS, L., JANNSEN, H. G., BORM, P. J. & VOGELS, G. D. (1983*c*). A novel one-carbon carrier (cTHMP) isolated from *Methanobacterium thermoautotrophicum* and derived from methanopterin. *European Journal of Biochemistry*, **130**, 545–52.

KELTJENS, J. T., HUBERTS, M. J., LAARHOVEN, W. H. & VOGELS, G. D. (1983*d*). Structural elements of methanopterin, a novel pterin present in *Methanobacterium thermoautotrophicum. European Journal of Biochemistry*, **130**, 537–44.

KELTJENS, J. T., WHITMAN, W. B., CAERTELING, C. G., VAN KOOTEN, A. M., WOLFE, R. S. & VOGELS, G. D. (1982). Presence of coenzyme M derivatives in the prosthetic group (coenzyme MF₄₃₀) of methylcoenzyme M reductase from *Methanobacterium thermoautotrophicum. Biochemical and Biophysical Research Communications*, **108**, 495–503.

KENEALY, W. R. & ZEIKUS, J. G. (1982). One-carbon metabolism in methanogens:

evidence for synthesis of a two-carbon cellular intermediate and unification of catabolism and anabolism in *Methanosarcina barkeri*. *Journal of Bacteriology*, **151**, 932–41.

KLINGENBERG, M. & SCHOLLMEYER, P. (1960). Zur Reversibilität der oxydativen Phosphorylierung. *Biochemische Zeitschrift*, **333**, 335–50.

KNAFF, D. B. (1978). Reducing potentials and the pathway of NAD^+ reduction. In *The Photosynthetic Bacteria*, ed. R. K. Clayton & W. R. Sistrom, pp. 629–40. New York: Plenum Press.

KNAPPE, J., BLASCHKOWSKI, H. P., GRÖBNER, P. & SCHMITT, T. (1974). Pyruvate formate-lyase of *Escherichia coli*: the acetyl-enzyme intermediate. *European Journal of Biochemistry*, **50**, 253–63.

KNAPPE, J., BLASCHKOWSKI, H. P. & EDENHARDER, R. (1972). Enzyme-dependent activation of pyruvate formate-lyase of *Escherichia coli*. In *Metabolic Interconversion of Enzymes*, ed. O. Wieland, E. Helmreich & H. Holzer, pp. 319–29. Berlin: Springer Verlag.

KONINGS, W. N. & BOOTH, I. R. (1981). Do the stoichiometries of ion-linked transport systems vary? *Trends in Biochemical Sciences*, **6**, 257–62.

KONINGS, W. N. & VELDKAMP, H. (1983). Energy transduction and solute transport mechanisms in relation to environments occupied by microorganisms. In *Microbes in their Natural Environments*, ed. J. H. Slater, R. Whittenbury & J. W. T. Wimpenny, *Society for General Microbiology Symposium 34*, pp. 153–86. Cambridge University Press.

KRISTJANSSON, J. K., SCHÖNHEIT, P. & THAUER, R. K. (1982). Different K_s values for hydrogen of methanogenic bacteria and sulfate reducing bacteria: an explanation for the apparent inhibition of methanogenesis by sulfate. *Archives of Microbiology*, **131**, 278–82.

KRÖGER, A. & WINKLER, E. (1981). Phosphorylative fumarate reduction in *Vibrio succinogenes*: stoichiometry of ATP synthesis. *Archives of Microbiology*, **129**, 100–4.

KRÜGER, H.-J., HUYNH, B. H., LJUNGDAHL, P. O., XAVIER, A. V., DERVARTANIAN, D. V., MOURA, I., PECK JR, H. D., TEIXEIRA, M., MOURA, J. J. G. & LEGALL, J. (1982). Evidence for nickel and a three-iron center in the hydrogenase of *Desulfovibrio desulfuricans*. *Journal of Biological Chemistry*, **257**, 14620–3.

KRZYCKI, J. A., WOLKIN, R. H. & ZEIKUS, J. G. (1982). Comparison of unitrophic and mixotrophic substrate metabolism by an acetate-adapted strain of *Methanosarcina barkeri*. *Journal of Bacteriology*, **149**, 247–54.

KRZYCKI, J. & ZEIKUS, J. G. (1980). Quantification of corrinoids in methanogenic bacteria. *Current Microbiology*, **3**, 243–5.

KÜHN, W., FIEBIG, K., HIPPE, H., MAH, R. A., HUSER, B. A. & GOTTSCHALK, G. (1983). Distribution of cytochromes in methanogenic bacteria. *Journal of Bacteriology*, in press.

KÜHN, W., FIEBIG, K., WALTHER, R. & GOTTSCHALK, G. (1979). Presence of a cytochrome b_{559} in *Methanosarcina barkeri*. *FEBS Letters*, **105**, 271–4.

KÜHN, W. & GOTTSCHALK, G. (1983). Characterization of cytochromes occurring in *Methanosarcina* species. *European Journal of Biochemistry*, in press.

LANYI, J. K. (1979). The role of Na^+ in transport processes of bacterial membranes. *Biochimica et Biophysica Acta*, **559**, 377–97.

LEGALL, J., LJUNGDAHL, P. O., MOURA, I., PECK JR, H. D., XAVIER, A. V., MOURA, J. J. G., TEIXERA, M., HUYNH, B. H. & DERVARTANIAN, D. V. (1982). The presence of redox-sensitive nickel in the periplasmic hydrogenase from *Desulfovibrio gigas*. *Biochemical and Biophysical Research Communications*, **106**, 610–16.

LEIGH, J. A. & WOLFE, R. S. (1983). Carbon dioxide reduction factor and methanopterin, two coenzymes required for CO_2 reduction to methane by extracts of *Methanobacterium*. *Journal of Biological Chemistry*, **258**, 7536–40.

LEONHARDT, U. & ANDREESEN, J. R. (1977). Some properties of formate dehydrogenase accumulation and incorporation of [185]W-tungsten into proteins of *Clostridium formicoaceticum*. *Archives of Microbiology*, **115**, 277–84.

LIU, C. L., HART, N. & PECK JR, H. D. (1982). Inorganic pyrophosphate: energy source for sulfate-reducing bacteria of the genus *Desulfotomaculum*. *Science*, **217**, 363–4.

LJUNGDAHL, L. G. (1976). Tungsten, a biologically active metal. *Trends in Biochemical Sciences*, **1**, 63–5.

LJUNGDAHL, L. G. & ANDREESEN, J. R. (1975). Tungsten, a component of active formate dehydrogenase from *Clostridium thermoaceticum*. *FEBS Letters*, **54**, 279–82.

LJUNGDAHL, L. G. & ANDREESEN, J. R. (1978). Formate dehydrogenase, a selenium–tungsten enzyme from *Clostridium thermoaceticum*. In *Methods in Enzymology*, vol. 53, ed. S. Fleischer & L. Parker, pp. 360–72. New York & London: Academic Press.

McCORMICK. J. R. D. & MORTON, G. O. (1982). Identity of cosynthetic factor 1 of *Streptomyces aureofaciens* and fragment FO from coenzyme F_{420} of *Methanobacterium* species. *Journal of the American Chemical Society*, **104**, 4014–15.

McINERNEY, M. J., BRYANT, M. P., HESPELL, R. B. & COSTERTON, J. W. (1981). *Syntrophomonas wolfei* gen. nov. sp. nov., an anaerobic syntrophic, fatty acid-oxidizing bacterium. *Applied and Environmental Microbiology*, **41**, 1029–39.

MAH, R. A. (1980). Isolation and characterization of *Methanococcus mazei*. *Current Microbiology*, **3**, 321–6.

MALONEY, P. C. (1982). Energy coupling to ATP synthesis by the proton-translocating ATPase. *Journal of Membrane Biology*, **67**, 1–12.

MALONEY, P. C. (1983). Relationship between phosphorylation potential and electrochemical H^+ gradient during glycolysis in *Streptococcus lactis*. *Journal of Bacteriology*, **153**, 1461–70.

MARGULIS, L. (1981). *Symbiosis in Cell Evolution. Life and its Environment on the Early Earth*. San Francisco: Freeman.

MELL, H., BRONDER, M. & KRÖGER, A. (1982). Cell yields of *Vibrio succinogenes* growing with formate and fumarate as sole carbon and energy sources in chemostat culture. *Archives of Microbiology*, **131**, 224–8.

MICHELS, P. A. M., MICHELS, J. P. J., BOONSTRA, J. & KONINGS, W. N. (1979). Generation of an electrochemical proton gradient in bacteria by the excretion of metabolic end products. *FEMS Microbiology Letters*, **5**, 357–64.

MIRAGLIA, G. J. (1974). Pathogenic anaerobic bacteria. *CRC Critical Reviews in Microbiology*, **3**, 161–81.

MORRIS, J. G. (1977). Obligately anaerobic bacteria. *Trends in Biochemical Sciences*, **2**, 81–4.

MORRIS, J. G. (1983). Anaerobic fermentations: some new possibilities. In *Biotechnology, 48th Symposium of The Biochemical Society*, ed. C. F. Phelps & P. H. Clarke.

MOUNTFORT, D. O. & BRYANT, M. P. (1982). Isolation and characterization of an anaerobic syntrophic benzoate-degrading bacterium from sewage sludge. *Archives of Microbiology*, **133**, 249–56.

MOURA, J. J. G., MOURA, I., BRUSCHI, M., LE GALL, J. & XAVIER, A. V. (1980). A cobalt containing protein isolated from *Desulfovibrio gigas*, a sulfate reducer. *Biochemical and Biophysical Research Communications*, **92**, 962–70.

NETHE-JAENCHEN, R. & THAUER, R. K. (1983). Growth yields and saturation constant of *Desulfovibrio vulgaris* in chemostat culture. *Archives of Microbiology*, submitted.

NUCCITELLI, R. & HEIPLE, J. M. (1982). Summary of the evidence and discussion concerning the involvement of pH_i in the control of cellular functions. In *Intracellular pH: Its Measurement, Regulation and Utilization in Cellular Functions*, ed. R. Nuccitelli & D. W. Deamer, pp. 567–86. New York: Liss.

NYRÉN, P. & BALTSCHEFFSKY, M. (1983). Inorganic pyrophosphate-driven ATP-synthesis in liposomes containing membrane-bound inorganic pyrophosphatase and F_0–F_1 complex from *Rhodospirillum rubrum*. *FEBS Letters*, **155**, 125–30.

OTTO, R., TEN BRINK, B., VELDKAMP, H. & KONINGS, W. N. (1983). The relation between growth rate and electrochemical proton gradient of *Streptococcus cremoris*. *FEMS Microbiology Letters*, **16**, 69–74.

OTTO, R., HUGENHOLTZ, J., KONINGS, W. N. & VELDKAMP, H. (1980a). Increase of molar growth yield of *Streptococcus cremoris* for lactose as a consequence of lactate consumption by *Pseudomonas stutzeri* in mixed culture. *FEMS Microbiology Letters*, **9**, 85–8.

OTTO, R., SONNENBERG, A. S. M., VELDKAMP, H. & KONINGS, W. N. (1980b). Generation of an electrochemical proton gradient in *Streptococcus cremoris* by lactate efflux. *Proceedings of the National Academy of Sciences, USA*, **77**, 5502–6.

PADAN, E., ZILBERSTEIN, D. & SCHULDINER, S. (1981). pH homeostasis in bacteria. *Biochimica et Biophysica Acta*, **650**, 151–66.

PECHER, A., BLASCHKOWSKI, H. P., KNAPPE, K. & BÖCK, A. (1982). Expression of pyruvate–formate lyase of *Escherichia coli* from the cloned structural gene. *Archives of Microbiology*, **132**, 365–71.

PECK JR, H. D. & LEGALL, J. (1982). Biochemistry of dissimilatory sulphate reduction. *Philosophical Transactions of the Royal Society of London, Series B*, **298**, 443–66.

PECK JR, H. D., LIU, C.-L., VARMA, A. K., LJUNGDAHL, L. G., SZULCZYNSKI, M., BRYANT, F. & CARREIRA, L. (1982). The utilization of inorganic pyrophosphate, tripolyphosphate and tetrapolyphosphate as energy sources for the growth of anaerobic bacteria. In *Basic Biology of New Developments in Biotechnology*, ed. A. Hollaender, B. Laskin & C. Rogers, pp. 317–48. New York & London: Plenum Press.

PERSKI, H.-J., MOLL, J. & THAUER, R. K. (1981). Sodium dependence of growth and methane formation in *Methanobacterium thermoautotrophicum*. *Archives of Microbiology*, **130**, 319–21.

PERSKI, H.-J., SCHÖNHEIT, P. & THAUER, R. K. (1982). Sodium dependence of methane formation in methanogenic bacteria. *FEBS Letters*, **143**, 323–6.

PFALTZ, A., JAUN, B., FÄSSLER, A., ESCHENMOSER, A., JAENCHEN, R., GILLES, H. H., DIEKERT, G. & THAUER, R. K. (1982). Zur Kenntnis des Faktors F_{430} aus methanogenen Bakterien: Struktur des porphinoiden Ligandsystems. *Helvetica Chimica Acta*, **65**, 828–65.

PFENNIG, N. & BIEBL, H. (1976). *Desulfuromonas acetoxidans* gen. nov. and sp. nov., a new anaerobic, sulfur-reducing, acetate-oxidizing bacterium. *Archives of Microbiology*, **110**, 3–12.

PFENNIG, N. & WIDDEL, F. (1981). Ecology and physiology of some anaerobic bacteria from the microbial sulfur cycle. In *Biology of Inorganic Nitrogen and Sulfur*, ed. H. Bothe & A. Trebst, pp. 169–77. Berlin, Heidelberg & New York: Springer Verlag.

PINE, M. J. & BARKER, H. A. (1956). Studies on methane fermentation. XII. The pathway of hydrogen in acetate fermentation. *Journal of Bacteriology*, **71**, 644–8.

PIRT, S. J. (1975). *Principles of Microbe and Cell Cultivation*. Oxford: Blackwell Scientific.

PIRT, S. J. (1982). Maintenance energy: a general model for energy-limited and energy-sufficient growth. *Archives of Microbiology*, **133**, 300–2.

PITSCH, C. & SIMON, H. (1982). The stereochemical course of the water elimination from (2R)-phenyllactate in the amino acid fermentation of *Clostridium sporogenes*. *Hoppe-Seyler's Zeitschrift für physiologische Chemie*, **363**, 1253–7.

POL, A., VAN DER DRIFT, C. & VOGELS, G. D. (1982). Corrinoids from *Methanosarcina barkeri:* structure of the α-ligand. *Biochemical and Biophysical Research Communications*, **108**, 731–7.

RAGSDALE, S. W., LJUNGDAHL, L. G. & DERVARTANIAN, D. V. (1982). EPR (1983). Properties of purified carbon monoxide dehydrogenase from *Clostridium thermoaceticum*, a nickel, iron–sulfur protein. *Journal of Biological Chemistry*, **258**, 2364–9.

RAGSDALE, S. W., LJUNGDAHL, L. G. & DERVARTANIAN, D. V. (1982). EPR evidence for nickel–substrate interaction in carbon monoxide dehydrogenase from *Clostridium thermoaceticum*. *Biochemical and Biophysical Research Communications*, **108**, 658–63.

REEVES, R. E. (1976). How useful is the energy in inorganic pyrophosphate? *Trends in Biochemical Sciences*, **1**, 53–5.

RIEBELING, V., THAUER, R. K. & JUNGERMANN, K. (1975). The internal-alkaline pH gradient, sensitive to uncoupler and ATPase inhibitor, in growing *Clostridium pasteurianum*. *European Journal of Biochemistry*, **55**, 445–53.

ROMESSER, J. A. & WOLFE, R. S. (1982). CDR factor, a new coenzyme required for carbon dioxide reduction to methane by extracts of *Methanobacterium thermoautotrophicum*. In *Archaebacteria*, ed. O. Kandler, pp. 271–6. Stuttgart & New New York: G. Fischer Verlag.

RÜHLEMANN, M., STUPPERICH, E. & FUCHS, G. (1983). Detection of acetyl coenzyme A as an early CO_2 assimilation intermediate in *Methanobacterium*. In preparation.

SCHERER, P. & SAHM, H. (1980). Growth of *Methanosarcina barkeri* on methanol or acetate in a defined medium. In *First International Symposium on Anaerobic Digestion*, ed. B. Wheatley, D. Stafford & V. Mason, pp. 45–7. Cardiff: Scientific Press.

SCHINK, B. & PFENNIG, N. (1982*a*). *Propionigenium modestum* gen. nov. sp. nov., a new strictly anaerobic, nonsporing bacterium growing on succinate. *Archives of Microbiology*, **133**, 209–16.

SCHINK, B. & PFENNIG, N. (1982*b*). Fermentation of trihydroxybenzenes by *Pelobacter acidigallici* gen. nov. sp. nov., a new strictly anaerobic, non-spore-forming bacterium. *Archives of Microbiology*, **133**, 195–201.

SCHÖNHEIT, P., KRISTJANSSON, J. K. & THAUER, R. K. (1982). Kinetic mechanism for the ability of sulfate reducers to out-compete methanogens for acetate. *Archives of Microbiology*, **132**, 285–8.

SCHÖNHEIT, P., MOLL, J. & THAUER, R. K. (1979). Nickel, cobalt, and molybdenum requirement for growth of *Methanobacterium thermoautotrophicum*. *Archives of Microbiology*, **123**, 105–7.

SEELY, R. J. & FAHRNEY, D. E. (1983). A novel diphospho-P,P'-diester from *Methanobacterium thermoautotrophicum*. *Journal of Biological Chemistry*, in press.

SEEWALDT, E. & STACKEBRANDT, E. (1982). Partial sequence of 16S ribosomal RNA and the phylogeny of *Prochloron*. *Nature, London*, **295**, 618–20.

SLEAT, R. & ROBINSON, J. P. (1983). Methanogenic degradation of sodium benzoate in profundal sediments from a small eutrophic lake. *Journal of General Microbiology*, **129**, 141–52.

SLONCZEWSKI, J. L., ROSEN, B. P., ALGER, J. R. & MACNAB, R. M. (1981). pH homeostasis in *Escherichia coli:* measurement by ^{31}P nuclear magnetic resonance of methylphosphonate and phosphate. *Proceedings of the National Academy of Sciences, USA*, **78**, 6271–5.

SPIVEY, M. J. (1978). The acetone/butanol/ethanol fermentation. *Process Biochemistry*, **13**, no. 11, 2–4, 25.

STADTMAN, T. C. (1978). Selenium-dependent clostridial glycine reductase. In *Methods in Enzymology*, vol. 53, ed. S. Fleischer & L. Packer, pp. 373–82. New York & London: Academic Press.

STADTMAN, T. C. (1980). Biological functions of selenium. *Trends in Biochemical Sciences*, **5**, 203–6.

STADTMAN, T. C. (1981). Selenoenzymes. In *Structural and Functional Aspects of Enzyme Catalysis*, ed. H. Eggerer & R. Huber, pp. 96–103. Berlin, Heidelberg & New York: Springer Verlag.

STOUTHAMER, A. H. (1979). The search for correlation between theoretical and experimental growth yields. In *Microbial Biochemistry*, vol. 21. ed. J. R. Quayle, pp. 1–47. Baltimore: University Park Press.

STUCKI, J. W. (1982). Thermodynamic optimizing principles in mitochondrial energy conversions. In *Metabolic Compartmentation*, ed. H. Sies, pp. 39–69. New York & London: Academic Press.

STUPPERICH, E. & FUCHS, G. (1983). Autotrophic acetyl coenzyme A synthesis *in vitro* from two CO_2 in *Methanobacterium*. *FEBS Letters*, **156**, 345–8.

STUPPERICH, E., HAMMEL, K. E., FUCHS, G. & THAUER, R. K. (1983). Carbon monoxide fixation into the carboxyl group of acetyl coenzyme A during autotrophic growth of *Methanobacterium*. *FEBS Letters*, **152**, 21–3.

TANNER, A. C. R., BADGER, S., LAI, C.-H., LISTGARTEN, M. A., VISCONTI, R. A. & SOCRANSKY, S. S. (1981). *Wolinella* gen. nov., *Wolinella succinogenes* (*Vibrio succinogenes* Wolin *et al.*) comb. nov., and description of *Bacteroides gracilis* sp. nov., *Wolinella recta* sp. nov., *Campylobacter concisus* sp. nov., and *Eikenella corrodens* from humans with periodontal disease. *International Journal of Systematic Bacteriology*, **31**, 432–45.

TARVIN, D. & BUSWELL, A. M. (1934). The methane fermentation of organic acids and carbohydrates. *Journal of the American Chemical Society*, **56**, 1751–5.

TAYLOR, C. D., McBRIDE, B. C., WOLFE, R. S. & BRYANT, M. P. (1974). Coenzyme M, essential for growth of a rumen strain of *Methanobacterium ruminantium*. *Journal of Bacteriology*, **120**, 974–5.

TAYLOR, C. D. & WOLFE, R. S. (1974). Structure and methylation of coenzyme M ($HSCH_2CH_2SO_3$). *Journal of Biological Chemistry*, **249**, 4879–85.

TAYLOR, G. T., KELLY, D. P. & PIRT, S. J. (1976). Intermediary metabolism in methanogenic bacteria (*Methanobacterium*). In *Symposium on Microbial Production and Utilization of Gases (H_2, CH_4, CO)*, ed. H. G. Schlegel, G. Gottschalk & N. Pfennig, pp. 173–80. Göttingen: E. Goltze.

TEMPEST, D. W. (1978). The biochemical significance of microbial growth yields: a reassessment. *Trends in Biochemical Sciences*, **3**, 180–4.

THAUER, R. K. (1972). CO_2-reduction to formate by NADPH. The initial step in the total synthesis of acetate from CO_2 in *Clostridium thermoaceticum*. *FEBS Letters*, **27**, 111–15.

THAUER, R. K. (1982). Dissimilatory sulphate reduction with acetate as electron donor. *Philosophical Transactions of the Royal Society of London, Series B*, **298**, 467–71.

THAUER, R. K., BRANDIS-HEEP, A., DIEKERT, G., GILLES, H.-H., GRAF, E. G., JAENCHEN, R. & SCHÖNHEIT, P. (1983). Drei neue Nickelenzyme aus anaeroben Bakterien. *Naturwissenschaften*, **70**, 60–4.

THAUER, R. K., DIEKERT, G. & SCHÖNHEIT, P. (1980). Biological role of nickel. *Trends in Biochemical Sciences*, **5**, 304–6.

THAUER, R. K., JUNGERMANN, K. & DECKER, K. (1977). Energy conservation in chemotrophic anaerobic bacteria. *Bacteriological Reviews*, **41**, 100–80.

THAUER, R. K., KIRCHNIAWY, F. H. & JUNGERMANN, K. A. (1972). Properties and function of the pyruvate–formate-lyase reaction in clostridia. *European Journal of Biochemistry*, **27**, 282–90.

THOMSON, F. M. & BOOTH, I. R. (1982). Control of butane-2,3-diol formation in *Klebsiella aerogenes*. *Biochemical Society Transactions*, **10**, 465–6.

TOUZEL, J. P. & ALBAGNAC, G. (1983). Isolation and characterization of *Methanococcus mazei* strain MC_3. *FEMS Microbiology Letters*, **16**, 241–5.

TUNG, K. K. & WOOD, W. A. (1975). Purification, new assay, and properties of coenzyme A transferase from *Peptostreptococcus elsdenii*. *Journal of Bacteriology*, **124**, 1462–74.

TURNER, D. C. & STADTMAN, T. C. (1973). Purification of protein components of the clostridial glycine reductase system and characterization of protein A as a selenoprotein. *Archives of Biochemistry and Biophysics*, **154**, 366–81.

UNDEN, G., BÖCHER, R., KNECHT, J. & KRÖGER, A. (1982). Hydrogenase from *Vibrio succinogenes*, a nickel protein. *FEBS Letters*, **145**, 230–4.

VARMA, A. K. & PECK JR, H. D. (1983). Utilization of short- and long-chain polyphosphates as energy sources for the anaerobic growth of bacteria. *FEMS Microbiology Letters*, **16**, 281–5.

VOGELS, G. D., KELTJENS, J. T., HUTTEN, T. J. & VAN DER DRIFT, C. (1982). Coenzymes of methanogenic bacteria. In *Archaebacteria*, ed. O. Kandler, pp. 258–64. Stuttgart & New York: G. Fischer Verlag.

WAGNER, R. & ANDREESEN, J. R. (1977). Differentiation between *Clostridium acidiurici* and *Clostridium cylindrosporum* on the basis of specific metal requirements for formate dehydrogenase formation. *Archives of Microbiology*, **114**, 219–24.

WAGNER, R. & ANDREESEN, J. R. (1979). Selenium requirement for active xanthine dehydrogenase from *Clostridium acidiurici* and *Clostridium cylindrosporum*. *Archives of Microbiology*, **121**, 255–60.

WESTERHOFF, H. V., HELLINGWERF, K. J. & VAN DAM, K. (1983). Thermodynamic efficiency of microbial growth is low but optimal for maximal growth rate. *Proceedings of the National Academy of Sciences, USA*, **80**, 305–9.

WESTERHOFF, H. V., LOLKEMA, J. S., OTTO, R. & HELLINGWERF, K. J. (1982). Thermodynamics of growth. Non-equilibrium thermodynamics of bacterial growth. The phenomenological and the mosaic approach. *Biochimica et Biophysica Acta*, **683**, 181–220.

WHITMAN, W. B. & WOLFE, R. S. (1980). Presence of nickel in factor F_{430} from *Methanobacterium bryantii*. *Biochemical and Biophysical Research Communications*, **92**, 1196–201.

WIDDEL, F. (1980). Anaerober Abbau von Fettsäuren und Benzoesäure durch neu isolierte Arten Sulfat-reduzierender Bakterien. Doctoral thesis, University of Göttingen.

WIDDEL, F. & PFENNIG, N. (1981). Studies on dissimilatory sulfate-reducing bacteria that decompose fatty acids. I. Isolation of new sulfate-reducing bacteria enriched with acetate from saline environments. Description of *Desulfobacter postgatei* gen. nov., sp. nov. *Archives of Microbiology*, **129**, 395–400.

WIEGEL, J., BRAUN, M. & GOTTSCHALK, G. (1981). *Clostridium thermoautotrophicum* species novum, a thermophile producing acetate from molecular hydrogen and carbon dioxide. *Current Microbiology*, **5**, 255–60.

WOLFE, R. S. & HIGGINS, I. J. (1979). Microbial biochemistry of methane – a study in contrasts. In *Microbial Biochemistry*, vol. 21, ed. J. R. Quayle, pp. 267–300. Baltimore: University Park Press.

WOOD, H. G., DRAKE, H. L. & HU, S. I. (1982). Studies with *Clostridium thermoaceticum* and the resolution of the pathway used by acetogenic bacteria that grow on carbon monoxide or carbon dioxide and hydrogen. In *Proceedings of the Biochemistry Symposium*, ed. E. E. Snell, pp. 29–56. Palo Alto: Annual Reviews Inc.

YAMAMOTO, I., SAIKI, T., LIU, S.-M. & LJUNGDAHL, L. G. (1983). Purification and properties of NADP-dependent formate dehydrogenase from *Clostridium thermoaceticum*, a tungsten–selenium–iron protein. *Journal of Biological Chemistry*, **258**, 1826–32.

YAMAZAKI, S. (1982). A selenium-containing hydrogenase from *Methanococcus vannielii*. *Journal of Biological Chemistry*, **257**, 7926–9.

ZEHNDER, A. J. B. & BROCK, T. D. (1979). Biological energy production in the apparent absence of electron transport and substrate level phosphorylation. *FEBS Letters*, **107**, 1–3.

ZEHNDER, A. J. B. & BROCK, T. D. (1980). Anaerobic methane oxidation: occurrence and ecology. *Applied and Environmental Microbiology*, **39**, 194–204.

ZEIKUS, J. G. (1980). Fate of lignin and related aromatic substrates in anaerobic environments. In *Lignin Biodegradation: Microbiology, Chemistry and Potential Applications*, vol. I, ed. T. K. Kirk, T. Higuchi & H. Chang, pp. 101–10. Boca Raton: CRC Press.

ZEIKUS, J. G., WELLSTEIN, A. L. & KIRK, T. K. (1982). Molecular basis for the biodegradative recalcitrance of lignin in anaerobic environments. *FEMS Microbiology Letters*, **15**, 193–7.

ZEIKUS, J. G. & WOLFE, R. S. (1972). *Methanobacterium thermoautotrophicus* sp. n., an anaerobic, autotrophic, extreme thermophile. *Journal of Bacteriology*, **109**, 707–13.

TRANSPOSABLE ELEMENTS, GENOME REORGANIZATION AND CELLULAR DIFFERENTIATION IN GRAM-NEGATIVE BACTERIA

JAMES A. SHAPIRO

University of Chicago, 920 East 58th Street, Chicago, IL 60637, USA

INTRODUCTION: GENOTYPE AND PHENOTYPE

The growth and division of individual eubacterial cells involve thousands of repeated macromolecular syntheses that are coordinated both temporally and spatially. The information for the sequences of the proteins which carry out and regulate these polymerizations is contained in DNA molecules whose collective length is of the order of 1000 times the length of a single bacterium (Cairns, 1963) and which must therefore exist in a highly compacted form within each cell (Worcel & Burgi, 1972; Pettijohn, 1976; Sjåstad *et al.*, 1982). It is apparent from these basic considerations that processes of DNA replication and protein synthesis (which occur contemporaneously) present a formidable problem of physical organization because the appropriate polymerases must have access to the condensed genome at the correct times and locations. The magnitude of this problem looms even greater when we think of the growth of bacterial populations on solid surfaces, for we know that each bacterial strain displays a characteristic colonial morphology and so reveals intercellular coordination of the growth process (Olds, 1974).

In other words, it is necessary to formulate a dynamic view of the genotype/phenotype relationship. We must try to understand how the information organized in constantly moving but highly compacted DNA molecules influences the characteristics of growing cell populations. The technology for examining the most basic level of genomic organization – the linear sequence of nucleotides in DNA molecules – is now highly sophisticated and has become widely known among microbiologists and students. Other papers in this volume will discuss some of the lessons we are learning about genome organization from studies of DNA primary structure. In this paper, I would like to address two other aspects of the genotype/phenotype question that are less well known: (1) the roles of transposable elements in the reorganization of bacterial genomes, and (2) how transposable elements can provide a technology for revealing multicellular patterns of genome expression in bacterial growth on agar medium.

AGENTS OF GENOME REORGANIZATION: THE *MUDLAC* SYSTEM

Bacterial cells have a variety of biochemical systems for reorganizing their DNA molecules. These include mutator systems for changing one or a few nucleotides, homologous and non-homologous general recombination mechanisms, site-specific recombination mechanisms and transposable elements (reviewed in Shapiro, 1982, 1983*a, b*, 1984). In bacteria, transposable elements may be defined as DNA segments bounded by defined (usually symmetrical) sequences that recombine with different genome 'target' regions by mechanisms that frequently result in two sequence duplications – one of the transposable element itself and one of a short oligonucleotide at the target site (reviewed in Calos & Miller, 1980; Kleckner, 1981). These replicative recombination events can generate several different kinds of genome reorganizations, including appearance of the transposable element at the target site (simple transposition), deletions adjacent to the transposable element, fusions of circular DNA molecules, inversions, and transpositions of DNA segments external to the transposable element (Cohen & Shapiro, 1980; Shapiro, 1979, 1982; Iida, Meyer & Arber, 1983; Toussaint & Resibois, 1983).

Although the generality and functionality of these different kinds of mechanisms for genome reorganization are only now becoming widely accepted among geneticists, they have been exploited by genetic engineers over the past 25 years for *in vivo* clonings and other genetic manipulations, often without benefit of *in vitro* technology (Miller, 1972; Bassford *et al.*, 1978; Shapiro, 1983*a*, 1984). One of the major accomplishments of these efforts was the development of methods utilizing the transposing bacteriophage Mu for constructing hybrid protein coding sequences (Casadaban, 1976; Bassford *et al.*, 1978; Casadaban & Cohen, 1979; Casadaban & Chou, 1983). These methods provide model answers to questions about the origin of novel coding sequences in evolution. Interestingly, they reveal unanticipated genetic capabilities of phage Mu. Fig. 1 illustrates the original Casadaban technique for generating fusions in *E. coli* which create hybrid β-galactosidase coding sequences by connecting all but the first 60 or so nucleotides of the *lacZ* cistron to the beginning of virtually any other cistron. While the Mu prophage was initially used only as a kind of 'portable homology' to direct the insertion of the *lacZ* sequence adjacent to other cistrons, it has since become evident that the Mu prophage plays a direct role in the formation of cistron fusions. Two lines of evidence support this statement: (1) hybrid cistrons often contain Mu sequences as linkers between the regions encoding the two domains of the hybrid β-galactosidase (Moreno *et al.*, 1980; Casadaban & Chou, 1983; M. Berman & S. Benson, personal communication); and (2) alterations in the regulation of Mu functions by the product of the Mu *c* cistron result in quantitative and kinetic changes in the appearance of clones containing *araB–lacZ* fusions (J. Shapiro, unpublished).

In the original Casadaban technique, the insertion of Mu in a particular cistron was the *first* step in a series of DNA reorganizations that gave rise to a hybrid coding sequence (Fig. 1). It is possible to invert this genealogy and create a Mu derivative already incorporating *lac* sequences so that insertion in a given cistron is the final step and automatically results in a genetic fusion (Casadaban & Cohen, 1979; Casadaban & Chou, 1983). These Mu derivatives are called Mu*dlac* (or 'Mud') phages, for Mu *d*efectives carrying *lac* sequences. There are two general classes of Mu*dlac* fusion phages: Mud I phages which create transcriptional fusions of the inserted cistron to an intact *lacZ* cistron, and Mud II phages which create translational fusions of the inserted cistron, some Mu nucleotides, and a *lacZ* sequence lacking translational initiation signals. The structure of such a Mud II phage is schematized in Fig. 2. The antibiotic resistance markers incorporated in these Mud phages facilitate the isolation of insertions without reference to the *lac* phenotype of the resulting fusion. Once a fusion has been isolated, the expression of β-galactosidase activity reflects the regulation of the target cistron. β-Galactosidase expression can be determined in mass populations by enzyme assays on cell extracts. However, it is often more convenient for screening many clones to visualize β-galactosidase activity directly on colonies growing on agar containing a chromogenic substrate, such as 5-bromo-3-chloro-3-indolyl-β-D-galactoside (XGal) which turns blue–green on hydrolysis (Miller, 1972). The pigment produced by XGal hydrolysis sticks to

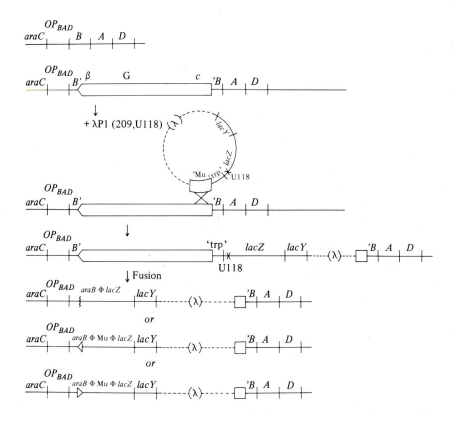

Fig. 1. Schematic representation of the DNA structures involved in the formation of *araB–lacZ* cistron fusions. *E. coli* chromosomal sequences are represented by continuous lines, lambda sequences by dashed lines, and Mu sequences by an open box. Apostrophes indicate interrupted or deleted sequences. The 'trp' region upstream of *lacZ* indicates *trp* operon material connected to a small segment of *lacO* by the W209 *trpA–lacO* deletion (Mitchell, Reznikoff & Beckwith, 1974). The orientation of the Mu prophage is indicated by the pointed and blunt ends of the box and this is oriented with respect to the Mu genetic map by indicating the relative positions of *c* (repressor cistron), invertible G region, and *β* region (Toussaint & Resibois, 1983). The presumptive steps in fusion formation are: (i) insertion of a Mu *cts*62 prophage into *araB*; (ii) lysogenization of the araB::Mu*cts*62 strain with λpl (209,U118) by reciprocal homologous exchange between the Mu prophage and the Mu *c* terminus fragment of λpl (Casadaban, 1976); (iii) removal of transcriptional and translational stop signals between *araB* sequences and *lacZ* sequences 3′ to the U118 *ochre* triplet to generate a functional *araB–Φ-lacZ* hybrid cistron. Three possible structures are shown resulting from step (iii): a simple in-phase deletion between *araB* and *lacZ*, a deletion between a site close to the *β* terminus of Mu and *lacZ*, and a more complex rearrangement leaving in some nucleotides from the *β* terminus of Mu but inverted with respect to their original orientation. Although no *araB–Φ-lacZ* fusion has yet been sequenced, sequences of other cistron fusions generated by this method have revealed analogous structures (Moreno *et al.*, 1980; M. Berman & S. Benson, personal communication), and the presence of a functional Mu *β* terminus has been demonstrated by recombination tests in two *araB–lacZ* fusions (Casadaban & Chou, 1983).

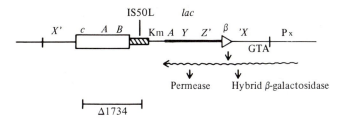

Fig. 2. The structure of the MudII1681 phage (not drawn to scale). The open box and triangle indicate Mu sequences. The heavy line indicates *lac* sequences deleted for the beginning of *lacZ*. The hatched box indicates the IS50 left of Tn5, and the thin line indicates the kanamycin resistance determinant of Tn5. *X'* and *'X* indicate a cistron interrupted by MudII1681 insertion, and Px is the promoter for this cistron. The Δ1734 bar indicates where a *Hind*III fragment has been deleted to remove the Mu *A* and *B* cistrons encoding transposition/replication functions. The right (β) end of Mu extends 116 nucleotides to a *Sau*IIIA site that has been joined to a *Bam*HI site in codon 8 of *lacZ* (Casadaban & Chou, 1983).

bacterial cells and cannot be extracted from cell material by detergent lysis. Thus, the deposition of pigment remains with the bacteria that have produced β-galactosidase, and the pigment only appears to diffuse in the agar medium when cells lyse and liberate enzyme. The intensity of staining is at least grossly proportional to the amount of β-galactosidase activity synthesized but is also subject to other factors, such as the rate of XGal transport into the bacteria. Although the physiological details of XGal staining are still unknown, we shall see that the resolution of this method is very great and that it can provide a great deal of information about the control of bacterial growth on agar surfaces.

MUDLAC AS A PROBE FOR PATTERN FORMATION IN THE GROWTH OF *PSEUDOMONAS PUTIDA* COLONIES

One of the great advantages of the Mud phage system is that it can be applied in a wide variety of gram-negative bacterial species. Some Mud phages are known to be capable of transposition in non-enteric bacteria (F. Richaud, B. de Castilho & M. Casadaban, personal communication), and they can be transferred on broad-host-range plasmids into many different strains. I took advantage of an IncP1 plasmid (pPH1) carrying the MudII1681 phage to introduce the Mud phage into a *Pseudomonas putida* strain carrying the CAM–OCT degradative plasmid. The resulting PPS2298 strain is useful for isolating Mud phage transpositions because the CAM–

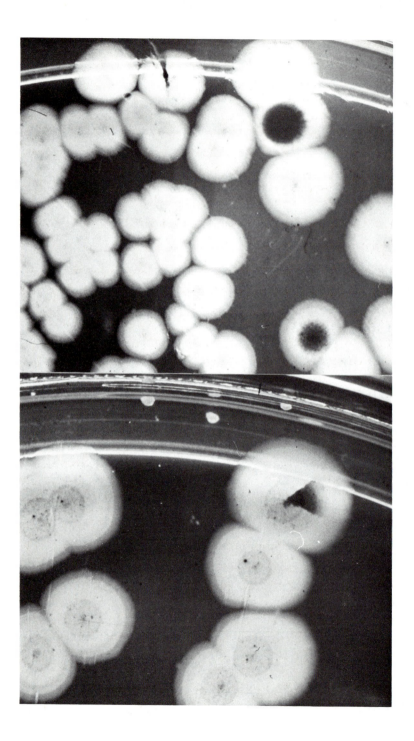

OCT plasmid transfers itself at high frequency but severely inhibits the transfer of IncP1 plasmids (Fennewald & Shapiro, 1979). Consequently, when PPS2298 clones were mated with a second *P. putida* recipient (PPS587), the resulting exconjugants selected for receipt of the MudII1681 kanamycin-resistance (Km^r) marker proved (in the vast majority) to harbour only CAM–OCT::MudII1681 plasmids. Some of the exconjugants stained intensely with XGal, but the majority showed very low or undetectable pigmentation.

Since the original purpose of introducing the MudII1681 phage into *P. putida* was to fuse the *alk* loci on CAM–OCT to *lacZ*, clones carrying alkane-inducible β-galactosidase activity were sought by plating independent subclonal cultures of the CAM–OCT::MudII1681 strains PPS2299b, PPS2306a and PPS2306b on XGal glucose agar and incubating in the presence of octane vapours. As illustrated in Fig. 3, more intensely pigmented sectors or colonies arose on these plates. These sectors and pigmented colonies were picked with sterile toothpicks and restreaked on XGal–glucose–octane agar. Surprisingly, these restreakings produced a mixture of some pigmented colonies and sectors with a majority of colonies that resembled those of the parental PPS2299 and PPS2306 strains. A second restreaking of the pigmented areas did produce the expected populations of colonies which mostly showed reproducible staining and were consequently assigned strain designations (Figs. 4–6). The genealogies of some of these strains are given in Table 1. Although none of the clones proved to harbour the desired *alk–lacZ* fusions, inspection of the pigmented colonies revealed a variety of specific patterns that illuminate the operation of morphogenetic mechanisms during *P. putida* colony formation. (Comparable patterns are visible in *E. coli* and *Agrobacterium tumefaciens* colonies produced by strains harbouring Mud phages; so these morphogenetic mechanisms are probably operative in most, if not all, gram-negative species.) As we shall see, the Mud phage system offers at least three great advantages in the study of colony morphogenesis: (1) it makes colony morphology visible with

Fig. 3. Parts of fields of PPS2306a1 (upper panel) and PPS2306b2 (lower panel) incubated 8 days on supplemented minimal glucose (0.4%) XGal agar in the presence of octane vapours. The upper panel shows sectors 1 (12 o'clock), 2 (2 o'clock) and 3 (5 o'clock), and the lower panel shows sector 6 (see Table 1). The pigmented PPS2306a1 colonies were about 3 mm in diameter, and the double pair of PPS2306b2 colonies at the lower left were each 5 mm in diameter. The photograph was originally taken on Ektachrome 64 and copied on Kodak Technical Pan 2415.

Table 1. *Genealogy of* Pseudomonas putida *strains*

	subcloning	sectoring	streaking	streaking
PPS2298a5				
× ⟶ PPS2299b	⌐ b1 ⟶	sector 9 ⟶	sector 91 ⟶	PPS2344
PPS587	├ b3 ⟶	sector 13 ⟶	sector 132 ⟶	PPS2357
	└ b4 ⟶	sector 11 ⟶	sector 113 ⟶	PPS2371
PPS2298d1				
× ⟶ PPS2306a	⟶ a1	sector 1 ⟶	sector 11 ⟶	PPS2331
PPS587		└ sector 2 ⟶	sector 21 ⟶	PPS2334
			sector 24 ⟶	PPS2338
PPS2298d2				
× ⟶ PPS2306b	⟶ b2	⌐ sector 5 ⟶	sector 52 ⟶	PPS2341a
PPS587			sector 53 ⟶	PPS2369
		└ sector 6 ⟶	sector 62 ⟶	PPS2348
			sector 63 ⟶	PPS2352

PPS2298 is *met* (CAM–OCT) (pPH1::MudII1681) and PPS5287 is *his⁻ trp⁻*. Most Met⁺ Kmʳ exconjugants lacked the gentamycin marker of pPH1 and retransferred Kmʳ (i.e. MudII1681) with CAM–OCT. Sectors 1 and 2 of PPS2306a1 and sector 6 of PPS2306b2 are shown in Fig. 3. PPS2369 appeared as an isolated dark colony in a field of lighter colonies (labelled PPS2341b).

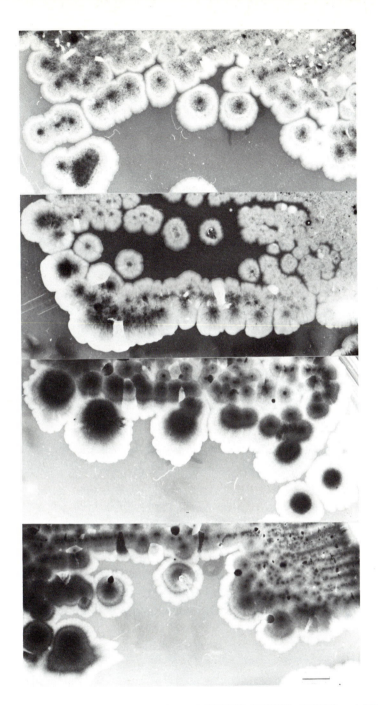

Fig. 4. Parts of fields of (from top to bottom) PPS2344, PPS2331, PPS2338 and PPS2348 incubated 7 days on supplemented minimal glucose XGal agar in the presence of octane vapours. The bar represents 2 mm.

The colonies shown in Figs. 4 to 12 were photographed with a Canon FD Macro lens on Kodak Technical Pan 2415 film through a No. 15 deep yellow filter to enhance the contrast of the blue pigmentation. The petri dishes were dark-field illuminated with a Bowens 'Texturelight' ring electronic flash unit placed 30 mm below the glass plate supporting the subject. Appropriate contrast in the negatives was achieved by developing the film 6 min in Kodak HC110 developer diluted 1:19 directly from the packaged concentrate.

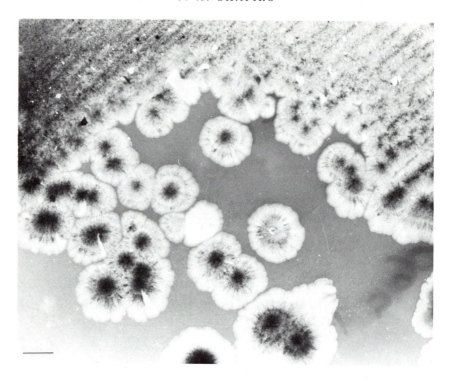

Fig. 5. A field of PPS2334 incubated as described in the legend to Fig. 4. The bar represents 2 mm.

unprecedented detail; (2) clonal changes can be seen directly in the stained colonies; and (3) the Mud phage provides both a physical and a genetic marker for probing how changes in DNA organization correlate with changes in colony morphogenesis.

The rest of this paper will present initial observations on the various patterns revealed with the Mud phage system and discuss some of the hypotheses that we can derive from those observations. The illustrations have generally been organized to emphasize clonal and subclonal relationships, but some show comparisons of different strains grown under similar conditions. Six of the original patterns observed on the second streaking of the initial PPS2299 and PPS2306 variants provide examples of different kinds of colony morphologies (Figs. 4–6). Note that subclones of the same initial PPS2306a1 or PPS2306b2 variant frequently display different patterns, such as PPS2334 and PPS2338. Indeed, variant no. 2 of PPS2306a1 (Fig. 3) gave rise to six different colony types. Some patterns from both PPS2299 and PPS2306 are reasonably stable on

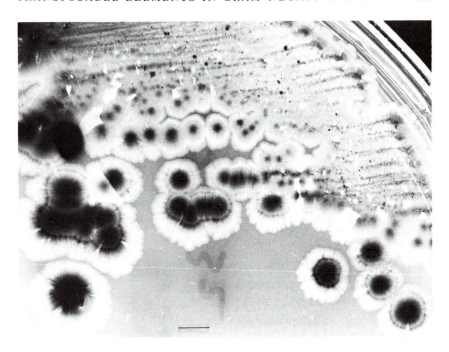

Fig. 6. A field of PPS2357 incubated as described in the legend to Fig. 4. The bar represents 2 mm.

restreaking (PPS2334, PPS2357) whereas others are very unstable and repeatedly throw off multiple colony types (PPS2341a and PPS2369). All *P. putida* strains carrying MudII1681 produce variant clones and sectors visible in at least 5% of the colonies, and it is likely that at least a portion of this genetic instability reflects continued expression of Mu DNA reorganization activities in the *Pseudomonas* background.

Pigmentation patterns, like other aspects of colonial development such as size and shape, are influenced by the composition of the growth medium and the proximity of the colonies to each other. Figs. 7 and 8 illustrate the patterns observed when strain PPS2357 is streaked on XGal agar containing different carbon sources. Organisms develop at about the same rate on glucose, citrate and casamino acids agar to yield well-isolated colonies of about 5 mm diameter in a week. Colonies on glycerol agar develop more slowly, rarely exceed a diameter of 2.5 mm even upon prolonged incubation, and generally have a more convex (less flattened) perimeter.

Sectoring is evident in many of the colonies illustrated here.

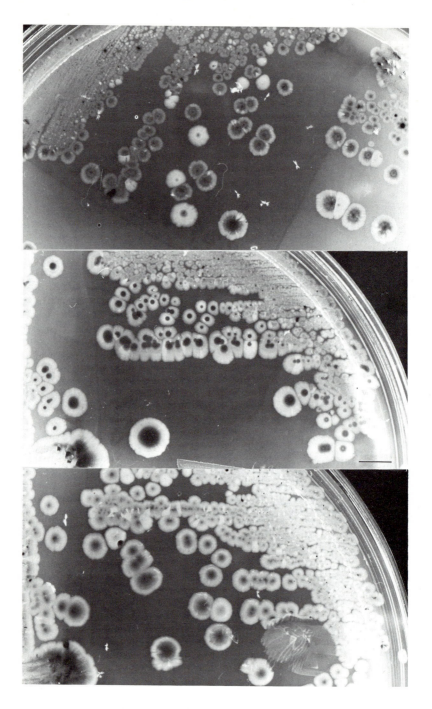

Fig. 7. Parts of fields of PPS2357 restreaked and incubated 4 days on supplemented minimal XGal agar containing the following carbon sources (top to bottom): casamino acids (0.2%), sodium citrate (0.25%) and glucose (0.4%). The bar represents 4 mm.

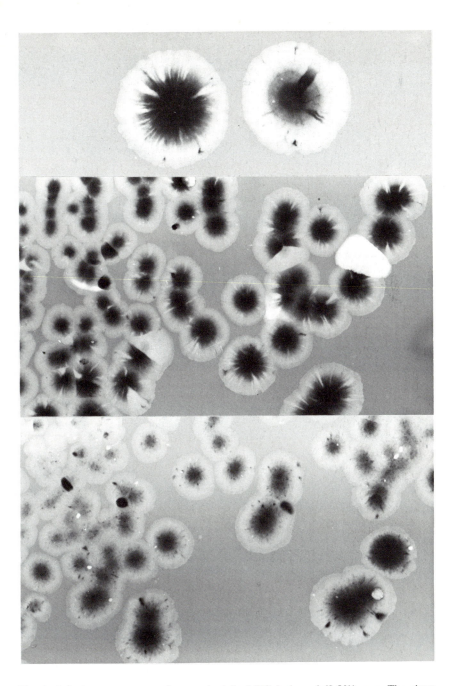

Fig. 8. Colonies grown on supplemented minimal XGal–glycerol (0.5%) agar. The plates were photographed 36 days after streaking, and the colonies averaged 2 mm in diameter. The top two panels are from a field of PPS2357, and the bottom panel is from a field of PPS2371.

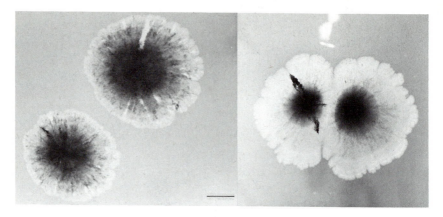

Fig. 9. Two pairs of colonies incubated on XGal–casamino acids agar. The pair on the left are from a field of PPS2348 and were photographed 36 days after streaking. The pair on the right are from a light sector of PPS2352 and were photographed 14 days after streaking. The bar represents 1 mm.

Restreaking of cells picked from different sectors from a single colony yields groups of colonies which often show different staining intensities or patterns. Experiments such as these confirm that visible sectors contain cells with different heritable properties. While most sectors produce a field of similar colony types, the population in some sectors may produce remarkably non-uniform collections of colony types. Fig. 10 shows a sectored PPS2369 colony and some of the progeny colonies which arose from the dark area at the top (sector A1) or from the lighter area at the bottom (sector A2). Sector A1 yielded a reasonably uniform population of dark colonies while sector A2 gave a variety of lighter and often highly variegated colonies.

The most common type of sector observed is the radially oriented wedge-shaped type visible in dark sector 6 of PPS2306b2 (Fig. 3) and the white sectors of PPS2334 (Fig. 5) and PPS2357 (Figs. 6, 7 and 8). It is interesting to note that growth on glycerol medium appears to exaggerate sectoring of PPS2357. The shape of these sectors is generally believed to result from the geometry of clonal expansion on agar surfaces and the time of appearance of a variant cell, but three observations suggest that the formation of sectors is a more involved process: (1) some sectors are not wedge-shaped, as for example the oval or pear-shaped dark sectors of PPS2348 (Fig. 4) and the irregular sectors of PPS2371 (Fig. 8) and PPS2341a (Fig. 11); (2) many sectors appear to have originated at some distance

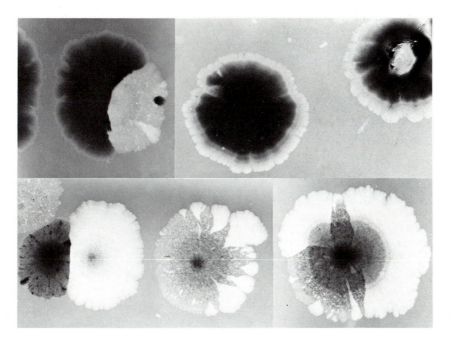

Fig. 10. A sectored PPS2369 colony and some of the progeny colonies obtained after restreaking. All colonies were incubated on XGal–casamino acids agar. The sectored PPS2369 colony in the upper left was photographed 13 days after streaking. The progeny colonies were photographed 14 days after streaking. The dark colonies in the upper right came from the dark (parental) sector of the PPS2369 colony. Note that one has been picked with a sterile toothpick. The lighter colonies in the bottom row descend from cells picked from an area inside the lighter sector that had no visible subsector. There is a dark ring in the PPS2369 colony which continues through the sector unchanged.

from the centre of the colony, so that it is likely the progenitor cell was only one of thousands in the developing colony and would not have given rise to a visible sector unless its progeny occupied a disproportionate share of the colonial surface; and (3) there are sectors which appear and start to radiate in a wedge-shaped manner but then shrink before they reach the edge of the colony (Figs. 9 and 10).

In contrast to the sectoring patterns are the circular and radial designs apparent in all the pigmented colonies. These designs consist of concentric rings, radial filaments, speckled disc regions, and complex mixtures of these basic elements. When examined carefully, all these patterns reveal a granularity in the pigmentation that suggests a non-uniform response of individual cells or groups of cells to the patterning mechanism. Inspection of these patterns

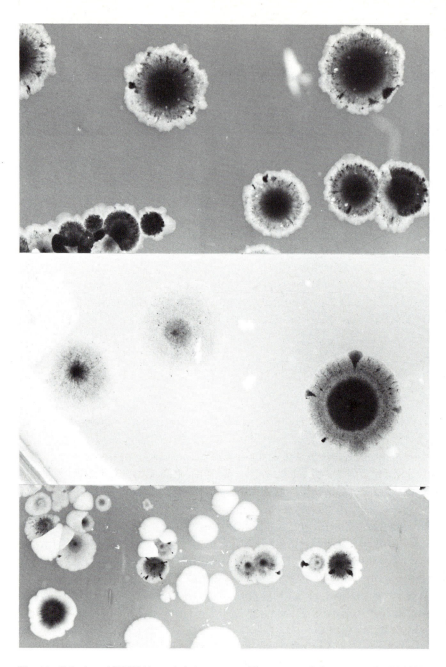

Fig. 11. Colonies of PPS2341a and their progeny. The top panel shows part of a PPS2341a field on XGal–glycerol agar photographed 36 days after streaking. The middle panel shows three PPS2341a colonies on XGal–casamino acids agar photographed 28 days after streaking. Note that this panel was printed lightly to bring out detail within the dark colony (denoted PPS2416). The bottom panel shows part of a field when cells from PPS2416 are restreaked on casamino acids agar containing Bluo-Gal (BRL Laboratories), a chromogenic β-galactosidase substrate that produces a deeper blue pigment than XGal. This field was photographed 5 days after streaking. Note the bizarre patterns of sectors and deep blue streaks in some of the colonies.

suggests that they are often non-heritable cellular differentiations. This conclusion has been confirmed for PPS2334 on glucose–octane agar and for PPS2357 on citrate agar by picking bacteria from both the darkly pigmented centres and lightly pigmented edges of individual colonies and restreaking them on XGal glucose or XGal citrate plates, when it is found that cells from either location produce indistinguishable colony types displaying the parental pigmentation pattern.

One important clue to the factors which contribute to pattern formation in these colonies comes from examining the relationship of sectors to the circular patterns in individual colonies. In many cases, the sectors do not alter the underlying rings which continue through a zone of greater or less pigmentation. Examples of this are sector 6 of PPS2306b2 (Fig. 3), one of the colonies from sector A2 of PPS2369 (lower right of Fig. 10), and the sectoring colony of PPS2341a on glycerol (lower right of the top panel of Fig. 11). A minority of observed sectors deranges the underlying circular patterns, as seen in one of the PPS2357 colonies on glycerol medium (top panel of Fig. 8). In the first kind of sector, it appears that the heritable variation has affected the hybrid β-galactosidase sequence so that it responds to the unchanged basic circular patterns in a quantitatively different manner. In the second case, we can hypothesize that the heritable change has not altered the β-galactosidase sequence but rather changed some component of the regulatory system which controls colony morphogenesis (and, consequently, β-galactosidase expression). In agreement with this interpretation is the observation that the PPS2357–glycerol sector not only changes the XGal staining but also the structure of the edge of the colony (Fig. 8).

A second important clue about morphogenetic mechanisms comes from the interactions between distinct colonies. These interactions can be seen because differences in staining intensities or the juxtaposition of patterns make visible the boundaries of the interacting colonies. Although there is sometimes ambiguity as to whether a variegating region represents two different colonies or a single sectoring colony, careful inspection usually reveals whether there is one or more than one focus for the circular patterns. In the latter case, it is likely that the variegating regions result from intersections of two separate colonies. Intersections are visible in the lower centre of the PPS2334 field where two pairs of colonies have grown together (Fig. 5). The distance between the foci of the two colonies in each pair is less than the long diameter of the colonies. This means that the two colonies must have come into contact and then continued expanding. The boundary of the two patterns is clearly visible in the upper pair, and it remains sharp across the entire intersection. Similar sharp intersections are visible in fields of mixed dark and light colonies. It thus appears that the bacteria from each colony respond only to the pattern of that colony and do not mix with the bacteria of a contiguous colony. In other words, bacteria respect the boundaries between adjacent colonies, even when they have proliferated in contact with each other. A different aspect of intercolonial contact appears in the field of PPS2357 (Fig. 6). Here several colonies intersect, and the boundaries are visible. These boundaries neatly terminate the intensely pigmented rings characteristic of PPS2357 on this plate, and it is interesting to note that the rings can be seen to remain out of register at some of the intersections. These observations indicate that the colony does not have to develop as a complete circle for arcs of the circular pattern to be laid down.

One intriguing observation (which does not require XGal staining) is that colonies avoid direct contact with each other once they have reached a certain size. In the field of PPS2357 grown on citrate agar, for example, there are examples of double fused colonies and also of adjacent colonies where there is a clear gap between their perimeters (Fig. 7; similar cases are visible in other figures as well). From the dimensions of the colonies along these gaps, it is clear that the colonies would have contacted each other if they had not been blocked in expansion in that direction.

The colony types produced by PPS2299 and PPS2306 derivatives are highly varied but by no means reveal the full extent of the *P. putida* developmental repertoire. Introduction of Mud phages into PPS587 on plasmids other than CAM–OCT reveals still other patterns, as shown by the photographs of PPS2455, PPS2457 and PPS2463 in Fig. 12. PPS2455 and PPS2457 both contain the MudI1734 element which is deleted for Mu transposition functions (Fig. 2) and so produces more (but not completely) stable patterns. PPS2463 has a very interesting pattern, different from all the others, because the XGal staining is most intense at the colony periphery. Moreover, the edges of PPS2463 colonies frequently show zones of staining whose shape indicates response to some diffusible signal coming from outside the colony. This signal appears to be the local concentration of kanamycin because PPS2463 colonies do not stain

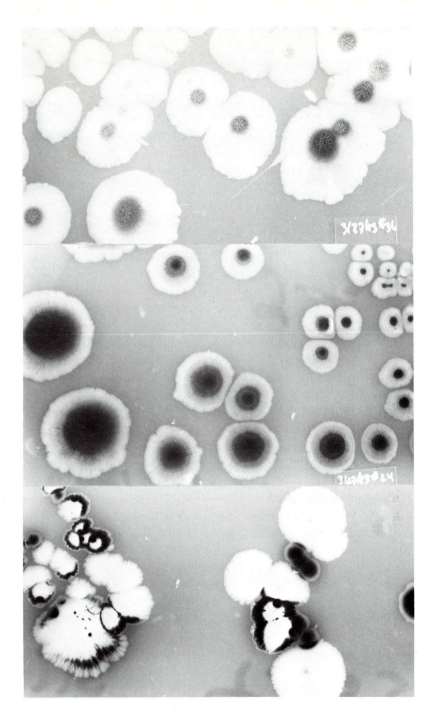

Fig. 12. Colonies of PPS587 carrying broad-host-range plasmids with Mud phage insertions. Independent R751::MudI1734 and R388::MudII1681 plasmids were isolated in *E. coli* and then transferred to PPS587 to generate (from top to bottom): PPS2455 (R751::MudI1734), PPS2457 (R751::MudI1734) and PPS2463 (R388::MudII1681). These strains were streaked on supplemented minimal XGal–casamino acids agar containing kanamycin (25 μg ml^{-1}) and had been incubated for 9 days when this photograph was taken.

in the absence of the antibiotic, and a kanamycin gradient on XGal agar can be visualized qualitatively by the distribution of blue stain in a PPS2463 field. It is interesting to compare the detailed structure of this chemically induced pigmentation pattern with those generated during colony development.

SOME IDEAS ABOUT GENOME EXPRESSION AND MORPHOGENESIS IN BACTERIAL COLONY FORMATION

Interpretation of the staining patterns presented here is complicated by the many parameters that can influence the amount and location of pigment in a colony. Intensity of pigmentation may reflect the levels of transcription, translation, protein processing, protein degradation, transport of enzyme or substrate, and local differences in cell population density. Distribution of pigment probably reflects both the positions of β-galactosidase expression during colony development and the results of migration of stained cells. Clearly, all these factors need to be sorted out by a detailed kinetic study of colony development.

The above uncertainties notwithstanding, I believe the colonies illustrated here do reflect the activity of specific morphogenetic mechanisms controlling bacterial growth on solid surfaces. The variety of reproducible patterns observed is incompatible with 'accidental' explanations, such as the generation of pH gradients within the colony that alter the intensity of pigmentation. Moreover, such gradients would not explain the granularity of the patterns. Just like the morphologies that allow microbiologists to identify different bacterial species, these staining patterns tell us that the positions and phenotypes of the component cells in a colony are highly organized. In other words, specific systems of cell aggregation and differentiation, like fruiting body formation by *Myxococcus* (Dworkin, 1966), are really typical of mechanisms operating in bacterial growth on solid surfaces rather than exceptional specializations. Since the XGal staining reveals both clonal and non-clonal forms of organization, it facilitates the formulation of working hypotheses about the morphogenetic systems operative in *Pseudomonas putida* colonies (and, insofar as *P. putida* is typical, in bacterial colonies in general). At the considerable risk of making foolish statements, I will try to sketch some of these hypotheses.

(1) Each colony has a mechanism for generating a field of

Fig. 13. Colonies of PPS587 on rich tryptone–yeast extract–salt agar after 2 weeks at room temperature. The large colony on the left measured 15 mm in diameter. The petri dish was illuminated by a single elevated flash unit inclined at 45° to the agar surface.

concentric circles that governs both the synthesis of specific proteins and cell position. Circular patterns are obvious in the shape of the colonial surface (Fig. 13) and so must govern the spatial accumulation of cells. This mechanism operates by intercellular signals transmitted over short distances either by cell contact or chemically so that local segments of the circular field are not disrupted by perturbations of colony development at other places. This hypothesis may be testable by genetic experiments (isolating mutants with defective circle generation) or by finding agents that disrupt the signals needed to construct a circular field. Some strains that produce irregularly shaped and highly polymorphic colonies, such as PPS2341a, may have defects in the mechanisms of colony morphogenesis.

(2) Each coding sequence has a characteristic response to the circular morphogenetic field that depends upon specific transcription signals, translational signals, protein processing signals, and upon its location in the genome. It seems logical to suppose that changes in transcription, translation and processing signals will give rise to sectors with deeper or fainter staining where the ring patterns are not disturbed.

I speculate that the major determinant of whether a pattern is concentric, radial, speckled or something more complicated is the location of the coding sequence in the genome (both in a linear and a three-dimensional sense). This idea cannot be tested with the existing Mud phage systems because each change in genomic location simultaneously results in new transcription, translation and processing signals. However, transposable elements that incorporate *lacZ* with its own promoter and translation initiation sequences should reveal whether insertions at different locations in various replicons produce patterns as diverse as the ones described here.

The idea that a locus encoding a specific activity can undergo structural change and consequently respond differently to a morphogenetic pattern is well known in eukaryotic developmental genetics (Stern, 1966; McClintock, 1965, 1967).

(3) There are mechanisms for maintaining the homogeneity of a clone by excluding neighbouring bacteria from outside the clone. These mechanisms account for the sharp boundaries between colonies that have come into contact and then proliferated side by side. I suspect that clonal exclusiveness, as in the tissues of higher organisms, depends largely on surface markers and that the specificity of these markers may be determined by non-genetic mechanisms (similar to those operating in reproduction of the *Paramecium* cortex: Sonneborn, 1970). Moreover, individual clones do not grow strictly in proportion to other clones in the colony. Some clones expand and others contract relative to each other, and this interclonal competition can vary at different positions. According to this view, visible sectoring would generally result from at least three different events: a genetic change affecting some discernible phenotype, establishment of clonal exclusiveness in cells of the novel genotype, and amplification of the clone relative to the parental cells to end up occupying a visible fraction of the colony surface. There must be several ways to achieve relative clonal amplification because we can see different sectorial shapes, especially near the edges of colonies. The operation of these clonal amplification mechanisms probably is the major determinant of the structure of the colony perimeter because strains with altered sectoring also show altered colony borders.

(4) We know that the generation of particular patterns in individual colonies depends on DNA structures because the introduction of different plasmids into PPS587 gives different phenotypes. However, there is probably also a non-DNA sequence component to determining colony morphology. The reason for invoking this

extragenetic factor is to explain clones such as PPS2341a and sector A2 of PPS2369 which produce collections of distinct colony types upon repurification. The non-DNA sequence components may be such things as cell surface determinants that control patterns of aggregation in colony development, cytoplasmic regulatory molecules that are unequally distributed at cell division, septum-determining structures in the envelope, or the spatial organization of DNA molecules (see Mendelson, 1982, for a discussion of related questions affecting cell division).

While these ideas will ultimately prove to be very naive, I offer them to emphasize the need to integrate the control of genome expression with the temporal and spatial aspects of bacterial growth. We have made enormous progress in studying the one-dimensional organization of DNA molecules and elucidating some of the factors which influence the biochemistry of protein synthesis, DNA replication and DNA reorganization. But complex inhomogeneous multicellular phenotypes like the colonies illustrated here should remind us that we still have a long way to go before we will produce satisfactory answers to basic questions about hereditary pattern formation. The value of the initial observations presented here does not lie in providing new answers but rather in showing that developmental biology is as vital a subject for bacteriologists as it is for botanists and zoologists. The utility of transposable elements as probes for morphogenesis is no accident, for we often only become aware of patterns when they change to new and unfamiliar forms, and genome reorganization creates new patterns. Indeed, it may well turn out that we will only understand the relationship of genome organization to morphogenesis when we achieve a more coherent view of genome reorganization.

I thank F. Richaud, B. de Castilho and M. Casadaban for the various Mud phages, Gerry Grofman for introducing me to Kodak Technical Pan 2415, BRL Laboratories for a gift of Bluo-Gal, and C. Stroman for manuscript preparation. This research was supported by a grant from the National Science Foundation (PCM–8200971).

REFERENCES

BASSFORD, P., BECKWITH, J., BERMAN, M., BRICKMAN, E., CASADABAN, M., GUARENTE, L., SAINT-GIRONS, I., SARTHY, A., SCHWARTZ, M. & SILHAVY, T. (1978). Genetic fusion of the *lac* operon: a new approach to the study of biological processes. In *The Operon*, ed. J. Miller & W. Reznikoff, pp. 245–61. Cold Spring Harbor: Cold Spring Harbor Laboratory.

CAIRNS, J. (1963). The chromosome of *Escherichia coli*. *Cold Spring Harbor Symposia on Quantitative Biology*, **28**, 43–6.

CALOS, M. P. & MILLER, J. A. (1980). Transposable elements. *Cell*, **20**, 579–95.

CASADABAN, M. (1976). Transposition and fusion of the *lac* genes to selected promoters in *Escherichia coli* using bacteriophages lambda and Mu. *Journal of Molecular Biology*, **104**, 541–55.

CASADABAN, M. & CHOU, J. (1983). *In vivo* formation of hybrid protein β-galactosidase gene fusions in one step with a new transposable Mu–*lac* transducing phage. *Proceedings of the National Academy of Sciences, USA*, in press.

CASADABAN, M. J. & COHEN, S. N. (1979). Lactose genes fused to exogenous promoters in one step using a Mu–*lac* bacteriophage: *in vivo* probe for transcriptional control sequences. *Proceedings of the National Academy of Sciences, USA*, **76**, 4530–3.

COHEN, S. N. & SHAPIRO, J. A. (1980). Transposable genetic elements. *Scientific American*, **242**(2), 40–9.

DWORKIN, M. (1966). Biology of the myxobacteria. *Annual Review of Microbiology*, **20**, 75–106.

FENNEWALD, M. & SHAPIRO, J. (1979). Transposition of Tn7 in *P. aeruginosa* and isolation of *alk*::Tn7 mutations. *Journal of Bacteriology*, **136**, 264–9.

IIDA, S., MEYER, J. & ARBER, W. (1983). Prokaryotic IS elements. In *Mobile Genetic Elements*, ed. J. A. Shapiro, pp. 159–221. New York: Academic Press.

KLECKNER, N. (1981). Transposable elements in prokaryotes. *Annual Review of Genetics*, **15**, 341–404.

McCLINTOCK, B. (1965). The control of gene action in maize. *Brookhaven Symposia in Biology*, **18**, 162–84.

McCLINTOCK, B. (1967). Genetic systems regulating gene expression during development. *Developmental Biology, Supplement 1*, 84–112.

MENDELSON, N. (1982). Bacterial growth and division: genes, structures, forces and clocks. *Microbiological Reviews*, **46**, 341–75.

MILLER, J. H. (1972). *Experiments in Molecular Genetics*. Cold Spring Harbor: Cold Spring Harbor Laboratory.

MITCHELL, D., REZNIKOFF, W. & BECKWITH, J. (1974). Genetic fusions defining *trp* and *lac* regulatory elements. *Journal of Molecular Biology*, **93**, 331–50.

MORENO, F., FOWLER, A., HALL, M., SILHAVY, T., ZABIN, I. & SCHWARTZ, M. (1980). A signal sequence is not sufficient to lead β-galactosidase out of the cytoplasm. *Nature, London*, **286**, 356–9.

OLDS, R. J. (1974). *Color Atlas of Microbiology*. Chicago: Year Book Medical Publishers.

PETTIJOHN, D. E. (1976). Prokaryotic DNA in nucleoid structure. *CRC Critical Reviews in Biochemistry*, **4**, 175–202.

SHAPIRO, J. (1979). A molecular model for the transposition and replication of bacteriophage Mu and other transposible elements. *Proceedings of the National Academy of Sciences, USA*, **76**, 1933–7.

SHAPIRO, J. A. (1982). Changes in gene order and gene expression. In *Research Frontiers in Aging and Cancer, National Cancer Institute Monograph 60*, pp. 87–116.

SHAPIRO, J. A. (1983*a*). Mobile genetic elements and reorganization of procaryotic genomes. In *Proceedings of the Fourth International Symposium on the Genetics of Industrial Microorganisms*, Kyoto, Japan, June 1982, ed. Y. Ikeda & T. Beppu, pp. 9–32. Tokyo: Kodansha.

SHAPIRO, J. A. (1983*b*). Variation as a genetic engineering process. In *Evolution of Molecules and Men*, ed. D. S. Bendall, J. C. Metcalfe & N. J. Jardine. Cambridge University Press, pp. 253–70.

SHAPIRO, J. A. (1984). Mechanisms of DNA reorganization in bacteria. *International Review of Cytology*, in press.

SJÅSTAD, K., FADNES, P., KRÜGER, P. G., LOSSIUS, I. & KLEPPE, K. (1982). Isolation, properties and nucleolytic degradation of chromatin from *Escherichia coli*: *Journal of General Microbiology*, **128**, 3037–50.

SONNEBORN, T. M. (1970). Gene action in development. *Proceedings of the Royal Society of London, Series B*, **176**, 347–66.

STERN, C. (1966). *Genetic Mosaics and Other Essays*. Cambridge, Mass.: Harvard University Press.

TOUSSAINT, A. & RESIBOIS, A. (1983). Phage Mu: transposition as a life-style. In *Mobile Genetic Elements*, ed. J. A. Shapiro, pp. 105–58. New York: Academic Press.

WORCEL, A. & BURGI, E. (1972). On the structure of the folded chromosome of *Escherichia coli*. *Journal of Molecular Biology*, **71**, 127–47.

GENE EXPRESSION IN MICROBES: THE LACTOSE OPERON MODEL SYSTEM

WILLIAM S. REZNIKOFF

Dept of Biochemistry, College of Agricultural and Life Sciences,
University of Wisconsin, Madison, WI 53706, USA

The nature of all organisms is dictated by an interplay between their genetic potential and the environment. This simple statement leads to the realization that achieving an understanding of how gene expression occurs and how that expression is regulated is a fundamental goal in biological research. Micro-organisms, in particular *Escherichia coli* and its viruses, have been extremely productive tools in this endeavour because they are amenable to relatively simple genetic, physiological and biochemical manipulations and because there have been a large number of scientists working with these tools.

This paper will use one bacterial model system, the lactose (*lac*) operon of *E. coli* (illustrated in Fig. 1), to describe our current understanding of the steps in gene expression and its regulation. Expression of the *lac* genes will be illustrated by defining how the *lac* mRNA is made (or more precisely how transcription initiation occurs) and how it is utilized to synthesize β-galactosidase, lactose permease and thiogalactoside-transacetylase. For analysing these steps in gene expression, the *lac* operon is a logical choice since it is one of the best-studied systems and the principles so elucidated are generally applicable to other systems. However, although the *lac*

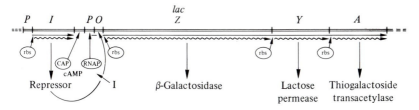

Fig. 1. The lactose operon. The transcription of the *lacZ*, *Y* and *A* genes results from transcription initiation by RNA polymerase (RNAP) at *lacP*. This event is regulated in a positive fashion by the catabolite gene activator protein (CAP) when it is complexed with cyclic AMP and in a negative fashion by the *lacI* gene product, the repressor. The repressor in turn is inactivated by the inducer (I). Translation (indicated by the wavy arrows) of each of the product proteins results from ribosome (rbs) binding to the mRNA translation initiation signal for each gene.

operon was once considered the paradigm of genetic regulatory systems, we now know that other gene systems are regulated by mechanisms not operative for *lac*. I will introduce some of these other mechanisms in summary form at the end of the paper.

TRANSCRIPTION INITIATION

The initiation of transcription occurs as a result of a series of four DNA-sequence-specific events during which RNA polymerase interacts with a specific segment of DNA termed the promoter (Chamberlin, 1974, 1976; Reznikoff & Abelson, 1978; McClure, 1980; Carpousis & Gralla, 1980; Munson & Reznikoff, 1981). These events are:

1. Loose binding of RNA polymerase to the promoter to form a 'closed complex'. This presumably results from the recognition of the promoter DNA sequence without disrupting the normal hydrogen bonds.
2. Isomerization of the RNA polymerase–promoter 'closed complex' to form an 'open complex'. This involves localized denaturation of the DNA helix.
3. Formation of the first mRNA phosphodiester bond.
4. A series of polymerization steps out to position 8 in which the elongation reaction competes at each step with a reaction that releases the oligonucleotide and recycles the RNA polymerase back to the initiation site. The sigma subunit is believed to be released after this series of steps.

The *lac* promoter,* which dictates the rate of all of these events, is a 50 base-pair (bp) DNA sequence, illustrated in Figs. 2 and 3. The promoter sequence has been defined by the following types of criteria (Maizels, 1973; Dickson *et al.*, 1975; Rosenberg & Court, 1979; Siebenlist, Simpson & Gilbert, 1980; Hawley & McClure, 1983): (1) the location of the precise initiation point for *lac* mRNA; (2) DNA sequence similarities with other known promoters; (3) the location of cis-dominant mutations which alter the ability of RNA

---

* The term 'promoter' generally refers to that portion of the DNA molecule recognized by RNA polymerase during the transcription initiation process; however, the *lac* promoter was originally defined by cis-dominant transcription-initiation-defective mutations which altered both the RNA polymerase and the catabolite gene activator protein (CAP) interaction sites. In this paper we will use the term to refer to only the RNA polymerase interaction site.

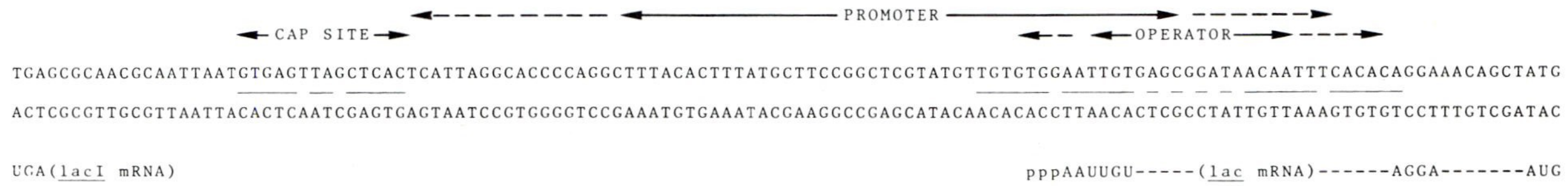

```
                                    ←─ CAP SITE ─→      ←────────── PROMOTER ──────────→        ←─ OPERATOR ─→
                          ←───────────────────→     ←────────────────────────────────→    ←───────────→

TGAGCGCAACGCAATTAATGTGAGTTAGCTCACTCATTAGGCACCCCAGGCTTTACACTTTATGCTTCCGGCTCGTATGTTGTGTGGAATTGTGAGCGGATAACAATTTCACACAGGAAACAGCTATG

ACTCGCGTTGCGTTAATTACACTCAATCGAGTGAGTAATCCGTGGGGTCCGAAATGTGAAATACGAAGGCCGAGCATACAACACACCTTAACACTCGCCTATTGTTAAAGTGTGTCCTTTCGATAC

UGA(lacI mRNA)                                                                       PPPAAUUGU------(lac mRNA)------AGGA-------AUG

                                                                                      UCA(lacI mRNA)
```

Fig. 2. The *lac* controlling elements. The DNA sequence of the *lac* controlling element region (Dickson *et al.*, 1975) is presented. The following points are indicated: the principal 5′ start point for *lac* mRNA, the *lacZ* translation initiation signals (AGGA-----AUG), the *lacI* translation termination signal (UGA), the CAP binding site, the promoter (RNA polymerase interaction site) and the operator. The boundaries of the CAP site (↔) were deduced from chemical protection studies (see Fig. 7). The boundaries of the promoter (↔) were deduced from genetic analyses (see

Fig. 3). DNase protection studies by the footprinting technique (Schmitz & Galas, 1979) indicated that the RNA polymerase 'covered' additional sequences (←--→) while more extended nuclease digestion results in the generation of a protected fragment extending from 24 bp before the mRNA start site to 19 bp downstream (Gilbert, 1976). The boundaries of the operator were determined from genetic studies and the minimal synthetic binding fragment ends (see Figs. 5 and 6). The footprinting protection results for the repressor–operator complex are also indicated (←--→).

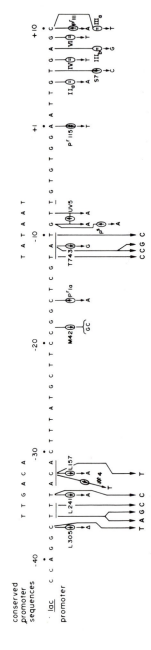

Fig. 3. The *lac* promoter. One strand of the *lac* promoter DNA sequence is presented with the bases numbered according to the conventions: +1 = mRNA start site, positive numbers = downstream, negative numbers = upstream. Above the sequence are the highly conserved −35 and −10 region sequences, and those *lacP* sequences that resemble these are underlined. Mutations which affect *lacP* activity are indicated below the sequence. Most of these are described in the text and Reznikoff *et al.* (1982). ↑ indicates that the mutation enhances *lacP* activity, ↓ indicates decreased

lacP activity, (=) indicates no effect on *lacP* activity (these are O[c] mutations) and ↕ indicates an ambiguous mutation described in more detail in the text. The mutation changes presented at the bottom of the figure were ↓ mutations isolated as ultraviolet-induced changes in the M13 vehicle mp8 (LeClerc & Istock, 1982) and #4 is a ↓ mutation isolated in the M13 vehicle mRZ361 (L. M. Munson & W. S. Reznikoff, unpublished). Not shown are any double mutations (either resulting from double hits or as single hits in the UV5 mutant background).

polymerase to interact with the promoter; (4) chemical and enzymatic probe studies.

The primary transcription initiation site has been defined for both *in vitro lac* transcripts (by direct 5'-end sequence analysis of transcription products: Maizels, 1973) and *in vivo lac* transcripts (by S1 mapping studies. M. Peterson, L. M. Munson & W. S. Reznikoff, unpublished) (see Figs. 2 and 3). Undoubtedly it is the events which lead to transcription initiation at this location which result in *lac* expression *in vivo* for the wild type and almost all mutant promoters. However, this is not the whole story. Recent studies have identified 'shadow' transcription initiation sites whose use is dictated by 'shadow' promoter sequences overlapping the standard *lac* promoter. These include a start site 13 bp downstream from the normal start site, which is 'activated' by the mutation P^r115 to be described later (Maquat & Reznikoff, 1980; M. Peterson & W. S. Reznikoff, unpublished). They also include a start site (called P2) 22 bp upstream from the normal start site (Reznikoff *et al.*, 1982; McClure, Hawley & Malan, 1982). This mRNA start is observed under certain 'permissive' conditions *in vitro*. These observations raise important questions, including what the physiological roles are for the 'shadow' promoters and whether they are a general feature of promoters. A possible physiological role for the P2 promoter will be discussed later. Whether they are a general feature of promoters is an unanswered question, although most promoters are embedded in A+T-rich regions and would, therefore, have a high probability of fortuitously having other promoter-like sequences nearby. The *gal* operon clearly has two overlapping promoters (Musso *et al.*, 1977).

There are three classes of mutations which define the *lac* promoter. These include mutations which decrease *lac* operon expression (indicated with ↓ in Fig. 3), mutations which increase *lac* operon expression (↑), and one ambiguous mutation (P^r115, ↕). The ↓ mutations were either isolated as cis-dominant mutations which pleiotropically reduced *lac* gene expression under conditions which normally provide high-level expression, or were *in vitro* constructs which were subsequently cloned into a promoter cloning vehicle and studied *in vivo*. The ↑ mutations (and P^r115) were isolated as mutations which enhanced *lac* expression under conditions of catabolite repression (in the absence of the positive stimulatory factors CAP and cyclic AMP) (Arditti, Grodzicker & Beckwith, 1973; Hopkins, 1974; Dickson *et al.*, 1975, 1977; Gilbert, 1976; Maquat, Thornton & Reznikoff, 1980; Reznikoff *et al.*, 1982; Mandecki & Reznikoff, 1982; LeClerc & Istock, 1982).

Over one hundred promoters recognized by *E. coli* RNA polymerase have been sequenced. By comparing these sequences after aligning them at their start point, it has become obvious that there are some highly conserved or 'canonical' sequences which all promoters approximate (Rosenberg & Court, 1979; Hawley & McClure, 1983). These are shown in Fig. 3 as related to *lacP*. These canonical sequences include a region approximately 35 bp before the start site (−35), a region approximately 10 bp before the start site (−10) and the region near the start site. As with almost all promoters, the *lac* promoter does not exactly fit the canonical sequences and, in particular, it is notably divergent in the −10 region. The general divergence of promoter sequences presumably explains why promoters are utilized at different levels and why the *lac* promoter is particularly inefficient in the absence of the positive stimulatory protein CAP to be described later.

There are two important caveats to this analysis:

(1) RNA polymerase does not 'read' As, Gs, Cs and Ts. It 'reads' the reactive groups found on these bases and/or it 'reads' the local DNA topology dictated by the sequence. In either case considerable ambiguity could exist in the recognized sequence.

(2) The existence of a 'canonical' sequence (usually the −10 region) is sometimes presented as evidence for the presence of a promoter in a particular sequence. In the absence of genetic or other biochemical evidence, this is not a reliable predictive factor.

The *lacP* mutations tend to fit the expectations of the sequence comparisons. That is, the mutations tend to be located in the −35 and −10 regions and in general those mutations which decrease the similarity to the canonical sequences decrease *lac* expression (↓) and those which increase the similarity increase *lac* expression (↑). There are, however, some particularly interesting mutations:

1. M42. This insertion (and deletions located at the same site: Stefano & Gralla, 1982) suggests that the precise spacing between the −35 and −10 regions is important in the recognition process.
2. P^r 115. This transversion of the bp at the start site appears to have two effects: it prevents initiation at or near the normal start site, and it generates a new −10 region which allows starts at +13. This latter effect means that it has uncovered a 'shadow' promoter.
3. S7 and P^r 111. These mutations indicate the sequences 10 bp

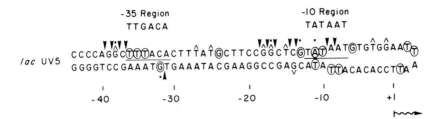

Fig. 4. Chemical probe studies of *lacP* UV5. This figure is similar to one published by Siebenlist *et al.* (1980) and summarizes the data presented in that article. The region which is unwound when the *lacP* UV5 is associated with RNA polymerase in the open complex (as determined by dimethylsulphate modification of the N-1 position of adenine) is indicated as a separation of the strands. Purines whose methylation, and 5-bromouridine substitutions whose protein cross-linking is inhibited (○) or enhanced (∧) by prior binding of polymerase are indicated. Sites where prior methylation of a purine (●) or prior ethylation of a phosphate (▼) inhibits polymerase binding are also represented.

upstream of the start site affect the interaction of RNA polymerase with *lacP*.

The ultimate goal of this type of analysis is to isolate all possible mutant variants at all sites in order to define which of the base pairs and possibly which reactive groups on these base pairs) are actually recognized during the transcription initiation process. The advent of synthetic polynucleotide techniques, promoter cloning vehicles, and targeted *in vitro* mutagenesis techniques certainly makes this a viable possibility.

A variety of *in vitro* enzymatic and chemical probe experiments essentially confirm the overall picture of promoter structure presented above (Schmitz & Galas, 1979; Siebenlist *et al.*, 1980). There are two basic experiments: protection experiments (in which perturbation of some agent's attack on the DNA by the prior binding of RNA polymerase is examined) and modification experiments (in which the agent is used to modify the promoter DNA and then this change is analysed for its effect on the RNA polymerase–promoter interaction).

The protection experiments are summarized in Figs. 2 and 4. The conclusion from these studies is that RNA polymerase sits on the DNA in the open complex such that it protects a region some 77 bp long with more intimate contacts (as suggested by the chemical probe experiments) lying from −39 bp to +4 bp and with many of the 'target' sites being located within the canonical sequence regions. Of particular interest is that RNA polymerase in the open complex facilitates dimethyl sulphate attack on the N-1 position of adenine at positions −9 to +2. Since the N-1 position is normally occupied in interstrand hydrogen bonds, this result is a direct confirmation that there is localized denaturation of the helix in the open complex and indicates the location of this denatured region.

The modification experiments are similar to mutation studies; they study what happens to the promoter activity when a portion of the sequence is removed by nuclease digestion or a potential reactive site is modified. The results of these experiments are shown in Fig. 4. Again we see that the basic conclusions from the genetic and sequence comparison studies are confirmed. One particularly interesting result not shown in Figs. 2 or 4 is that removing sequences upstream of −20 prevents formation of an RNA polymerase–promoter complex but qualitatively does not prevent its functioning if already formed (Gilbert, 1976). This suggests that the sequences upstream of this site are primarily used during the formation of the closed complex.

From this last observation, one is led to the hypothesis that the different steps in the transcription initiation process are dictated by different portions of the promoter sequence. This possibility is being studied by examining how different mutant templates affect some of the known biochemical steps as measured in *in vitro* transcription experiments (for instance see Hawley *et al.*, 1982). The UV5 mutation (located in the −10 region) acts primarily by enhancing open complex formation. In other systems, mutations in the −35 region alter the rate of closed complex formation (this is in agreement with the conclusion from the modification experiment mentioned above). The simplest interpretation is that there is a linear relationship between the sequence signals and the steps in transcription initiation. That is, the rate of closed complex formation is dictated by the −35 region, the rate of open complex formation by the −10 region and the rate of initiation by the sequence around the start site.

In fact this model is based upon very few data and there are some interesting complications. The transversion P^{r}115 clearly inhibits initiation at +1 and nearby sites such as −1 and thus presumably affects the initiation step. But it has not been determined whether it also alters the preceding reactions for +1 starts or not and, if it does so, whether this effect is due to an alteration in the sequence dictating the affected step or to the generation of a competing reaction. A more specific example of this latter possibility is exemplified by mutation L157. From filter binding studies, L157 does not appear to inhibit formation of either the closed or open complex (Reznikoff, 1976). Therefore, one might surmise that sequences out to −32 provide specific signals for the transcription initiation step. However, an alternative interpretation is that the RNA polymerase binding detected for this mutant DNA is not dictated by the *lacP* sequence but by an alternative binding reaction which might be activated by the L157 change (L157 is located in the −10 region for P2). If this were true, the L157 effect on *lacP* expression might be due to the enhancement of a competing reaction or to a combination of enhancing this competing reaction and altering some *lacP*-specific step.

Finally, although the properties of step 4 vary dramatically with different promoters no mutations have been found which alter this step for *lacP*.

THE REGULATION OF TRANSCRIPTION INITIATION

The frequency of transcription initiation in the *lac* operon is regulated; that is, mechanisms exist which enhance and repress its frequency relative to that dictated by the RNA polymerase–promoter interaction alone. As is well known, this was *the* model system used for discovering and analysing the negative regulation of transcription initiation. The *lac* negative regulation system is illustrated in Fig. 1. It involves three components: the *lac* repressor, a tetrameric protein encoded by the *lacI* gene; the *lac* operator, the target DNA sequence to which the *lac* repressor binds and thereby inhibits *lac* transcription initiation; and the inducer, β-galactoside compounds such as allolactose (the natural inducer) or isopropyl-β-D-thiogalactoside (a synthetic non-metabolizable inducer). When the inducer is bound to the repressor it causes an allosteric change in its configuration that reduces its affinity for *lacO* (for detailed descriptions of this system see Jacob & Monod, 1961; Beckwith & Zipser, 1970; Miller & Reznikoff, 1978).

The nature of the repressor–operator interaction has been the object of extensive genetic and biochemical investigations. The operator binding site on the repressor is believed to involve the N-terminal 60 amino acids of the repressor (Miller, 1978; Beyreuther, 1978; Weber & Geisler, 1978). All other well-studied repressors also have their operator binding site in their N-termini (Matthews *et al.*, 1982; von Wilcken-Bergmann & Müller-Hill,

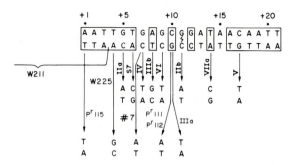

Fig. 5. Genetic analysis of the *lac* operator. The operator occupies the sequence from +3 to +19 (+1 corresponds to the *lac* mRNA start site). The O^c mutation changes are those described previously (Gilbert *et al.*, 1975; Maquat *et al.*, 1980) with the addition of the W211 and W225 changes (Yu Xian-Ming, L. M. Munson & W. S. Reznikoff, unpublished) and the #7 mutation (L. M. Munson & W. S. Reznikoff, unpublished). $P^r 115$ at +1 does not change the repressor–operator binding process, the deletion W211 decreases its affinity about 2-fold, while the deletion W225 (which is effectively a single base-pair substitution at +4) decreases its affinity about 11-fold (Reznikoff *et al.*, 1974*b*).

1982; Sauer *et al.*, 1982). High-affinity repressor–operator binding probably involves two subunits of the repressor held in the correct juxtaposition and each interacting with a portion of the operator (Kania & Brown, 1976; Weber & Geisler, 1978). By analogy with models derived from X-ray crystallographic analysis of other repressor proteins (such as λcI protein), we might expect that the *lac* repressor N-terminus forms an α-helix which participates in base-pair-specific interactions in the broad groove of the DNA helix (Matthews *et al.*, 1982).

The operator has been defined by cis-dominant mutations (shown in Fig. 5) and by a series of chemical and enzymatic probe experiments similar to those described for the promoter (shown in Figs. 2 and 6) (Smith & Sadler, 1971; Gilbert & Maxam, 1973; Gilbert *et al.*, 1975; Gilbert, Maxam & Mirzabekov, 1976; Ogata & Gilbert, 1977; Bahl *et al.*, 1977; Ogata & Gilbert, 1978; Schmitz & Galas, 1979; Maquat *et al.*, 1980; Yu Xian-Ming, L. M. Munson & W. S. Reznikoff, unpublished). In addition, synthetic constructs of the operator sequence containing base analogues at defined locations have been made and tested for their repressor affinities. These experiments have precisely defined some of the reactive groups in the operator that are recognized by the repressor (Caruthers, 1980). For instance, as shown in Fig. 5, the 5-methyl group of the thymine at position 8 in the operator has been shown to be critical because an A : U or a G : C substitution at this site

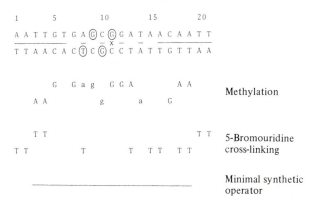

Fig. 6. Probe studies of the *lac* operator. Possible contact points between the repressor and the operator have been examined by dimethyl sulphate protection studies. Points of methylation inhibition and enhancement by bound repressor are indicated with upper- and lower-case letters respectively (Ogata & Gilbert, 1977, 1978). Positions at which 5-bromouridine can be cross-linked to the repressor in the presence of ultraviolet light are shown with letter 'T's. It should be noted that the 'T's in positions 1, 2, 20 and 21 probably do not represent specific contacts since a synthetic operator containing only base-pairs 3 to 19 binds repressor with normal affinity (the so-called minimal synthetic operator: Bahl *et al.*, 1977), and based upon the mutation analysis described in Fig. 5. Specific repressor–operator contacts have been identified through studies on the effect on repressor–operator binding of including base analogues at defined sites (Caruthers, 1980). These include the 5-methyl group on #8 thymine, and the 2-amino group on guanines #9, 10 and 11 (see circled bases on sequence).

decreases repressor–operator affinity by over 10-fold while a G:5-methyl C substitution does not.

The following questions are amongst those which remain to be elucidated with regard to the repressor–operator interaction: how can the repressor bind to the operator much faster than would be predicted from a collision reaction between the two molecules? How does the repressor block transcription initiation? What are the precise details of the repressor–operator complex?

The fact that the repressor binds to the operator faster than would be expected suggests that the true target for initial binding is the entire DNA molecule and that the operator is located by a one-dimensional diffusion process or by some intramolecular transfer reaction (Riggs, Bourgeois & Cohn, 1970; Barkley & Bourgeois, 1978). As does the promoter, the operator has pseudo-operator sites located close to it (Reznikoff, Winter & Hurley, 1974*b*; Gilbert *et al.*, 1975). It is conceivable that these act to 'trap' the diffusing repressor in the general environment of the operator.

The location of the operator relative to the promoter clearly suggests a mechanism for how the repressor might act in preventing *lac* transcription: the repressor would block RNA polymerase binding because the two binding events should be mutually exclusive. In fact an analysis by J. Majors (1975*a*) of the kinetics of *lac* transcription initiation *in vitro* following removal of the repressor from the template suggests that this is the operative mechanism. That is, he found that RNA polymerase added in the presence of repressor initiated transcription upon addition of IPTG with the same kinetics as if it had been added along with the IPTG. However, direct analysis of RNA polymerase binding in the presence and absence of the repressor by the 'footprinting' technique is not consistent with this conclusion (Schmitz & Galas, 1979). Addition of RNA polymerase to *lacP–O* DNA that already has bound repressor results in additional sequence protection and the additional protected sequence resembles that protected by RNA polymerase alone. How can this discrepancy be resolved? It is possible that the RNA polymerase binding in the presence of the repressor represents formation of the closed complex. This might not be apparent in the transcription assay unless RNA polymerase is added in limiting quantities.

Although *in vitro* repressor–synthetic operator binding studies have suggested some important aspects of the repressor–operator recognition reaction, a complete analysis of this event probably awaits an X-ray crystallographic analysis of repressor–operator complexes.

Only a few other negatively controlled systems have been studied in similar detail but there are some generalizations which can be made. The operator binding site is encoded in the N-terminus of the repressor protein. The target site is usually a symmetric sequence. But the target site and the promoter sequence are at different relative locations.

The *lac* operon is also subjected to positive control of transcription initiation and this is the molecular basis for the phenomenon of catabolite repression. The system is illustrated in Fig. 1 and is reviewed in several articles (Reznikoff & Abelson, 1978; de Crombrugghe & Pastan, 1978; Ullman & Danchin, 1983; de Crombrugghe, Busby & Buc, 1983). Transcription initiation at the *lac* promoter is activated by the binding of the catabolite gene activator protein (CAP) complexed with cyclic AMP. Those culture conditions which cause catabolite repression of *lac* (and other systems) do so by lowering the intracellular concentration of cyclic AMP and thus of activated CAP. In order to activate *lacP*, the CAP–cyclic

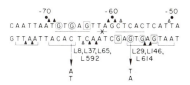

Fig. 7. The sequence of the *lac* CAP site together with the proposed symmetry element and the known *lac* class-I promoter mutation changes (Dickson *et al.*, 1977). The effect on methylation by dimethyl sulphate of the guanine residues of the *lac* CAP site by prior binding of the CAP–cyclic AMP complex is indicated by circles and squares: ○, methylation is blocked for the indicated residues; □, methylation is enhanced (Majors, 1977). No effect on the methylation of adenine residues was detected. Contacts between CAP and the phosphate groups on *lacP* DNA, as determined by which phosphotriester bonds block CAP–cyclic AMP binding in the ethyl nitrosourea experiment, are indicated as follows: ▲, completely missing species; △, partially missing species. This figure is similar to that presented in Reznikoff & Abelson (1978).

AMP complex must bind to a site shown in Figs. 1 and 7 which has been defined by cis-dominant Lac⁻ mutations as well as by chemical probe experiments (Scaife & Beckwith, 1966; Hopkins, 1974; Dickson *et al.*, 1977; Majors, 1977). The CAP site mutations are distinguished from *lac* promoter (RNA polymerase binding site) mutations by their failure to affect *lac* expression *in vivo* in the absence of CAP and by the decrease they cause in CAP binding affinity to *lacP–O* DNA *in vitro* (Majors, 1975*b*). The restricted genetic analysis of the CAP target site to date limits our ability to define its complete structure.

Detailed *in vitro* transcription assays have indicated that the CAP–cyclic AMP complex activates *lac* transcription by facilitating formation of the closed complex, the first step in the RNA polymerase–promoter interaction (McClure *et al.*, 1982). But the question remains as to how CAP acts to stimulate closed complex formation. Three models which are not mutually exclusive have been proposed.

1. CAP acts by perturbing the configuration of the DNA at the promoter (Dickson *et al.*, 1975). This model is consistent with the hypothesis from X-ray crystallographic studies on CAP that it binds to left-handed DNA (McKay & Steitz, 1981) and might be compatible with the fact that the location of CAP binding sites differ *vis-à-vis* promoter regions in different systems.

2. CAP acts by interacting with RNA polymerase and somehow stabilizing its binding to the promoter (Gilbert, 1976). Genetic evidence suggests that this is the mechanism by which the λcI protein activates RNA polymerase interaction with λP_{rm} (Guarante

et al., 1982). A mutation in the cI protein has been isolated which fails to activate λP_{rm} but has no effect on the protein's binding to the DNA.

3. CAP acts primarily as a repressor and only indirectly as an activator. That is, CAP represses RNA polymerase's interaction with P2 (Reznikoff *et al.*, 1982; McClure *et al.*, 1982). RNA polymerase bound to P2 normally inhibits RNA polymerase's binding to *lacP* and thus CAP's repression of RNA polymerase–P2 binding would activate *lacP*. *In vitro* assays suggest that this occurs to a certain extent; however, this model predicts that it should be possible to isolate mutations in the P2 promoter sequence which activate *lacP* expression by decreasing P2 activity. None have been found to date.

A number of CAP-stimulated systems have been analysed (see de Crombrugghe *et al.*, 1983 for a recent review). There does not seem to be a single simple picture emerging from these studies. CAP does not always interact with symmetrical sequences nor are the CAP site sequences in a standard location relative to the promoter. In one case (CAP stimulation of *araBAD* expression), CAP appears to act indirectly through an interaction with a second positively acting protein encoded by *araC*.

Other positively regulated systems such as *ara* and *mal* have also been studied. The most extensively studied system involves stimulation of λP_{rm} expression by λcI protein. The detailed structures of the protein and its target sites are known and evidence exists which suggests a mode of action for this protein (Guarente *et al.*, 1982).

TRANSLATION INITIATION

The utilization of *lac* mRNA to synthesize β-galactosidase (and permease and transacetylase) is determined by the stability of the mRNA and the formation of the translation initiation complex. The formation of this complex involves the binding of the 30S ribosome subunit and fMet–tRNA$_f^{Met}$ to the mRNA ribosome binding site. As has the promoter, the ribosome binding site has been studied by (1) determination of the translation start site, (2) sequence comparison studies and (3) mutation analysis (for recent reviews on this subject see Steitz, 1979; Gold *et al.*, 1981; Kozak, 1983).

The translation start site of *lacZ* was determined by Zabin & Fowler (1972) who showed that the initial N-terminal sequence of

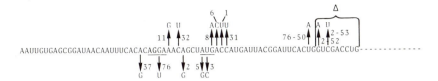

Fig. 8. lacZ translation initiation mutations. The 5'-end of the mRZ361 lac mRNA is shown. The proposed *lacZ* Shine–Dalgarno sequence and its AUG are underlined. Mutations isolated in mRZ361 which enhance *lacZα* expression are shown above the sequence and those that decrease *lacZα* expression are shown below the line. Mutations 2–52 and 2–53 are second-site revertants of mutation 2, and 76–50 is a second-site revertant of mutation 76 (L. M. Munson & W. S. Reznikoff, unpublished). Portions of the sequence downstream from the AUG are not found in normal *lacZ* but are products of the introduced restriction sites.

β-galactosidase was fMet-Thr-Met. Thus the translation start site must be the AUG underlined in Fig. 8 (see also Fig. 2).

Translation initiation sites as a general rule are preceded by a polypurine sequence spaced five to nine bases before the AUG. This was first recognized by Shine & Dalgarno (1974) and was postulated to play a role in translation initiation by virtue of the fact that this sequence is complementary to the 3'-end of the 16S RNA. The proposed 'Shine–Dalgarno' sequence in the *lacZ* translation initiation site is indicated in Fig. 8. Other than this sequence – the AUG initiation codon and the spacing between the Shine–Dalgarno sequence and the AUG – simple sequence comparisons have not led to elucidation of any other aspects of translation initiation signals; however, there probably are other determinants in the RNA. This has been predicted by some biochemical experiments (such as the reduced ability to form initiation complexes to RNA molecules trimmed to just the Shine–Dalgarno–AUG structure) and by genetic analyses. These other features fall into two categories: short-range RNA sequence features, and RNA secondary structural possibilities. Recent studies on *lacZ* translation initiation that relate to each of these mRNA determinants are discussed below.

Many of the mutations which affect *lacZ* translation initiation were isolated as mutants with altered levels of *lacZα* peptide synthesis in a derivative of the M13 mp8 cloning vehicle (L. M. Munson & W. S. Reznikoff, unpublished). These are shown in Fig. 8. Mutations whose properties are explainable by an alteration in the Shine–Dalgarno sequence or the AUG codon include numbers 76, 11 (enhances the Shine–Dalgarno sequence), 5 and 3. Mutation 5, which changes the AUG to GUG, is particularly interesting because it suggests that the GUG codon is less efficient as a

210   W. S. REZNIKOFF

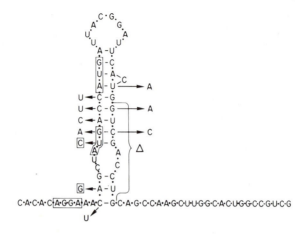

Fig. 9. A possible secondary structure for mRZ361 *lac* mRNA which occupies the AUG. Mutations which enhance *lacα* expression in general destabilize this structure. A few mutations which decrease *lacα* expression are noted in boxes. (L. M. Munson & W. S. Reznikoff, unpublished.)

translation initiation codon than AUG *in vivo*, although *in vitro* assays did not indicate this (Marcker, Clark & Anderson, 1966) (the same observation has recently been made for the T4 gene 32 cistron: L. Gold, personal communication). Other mutations, such as numbers 2 and 32, are located between the Shine–Dalgarno sequence and the AUG codon and might imply that this sequence is important in determining the frequency of translation initiation. For instance Gold *et al.* (1981) suggested from a computer analysis of ribosome binding site sequences that guanine residues were discriminated against in this region (mutation 2 is an adenine to guanine change which decreases expression) while uridine residues were favoured (mutation 32 is a cytosine to uridine change which enhances expression). However, an alternative explanation for the effect of these mutations will be presented below.

   The particular version of mp8 used for these studies encodes *lac* mRNA sequences with the potential for forming interesting secondary structures in the RNA (these secondary-structure possibilities are a fortuitous result of the cloning sites introduced into this particular mp8 derivative and are not characteristic of the normal *lacZ* gene). One such possibility is presented in Fig. 9. (This structure includes the *lacZ* AUG; alternative structures including both the *lacZ* AUG and the Shine–Dalgarno sequence can be drawn (L. M. Munson, personal communication).) It is a common assumption that secondary structures which occupy the Shine–Dalgarno sequence or the AUG in intrastrand hydrogen-bonding arrangements will inhibit translation initiation (Steitz, 1979; Gold *et al.*, 1981; Kozak, 1983). In fact a variety of mutations have been isolated which alter mp8 *lacZα* expression that can be most easily explained by virture of their effect on the proposed secondary structures. For instance point mutations 8, 6, 1, 31, 76–50, 2–52 and 2–53, and the deletion *Δ*, which all enhance *lacZα* expression, would be predicted to destabilize the proposed structures (see Figs. 8 and 9). As can be seen in Fig. 9, the effect of mutations 2 and 32 mentioned above may also be explained by their effect on the proposed secondary structure.

Another potential secondary structure that is encoded by *lac* region sequences which may have a special role in inhibiting *lacZ* translation initiation will be discussed below.

AVOIDING SPURIOUS GENE EXPRESSION

The whole model system of genes having specific transcription initiation signals whose expression may be regulated implies that genes have mechanisms for minimizing spurious expression resulting from read-through transcription. The *lac* operon has four mechanisms for accomplishing this goal:

1. The *lac* operon is preceded by the *lacI* gene (see Fig. 1) whose promoter functions very inefficiently (Calos, 1978; Miller, 1978; Steege, 1977); thus read-through transcription should be an infrequent event unless promoters further upstream can program read-through transcripts. An exception would be the case of some deletions.

2. A fraction of transcripts reading through *lacI* appear to terminate within the *lac* controlling elements even though no obvious termination signal sequences are present. This was found through *in vitro* analyses (Horowitz & Platt, 1982) and some *in vivo* *lacI* transcripts were found to have 3′-ends in this region (Cone, Sellitt & Steege, 1983).

3. The *lac* repressor–operator complex can block approximately 30–90% of read-through transcription (Reznikoff *et al.*, 1969).

4. In addition to all of the mechanisms listed above, there exists a 'fail-safe' mechanism for limiting the impact of read-through transcription. Translation initiation of *lacZ* is inefficient on all read-through transcripts carrying mRNA sequences encoded by DNA preceding the normal *lac* mRNA start site. This has been shown for

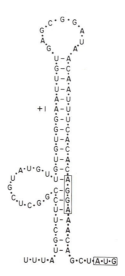

Fig. 10. Proposed *lac* read-through mRNA secondary structure. A possible read through mRNA secondary structure which occupies the Shine–Dalgarno sequence is shown (Yu Xian-Ming, L. M. Munson & W. S. Reznikoff, unpublished). The normal *lac* mRNA start site is indicated with a +1. The *lacZ* Shine–Dalgarno and AUG sequences are enclosed in boxes.

trp–lac transcripts (Reznikoff *et al.*, 1974*b*) and λ–*lac* transcripts (Mercereau-Puijalon & Kourilsky, 1976). The molecular basis for this inefficient translation initiation probably relates to the fact that these transcripts can form a secondary structure which occupies the *lacZ* Shine–Dalgarno sequence (Fig. 10; Yu Xian-Ming, L. M. Munson & W. S. Reznikoff, unpublished). Two striking features of this proposed mechanism are that it provides a *raison d'être* for the 35 bp symmetry element in the *lac* controlling elements and that read-through transcripts of the *gal* controlling elements show the same two properties (inhibition of translation initiation for the first gene product (Merril *et al.*, 1978) and an RNA secondary structure occupying the Shine–Dalgarno sequence (Merril, Gottesman & Adhya, 1981). One wonders whether this fail-safe mechanism might not exist for other gene systems.

SOME FACTORS CONTROLLING GENE EXPRESSION IN OTHER SYSTEMS NOT APPARENT FOR *lac*

Below I will briefly summarize three additional mechanisms known to influence gene expression in other bacterial systems, which are not thought to be operative for *lac*.

Attenuation. The expression of many biosynthetic operons such as *his*, *thr*, *leu* and *trp* (partially) is determined by the frequency with which RNA polymerase extends the transcript beyond a transcription termination site that precedes the structural genes. The termination event is determined by the secondary structure of the preceding mRNA, whose formation (or lack thereof) is in turn determined by the rate of progress of the ribosomes along the message. This mechanism is described in detail in a variety of reviews such as Yanofsky (1981).

Regulation. The expression of ribosomal protein genes is inhibited by the binding of specific ribosomal proteins to specific target translation initiation sites thereby blocking the binding of the 30S subunit. Thus ribosomal proteins are only synthesized when free ribosomal proteins are not present in the cytoplasm (Yates, Arfsten & Nomura, 1980).

Translational coupling. In some operons, the translation initiation of a downstream gene is dependent upon the translation of the immediately preceding gene. Two models have been proposed to explain this. One possibility is that the upstream mRNA when 'naked' assumes a conformation which inhibits ribosome loading at the internal AUG and this inhibiting conformation is disrupted by upstream translation. Alternatively ribosome loading at the internal AUG is inefficient; rather ribosomes do not disengage from the preceding nonsense codon but instead stay associated with the mRNA so that they can initiate translation at the internal AUG. This appears to be the case for *trpD* and ribosomal protein genes but not for *lacY* or *lacA*. The evidence for this is that *lacZ* nonsense mutation polarity on *lacY* and *lacA* expression can be suppressed in SuA mutants (which have a reduced frequency of transcription termination) and thus this polarity can be relieved by increasing transcription reading through to *lacY* without enhancing *lacZ* C-terminus translation, and that the intercistronic regions separating *lacZ* and *Y* and *Y* and *A* do not have the tight coupling typical of the *trpE–D* and ribosomal protein gene intercistronic regions (Beckwith, 1963; Oppenheim & Yanofsky, 1980; Buchel, Gronenborn & Müller-Hill, 1980).

We have learnt enough about the molecular mechanisms of prokaryotic gene expression to no longer be satisfied with simple line drawings such as that in Fig. 1. But there is still a great deal left to learn. It is hoped that exhaustive studies of a few model systems, such as *lac*, will lead to an elucidation of common mechanisms; but, although such mechanisms will be found in many cases, interesting differences between systems can be expected.

I would like to thank Lianna Munson, Yu Xian-Ming and Martha Peterson for stimulating conversations and for providing important information prior to its publication. Research work described here that was done in the author's laboratory was supported by grants from the National Institutes of Health (GM19670), the National Science Foundation (PCM7910686) and the 3M Foundation.

REFERENCES

ARDITTI, R., GRODZICKER, T. & BECKWITH, J. (1973). Cyclic adenosine monophosphate-independent mutants of the lactose operon of *Escherichia coli*. *Journal of Bacteriology*, **114**, 652–5.

BAHL, C. P., WU, R., STAWINSKY, J. & NARANG, S. (1977). Minimal length of the lactose operator sequence for the specific recognition by the lactose repressor. *Proceedings of the National Academy of Sciences, USA*, **74**, 966–70.

BARKLEY, M. D. & BOURGEOIS, S. (1978). Repressor recognition of operator and effectors. In *The Operon*, ed. J. H. Miller & W. S. Reznikoff, pp. 177–200. Cold Spring Harbor Laboratory.

BECKWITH, J. R. (1963). Restoration of operon activity by suppressors. *Biochimica et Biophysica Acta*, **76**, 162–4.

BECKWITH, J. R. & ZIPSER, D. (1970). *The Lactose Operon*. Cold Spring Harbor Laboratory.

BEYREUTHER, K. (1978). Chemical structure and functional organization of *lac* repressor from *Escherichia coli*. In *The Operon*, ed. J. H. Miller & W. S. Reznikoff, pp. 123–54. Cold Spring Harbor Laboratory.

BUCHEL, D. E., GRONENBORN, B. & MÜLLER-HILL, B. (1980). Sequence of the lactose permease gene. *Nature, London*, **283**, 541–5.

CALOS, M. (1978). DNA sequence for a low-level promoter of the *lac* repressor gene and an 'up' promoter mutation. *Nature, London*, **274**, 762–5.

CARPOUSIS, A. J. & GRALLA, J. D. (1980). Cycling of ribonucleic acid polymerase to produce digonucleotides during initiation *in vitro* at the *lac* UV5 promoter. *Biochemistry*, **19**, 3245–50.

CARUTHERS, M. H. (1980). Deciphering the protein–DNA recognition code. *Accounts of Chemical Research*, **13**, 155–60.

CHAMBERLIN, M. (1974). The selectivity of transcription. *Annual Review of Biochemistry*, **43**, 721–55.

CHAMBERLIN, M. (1976). RNA polymerase – an overview. In *RNA Polymerase*, ed. R. Losick & M. Chamberlin, pp. 17–67. Cold Spring Harbor Laboratory.

CONE, R. C. SELLITT, M. A. & STEEGE, D. A. (1983). *Lac* repressor mRNA transcription terminates *in vivo* in the *lac* control region. *Journal of Biological Chemistry*, in press.

DE CROMBRUGGHE, B., BUSBY, S. & BUC, H. (1983). Activation of transcription by the cyclic AMP receptor protein. In *Biological Regulation and Development*, vol. III-B, ed. K. Yamamoto. New York: Plenum Press (in press).

DE CROMBRUGGHE, B. & PASTAN, I. (1978). Cyclic AMP, the cyclic AMP receptor protein, and their dual control of the galactose operon. In *The Operon*, ed. J. H. Miller & W. S. Reznikoff, pp. 303–24. Cold Spring Harbor Laboratory.

DICKSON, R. C., ABELSON, J., BARNES, W. M. & REZNIKOFF, W. S. (1975). Genetic regulation: the *lac* control region. *Science*, **187**, 27–35.

DICKSON, R. C., ABELSON, J., JOHNSON, P., REZNIKOFF, W. S. & BARNES, W. M. (1977). Nucleotide sequence changes produced by mutations in the *lac* promoter of *Escherichia coli*. *Journal of Molecular Biology*, **111**, 65–75.

GILBERT, W. (1976). Starting and stopping sequences for the RNA polymerase. In *RNA Polymerase*, ed. R. Losicle & M. Chamberlin, pp. 193–206. Cold Spring Harbor Laboratory.

GILBERT, W., GRALLA, J., MAJORS, J. & MAXAM, A. (1975). Lactose operator sequences and the action of *lac* repressor. In *Protein–Ligand Interactions*, ed. H. Sund & G. Blauer, pp. 193–210. Berlin: Walter de Gruyter.

GILBERT, W. & MAXAM, A. (1973). The nucleotide sequence of the *lac* operator. *Proceedings of the National Academy of Sciences, USA*, **70**, 3581–4.

GILBERT, W., MAXAM, A. & MIRZABEKOV, A. (1976). Contacts between the *lac* repressor and DNA revealed by methylation. In *Control of Ribosome Synthesis, Alfred Benson Symposium IX*, ed. N. O. Kjeldgaard & O. Maaløc, pp. 139–48. New York: Academic Press.

GOLD, L., PRIBNOW, D., SCHNEIDER, T., SHINEDLING, S., SINGER, B. S. & STORMO, G. (1981). Translational initiation in prokaryotes. *Annual Review of Microbiology*, **35**, 365–403.

GUARENTE, L., NYE, J. S., HOCHSCHILD, A. & PTASHNE, M. (1982). Mutant λ phage repressor with a specific defect in its positive control function. *Proceedings of the National Academy of Sciences, USA*, **79**, 2236–9.

HAWLEY, D. K. & McCLURE, W. R. (1983). Compilation and analysis of *Escherichia coli* promoter DNA sequences. *Nucleic Acids Research*, **11**, 2237–55.

HAWLEY, D. K., MALAN, T. P., MULLIGAN, M. E. & McCLURE, W. R. (1982). Intermediates on the pathway to open-complex formation. In *Promoters: Structure and Function*, ed. R. L. Rodrigum & J. Chamberlin, pp. 54–68. New York: Praeger.

HOPKINS, J. D. (1974). A new class of promoter mutations in the lactose operon of *Escherichia coli*. *Journal of Molecular Biology*, **87**, 715–24.

HOROWITZ, H. & PLATT, T. (1982). A termination site for *lacI* transcription is between the CAP site and the *lac* promoter. *Journal of Biological Chemistry*, **257**, 11740–6.

JACOB, F. & MONOD, J. (1961). Genetic regulatory mechanisms in the synthesis of proteins. *Journal of Molecular Biology*, **3**, 318–56.

KANIA, J. & BROWN, D. T. (1976). The functional repressor parts of a tetrameric *lac* repressor–β-galactosidase chimera are organized as dimers. *Proceedings of the National Academy of Sciences, USA*, **73**, 3529–33.

KOZAK, M. (1983). Comparison of initiation of protein synthesis in procaryotes, eucaryotes, and organelles. *Microbiological Reviews*, **47**, 1–45.

LECLERC, J. E. & ISTOCK, N. L. (1982). Specificity of UV mutagenesis in the *lac* promoter of M13 *lac* hybrid phage DNA. *Nature, London*, **297**, 596–8.

McCLURE, W. R. (1980). Rate-limiting steps in RNA chain initiation. *Proceedings of the National Academy of Sciences, USA*, **77**, 5634–8.

McCLURE, W. R., HAWLEY, D. K. & MALAN, T. P. (1982). The mechanism of RNA polymerase activation on the λP_{rm} and $lacP^+$ promoters. In *Promoters: Structure and Function*, ed. R. L. Rodriguez & M. J. Chamberlin, pp. 111–20. New York: Praeger.

McKAY, D. & STEITZ, T. (1981). Structure of catabolite gene activator protein at 2.9 Å resolution suggests binding to left-handed B-DNA. *Nature, London*, **290**, 744–9.

MAIZELS, N. (1973). The nucleotide sequence of the lactose messenger ribonucleic acid transcribed from the UV5 promoter mutant of *E. coli*. *Proceedings of the National Academy of Sciences, USA*, **70**, 3585–9.

MAJORS, J. (1975a). Initiation of *in vitro* mRNA synthesis from the wild-type *lac* promoter. *Proceedings of the National Academy of Sciences, USA*, **72**, 4394–8.

MAJORS, J. (1975b). Specific binding of CAP factor to *lac* promoter DNA. *Nature, London*, **256**, 672–4.

MAJORS, J. (1977). Control of the *E. coli lac* operon at the molecular level. PhD thesis, Harvard University.

MANDECKI, W. & REZNIKOFF, W. S. (1982). A *lac* promoter with a changed distance between −10 and −13 regions. *Nucleic Acids Research*, **10**, 903–12.

MAQUAT, L. E. & REZNIKOFF, W. S. (1980). *Lac* promoter mutation P^r115 generates a new transcription initiation point. *Journal of Molecular Biology*, **139**, 551–6.

MAQUAT, L. E., THORNTON, K. & REZNIKOFF, W. (1980). *lac* promoter mutations located downstream from the transcription start site. *Journal of Molecular Biology*, **139**, 537–49.

MARCKER, K. A., CLARK, B. F. C. & ANDERSON, J. S. (1966). N-Formyl-methionyl-sRNA and its relation to protein biosynthesis. *Cold Spring Harbor Symposia on Quantitative Biology*, **31**, 279–85.

MATTHEWS, B. W., OHLENDORF, D. H., ANDERSON, W. F. & TAKEDA, Y. (1982). Structure of the DNA-binding region of *lac* repressor inferred from its homology with *cro* repressor. *Proceedings of the National Academy of Sciences, USA*, **79**, 1428–32.

MERCEREAU-PUIJALON, O. & KOURILSKY, P. (1976). Escape synthesis of β-galactosidase under the control of bacteriophage lambda. *Journal of Molecular Biology*, **108**, 733–52.

MERRIL, C. R., GOTTESMAN, M. E. & ADHYA, S. L. (1981). *Escherichia coli gal* operon proteins made after prophage lambda induction. *Journal of Bacteriology*, **147**, 875–87.

MERRIL, C., GOTTESMAN, M., COURT, D. & ADHYA, S. (1978). Disco-ordinate expression of the *Escherichia coli gal* operon after prophage lambda induction. *Journal of Molecular Biology*, **118**, 241–5.

MILLER, J. H. (1978). The *lacI* gene: its role in *lac* operon control and its use as a genetic system. In *The Operon*, ed. J. H. Miller & W. S. Reznikoff, pp. 31–88. Cold Spring Harbor Laboratory.

MILLER, J. H. & REZNIKOFF, W. S. (eds.) (1978). *The Operon*. Cold Spring Harbor Laboratory.

MUNSON, L. M. & REZNIKOFF, W. S. (1981). Abortive initiation and long ribonucleic acid synthesis. *Biochemistry*, **20**, 2081–5.

MUSSO, R., DiLAURO, R., ADHYA, S. & DE CROMBRUGGHE, B. (1977). Dual control for transcription of the galactose operon by cyclic AMP and its receptor protein at two interspersed promoters. *Cell*, **12**, 847–54.

OGATA, R. & GILBERT, W. (1977). Contacts between the *lac* repressor and thymines in the *lac* operator. *Proceedings of the National Academy of Sciences, USA*, **74**, 4973–6.

OGATA, R. & GILBERT, W. (1978). An amino-terminal fragment of *lac* repressor binds specifically to *lac* operator. *Proceedings of the National Academy of Sciences, USA*, **75**, 5851–4.

OPPENHEIM, D. S. & YANOFSKY, C. (1980). Translational coupling during expression of the tryptophan operon of *Escherichia coli*. *Genetics*, **95**, 785–95.

REZNIKOFF, W. S. (1976). Formation of the RNA polymerase–*lac* promoter open complex. In *RNA Polymerase*, ed. R. Losick & M. Chamberlin, pp. 441–454. Cold Spring Harbor Laboratory.

REZNIKOFF, W. S. & ABELSON, J. N. (1978). The *lac* promoter. In *The Operon*, ed. J. H. Miller & W. S. Reznikoff, pp. 221–43. Cold Spring Harbor Laboratory.

REZNIKOFF, W. S., MAQUAT, L. E., MUNSON, L. E., JOHNSON, R. C. & MANDECKI, W. (1982). The *lac* promoter: analysis of structural signals for transcription initiation and identification of a new sequence-specific event. In *Promoters: Structure and Function*, ed. R. L. Rodriguez & M. J. Chamberlin, pp. 80–95. New York: Praeger.

REZNIKOFF, W. S., MICHELS, C. A., COOPER, T. G., SILVERSTONE, A. E. & MAGASONIK, B. (1974a). Inhibition of *lacZ* gene translation initiation in *trp–lac* fusion strains. *Journal of Bacteriology*, **117**, 1231–9.

REZNIKOFF, W., MILLER, J. H., SCAIFE, J. G. & BECKWITH, J. R. (1969). A mechanism for repressor action. *Journal of Molecular Biology*, **43**, 201–13.

REZNIKOFF, W. S., WINTER, R. B. & HURLEY, C. K. (1974b). The location of the repressor binding sites in the *lac* operon. *Proceedings of the National Academy of Sciences, USA*, **71**, 2314–18.

RIGGS, A. D., BOURGEOIS, S. & COHN, M. (1970). The *lac* repressor–operator interaction. III. Kinetic studies. *Journal of Molecular Biology*, **53**, 401–17.

ROSENBERG, M. & COURT, D. (1979). Regulatory sequences involved in the promotion and termination of RNA transcription. *Annual Review of Genetics*, **13**, 319–53.

SAUER, R. T., YOCUM, R. R., DOOLITTLE, R. F., LEWIS, M. & PABO, C. O. (1982). Homology among DNA-binding proteins suggests use of a conserved super-secondary structure. *Nature, London*, **298**, 447–51.

SCAIFE, J. & BECKWITH, J. (1966). Mutational alteration of the maximal level of *lac* operon expression. *Cold Spring Harbor Symposia on Quantitative Biology*, **31**, 403–8.

SCHMITZ, A. & GALAS, D. J. (1979). The interaction of RNA polymerase and *lac* repressor with the *lac* control region. *Nucleic Acids Research*, **6**, 111–37.

SHINE, J. & DALGARNO, L. (1974). The 3′-terminal sequence of *E. coli* 165 ribosomal RNA: complementarity to nonsense triplets and ribosome binding sites. *Proceedings of the National Academy of Sciences, USA*, **71**, 1342–6.

SIEBENLIST, U., SIMPSON, R. B. & GILBERT, W. (1980). *E. coli* RNA polymerase interacts homologously with two different promoters. *Cell*, **20**, 269–81.

SMITH, T. F. & SADLER, J. R. (1971). The nature of lactose operator constitutive mutations. *Journal of Molecular Biology*, **59**, 273–305.

STEEGE, D. A. (1977). 5′-Terminal nucleotide sequence of *Escherichia coli* lactose repressor mRNA: features of translational initiation and reinitiation sites. *Proceedings of the National Academy of Sciences, USA*, **74**, 4163–7.

STEFANO, J. E. & GRALLA, J. D. (1982). Specific and non-specific interactions at mutant *lac* promoters. In *Promoters: Structure and Function*, ed. R. L. Rodriguez & M. J. Chamberlin, pp. 69–79. New York: Praeger.

STEITZ, J. A. (1979). Genetic signals and nucleotide sequences in messenger RNA. In *Biological Regulation and Development*, vol. 1, *Gene Expression*, ed. R. F. Goldberger, pp. 349–99. New York: Plenum Press.

ULLMAN, A. & DANCHIN, A. (1983). Role of cyclic AMP in bacteria. *Advances in Cyclic Nucleotides Research*, **15**, in press.

WEBER, K. & GEISLER, N. (1978). *Lac* repressor fragments produced *in vivo* and *in vitro*: an approach to the understanding of the interaction of repressor and DNA. In *The Operon*, ed. J. H. Miller & W. S. Reznikoff, pp. 155–76. Cold Spring Harbor Laboratory.

VON WILCKEN-BERGMANN, B. & MÜLLER-HILL, B. (1982). Sequence of *galR* gene indicates a common evolutionary origin of *lac* and *gal* repressor in *Escherichia coli*. *Proceedings of the National Academy of Sciences, USA*, **79**, 2427–31.

YANOFSKY, C. (1981). Attenuation in the control of expression of bacterial operons. *Nature, London*, **289**, 751–8.

YATES, J. L., ARFTSEN, A. E. & NOMURA, M. (1980). *In vitro* expression of *Escherichia coli* ribosomal protein genes: autogenous inhibition of translation. *Proceedings of the National Academy of Sciences, USA*, **77**, 1837–41.

ZABIN, I. & FOWLER, A. V. (1972). The amino acid sequence of β-galactosidase. III. Sequences of amino- and carboxyl-terminal tryptic peptides. *Journal of Biological Chemistry*, **247**, 5432–5.

NEW MICROBES FROM OLD HABITATS?

STANLEY T. WILLIAMS, MICHAEL GOODFELLOW AND JILL C. VICKERS

Dept of Botany, University of Liverpool, Liverpool L69 3BX, and Dept of Microbiology, University of Newcastle upon Tyne, Newcastle upon Tyne NE1 7RU, UK

The science of microbiology was initiated and is sustained by the isolation of microbes from natural habitats and their establishment in laboratory culture. Since the pioneering studies of Koch and other workers in the nineteenth century, countless man-hours have been devoted to this task. Our increased knowledge of the ecology, physiology and structure of micro-organisms has, inevitably, led to significant improvements in the objectivity and efficiency of isolation procedures, but problems remain. Thus, the element of chance can still play an important part in the detection of new microbes. Indeed, to some extent the ability to isolate a novel organism can still be as much an art as a science. Relatively few microbiologists would regard the isolation of new microbes from natural environments as a prime aim of their research. The majority are understandably consumed by the challenges and opportunities presented by investigation of microbes already in laboratory culture, of which there is no shortage.

Most new microbes are isolated by either microbial ecologists or applied microbiologists seeking novel, useful bioactive substances. The former seldom set out with the aim of obtaining new microbes but the possibility of their detection is inherent in any ecological study involving the isolation and enumeration of the microbial populations in a habitat. Our knowledge of microbes in the environment is somewhat akin to that of a sociologist whose theories are based primarily on observation of the behaviour of a prisoner in a cell. An increasing number of ecologists, frustrated by the problems of studying microbes *in vivo*, have justifiably concentrated on the reactions of isolates already incarcerated in controlled laboratory conditions. The benefits of studying 'cells containing more than one prisoner' have been emphasized (Bull & Slater, 1982). Others are understandably more concerned with the *processes* in natural environments than with the qualitative nature of the microbial populations. Probably most *direct* effort to isolate novel microbes is currently expended in the search for secondary metabolites of potential pharmaceutical or industrial value. The efficiency and objectivity of such programmes are, however, often limited by lack of knowledge about the distribution of potential producers and the natural roles, if any, of their secondary metabolites.

We have now entered the age of genetic engineering, with the exciting potential for the creation of new, useful microbes in the laboratory by applying 'Orwellian' techniques to microbial populations. It is, therefore, an appropriate time to evaluate the current and future relevance of the 'free' microbes in the environment as a source of new isolates for both scientific and biotechnological purposes. The almost limitless diversity of micro-organisms and their habitats precludes any attempt to present a comprehensive review of the principles and techniques of isolation, which has anyway been most admirably achieved by Starr *et al.* (1981). This paper, therefore, inevitably reflects our personal views and experience, supported by the results of others who have studied different microbes and environments. Most emphasis has been placed on non-pathogens, although it is recognized that there is still a steady stream of reports of new plant and animal pathogens. We all like to hope that most human pathogenic microbes have been characterized; it is reassuring that *Legionella pneumophila* (McDade *et al.*, 1977) was first isolated, though not identified, in 1947 (McDade, Brenner & Bozeman, 1979)!

THE DEFINITION OF NOVELTY

What is a 'new' microbe? It is pertinent to ask this question before proceeding further, but difficult to provide a concise answer. A glib reply is that novelty is in the eye of the beholder, as definition of a novel isolate clearly depends on the aims and interests of the investigator. As these vary widely, so do the criteria invoked to justify novelty. Thus, the isolation of *Sulfolobus acidocaldarius* (Brock *et al.*, 1972) during ecological studies of volcanic springs provided an organism with a highly unusual combination of ecological, chemical, physiological and structural attributes, qualifying it for membership of the archaebacteria, which may be regarded as a new division of the prokaryotes (Gibbons & Murray, 1978). Other archaebacteria, such as *Methanothermus* (Stetter *et al.*, 1981) and *Thermoproteus* (Zillig *et al.*, 1981) from 'exotic' environments form

Table 1. *Numbers of species and subspecies in selected bacterial genera*

Genus	No. of species and subspecies
Streptomyces	378
Pseudomonas	88
Clostridium	75
Mycoplasma	59
Lactobacillus	44
Mycobacterium	43
Bacillus	33
Flavobacterium	18
Cellulomonas	6
Photobacterium	4
Mycoplana	2
Beggiatoa	1
Gallionella	1
Nitrobacter	1

Compiled from Skerman *et al.* (1980).

the nucleus of new families. At the other extreme, many newly isolated streptomycete species, such as *Streptomyces sulfonofaciens* (Miyadoh *et al.*, 1983), differ from previously described species by little other than their ability to produce a novel secondary metabolite.

Guidelines for the minimum requirements necessary for the definition and description of species have been produced (e.g. Sneath, 1977; Trüper & Krämer, 1981) and the Approved Lists of Bacterial Names (Skerman, McGowan & Sneath, 1980) led to the demise of many poorly described species which were cluttering the literature. However, the criteria for recognition of species and higher taxa vary considerably between different taxonomists and microbial groups. As stated by Cowan (1965), the philosophy of taxonomists is reflected in the subdivisions they create ('clumpers' and 'splitters') and the group on which they work. Some idea of the disparity in the number of subdivisions within bacterial genera is given in Table 1. Definition of fungal taxa is largely based on morphological criteria and over 4000 genera have been described (Ainsworth, 1968; Ainsworth, Sparrow & Sussman, 1973). The number of species in a genus is undoubtedly to some extent a reflection of its inherent natural diversity, but also influenced by the aims of the taxonomist and the criteria adopted to define the species. This is well illustrated by the genus *Streptomyces*. Over 3000 species have been named, many in the patent literature, on the basis of their ability to produce novel secondary metabolites (Trejo, 1970). In the eighth edition of *Bergey's Manual of Determinative Bacteriology*, Pridham & Tresner (1974) recognized 463 validly described species, defined by a limited number of standardized criteria. A recent numerical phenetic classification of 394 named species using 139 unit characters (Williams *et al.*, 1983a) delimited 77 clusters, most of which were regarded as species. Another good example based on a different strategy involves the cyanobacteria. Classification based on 'phycological observations' of field material produced about 170 genera and over 1000 species (Rippka *et al.*, 1981*b*). In contrast, a taxonomic study of axenic isolates under laboratory conditions led to the recognition of 22 genera (Rippka *et al.*, 1979).

A sound classification is clearly needed to prove novelty in its broadest sense, although its relevance to the search for microbes producing a novel metabolite is not always appreciated. It is, however, of equal importance in minimizing the duplication of effort involved in the re-discovery of existing micro-organisms and their properties. Unfortunately, some 'new' taxa described owe more to the taxonomic ignorance of their creators than to the inherent novelty of the microbe. Taxonomists themselves must shoulder at least some of the blame for this situation. Many microbial taxonomists are concerned with the improvement and validation of classifications, often by application of modern, sophisticated techniques of chemical and numerical analysis (Norris, 1980). Such work is of fundamental importance. However, the logical subsequence to classification is identification (Sneath, 1978), which is the useful end-product of taxonomy (Cowan, 1965).

It is the identification scheme which is of particular value to non-taxonomists, providing them, ideally, with a relatively rapid but informative system for comparing new isolates with existing taxa. Unfortunately, such systems are not available for all microbial groups. In some cases an apparently workable identification scheme is produced which is derived from a classification of low information content. Thus the many attempts to classify streptomycete species on the basis of their pigmentation and morphology have resulted in identification schemes analogous to a library catalogue system based on the size, shape and binding colour of the books. The most visible attribute of numerical phenetic classifications is their high information content (see Williams *et al.*, 1983*a*, for an extreme example). Such data provide an ideal basis for construction of computerized identification matrices (Sneath & Sokal, 1973; Hill, 1974), but relatively few numerical classifications (e.g. Wayne *et al.*, 1980; Williams *et al.*, 1983*b*) have been developed into probabilistic identification systems. Sneath (1978) concluded that many areas of microbiology, such as ecology and environmental studies, are seriously hampered by lack of good identification systems. He also pointed out that identification had received little study in its own right and often consisted of a hotch-potch of techniques without much reasoned design. Such a situation contrasts markedly with that faced by the diagnostic medical or veterinary microbiologist. It is to be hoped that the lot of the microbiologist studying isolates from other environments will be gradually improved by the development of more automated identification procedures (Hedén *et al.*, 1976), rapid improvements in data storage and retrieval systems, and by the production of more useful, reliable schemes by taxonomists.

APPROACHES TO ISOLATION

Detailed reviews of the principles applied to the isolation of bacteria (Stolp & Starr, 1981), actinomycetes (Williams & Wellington, 1982*a, b*) and fungi (Booth, 1971) have been presented. Our prime aim here is to examine the approaches made to isolation and in so doing to give some impression of the extent to which they truly reflect the diversity of microbes occurring in natural environments.

Most isolation strategies rely on prior knowledge of the nutritional and growth requirements of the microbes being sought (Stolp & Starr, 1981). Therefore, they are inevitably somewhat conservative in their aims and design. It is clearly difficult to devise objective procedures for the isolation of unknown microbes or those producing completely novel metabolites. Even when a microbe has been detected in a habitat by direct observation, it can evade all attempts to isolate it. Stolp & Starr (1981) listed 21 bacterial groups each containing some members which are known to occur in natural environments but which have not yet been cultured axenically on laboratory media. Comparison of the results of direct observational and isolation procedures suggests that in habitats such as the soil (Gray & Williams, 1971) and the sea (Kriss, 1963; Baross & Morita, 1978) many more microbes are present than are isolated, although the possibility that some of the cells observed are inviable cannot be discounted. Electron microscopy of soil samples (Nikitin, 1973) and the rhizoplane (Foster & Rovira, 1978) revealed the presence of microbes which were morphologically different from any yet isolated.

All isolation procedures are selective although the degree of their selectivity cannot always be accurately assessed. Thus the possibility of unintentional selectivity is always present, a good example being the isolation of *Actinopolyspora halophila* from an unsterilized batch of medium used in studies of halobacteria (Gochnauer *et al.*, 1975). The distinction between so-called general and selective methods is in essence one of precision. While knowledge of the nutrient and growth requirements of microbes provides the basic rationale for their isolation, there is considerable variation in the efficiency of the procedures devised. This is largely a reflection of the degree of metabolic and physiological specialization of the microbes in question. Thus, for example, the principle of selective isolation of the chemolithotrophic hydrogen-oxidizing bacteria is simple, as only they can grow on a minimal medium in the presence of an atmosphere containing hydrogen, oxygen and carbon dioxide (Aragno & Schlegel, 1981). It is seldom possible to apply such clear-cut principles to the isolation of chemo-organotrophs. Here the relative paucity of specific nutrient and other growth requirements can lead to competition on isolation plates from unwanted microbes present in the diverse populations of rich habitats such as compost and sewage. It is difficult, for instance, to develop an isolation medium for saprophytic pseudomonads on which other chemo-organotrophs do not grow (Stolp & Gadkari, 1981). Such problems have led to the development of isolation techniques by trial and error, some of which have as much affinity with cookery as with microbiology. Sometimes there is no clear explanation for the effects of the selective factors applied, as in the dry heating of soil at 120 °C for 1 h to isolate *Microbispora* and *Streptosporangium* strains (Nonomura & Ohara, 1969). Nevertheless the ingenuity of such empirical approaches should not be underestimated, as they have often resulted in the isolation of novel microbes and facilitated the detection of known groups. A simple but logical modification can sometimes result in a marked improvement of an existing isolation procedure. Addition of novobiocin (25 μg ml^{-1}) to the isolation medium allows growth of *Thermoactinomyces* strains while inhibiting all associated bacteria capable of growth at 50 °C (Cross, 1968), thus facilitating the detection of these endospore-forming bacteria in a wide range of habitats (Cross, 1981).

However one views the current approaches to isolation, it is clear that they are still providing a steady stream of new taxa, with interesting properties, some recent examples of which are given in Table 2. This list serves not only to illustrate the diversity of the new isolates, but also to demonstrate that, given the ubiquitous distribution of microbes, novel taxa can be obtained from habitats which have not yet been intensively studied.

Five stages in the isolation of a microbe can be recognized (Table 3) and these will be discussed in turn, although they are not necessarily exclusive.

Selection and sampling of the habitat

The choice of a habitat for study obviously depends on the interests and aims of the investigator. Many ecological studies are primarily concerned with the distribution of known microbes within a habitat or niche, so detection of a novel isolate is likely to be inadvertent. Generally, habitats of indistinct physicochemical and nutritional characteristics (e.g. soil, water) support a high diversity of species, whereas those with strong selective pressures (e.g. salt lakes, hot springs) contain very few species (Schlegel & Jannasch, 1981). However, this trend may partially reflect a reluctance to examine extreme environments and to face the difficulties of cultivating their inhabitants. The attention paid to some habitats in both categories has been disproportionate. Thus, while populations in soil have received considerable study for many years, those in oceans have been comparatively neglected, resulting in a lack of recognition that distinct, unique species exist in this ecosystem (Baumann & Baumann, 1981). There are many other habitats which have received scant attention or have been totally ignored. While it would be unwise to assume that habitats such as the soil have yet yielded all their microbes, there is no doubt that when a neglected environment is 'opened up' it usually provides novel isolates. Studies of barren, lichen-free sites in the Dry Valleys of Antarctica showed that yeasts were apparently the only indigenous heterotrophs (Vishniac & Hempfling, 1979); several new species of *Cryptococcus* were recognized among the isolates (Baharaeen & Vishniac, 1982; Vishniac & Baharaeen, 1982). Thermophilic anaerobes in volcanic springs are now receiving more attention, and several novel taxa

Table 2. *Examples of some recently isolated novel genera and species*

Isolate	Source	Comment	References
Bacteria			
Acidiphilium cryptum[a]	Coal mine drainage	Gram-negative acidophile	Harrison (1981)
Agitococcus lubricus[a]	Fresh water	Lipolytic, twitching coccus	Franzmann & Skerman (1981)
Aquaspirillum magnetotacticum	Fresh water	Magnetic spirillum	Maratea & Blakemore (1981)
Caseobacter polymorphus[a]	Cheese	Coryneform	Crombach (1978)
Chitinophaga pinensis[a]	Pine litter	Chitionolytic myxobacter	Sangkhobol & Skerman (1981)
Clostridium papyrosolvens	Paper-mill effluent	Cellulolytic anaerobe	Madden *et al.* (1982)
Clostridium thermosulfurogenes	Volcanic pool	Thermophilic, forms elemental sulphur	Schink & Zeikus (1983)
Ectothiorhodospira abdelmalekii	Soda lakes	Halophilic, alkaliphilic phototroph	Imhoff & Trüper (1981)
Ensifer adhaerens[a]	Soil	Bacterial predator	Casida (1982)
Erythrobacter longus[a]	Seaweeds	Aerobe with bacteriochlorophyll *a*	Shiba & Simidu (1982)
Fusobacterium simiae	Dental plaque of monkey	Oral anaerobe	Slots & Potts (1982)
Lactobacillus amylovorus	Cattle waste-corn fermentations	Starch hydrolyser	Nakamura (1981)
Methanogenium thermophilicum	Sea water	Thermophilic methanogen	Rivard & Smith (1982)
Methanothermus fervidus	Volcanic spring	Extremely thermophilic methanogen	Stetter *et al.* (1981)
Mycobacterium sphagni	*Sphagnum* moor	Scotochromogenic, fast-growing	Kazda (1980)

Organism	Source	Description	Reference
Pasteurella testudinis	Desert tortoise	Parasite	Snipes & Biberstein (1982)
Propionispira arboris	Wetwood of trees	Dinitrogen-fixing anaerobe	Schink et al. (1982)
Pseudomonas mesophilica	Leaf surface of rye grass	Pink-pigmented	Austin & Goodfellow (1979)
Spiroplasma floricola	Tulip tree flowers	From nectar	Davis et al. (1981)
Staphylococcus auricularis	Human external ear	Skin saprophyte	Kloos & Schleifer (1983)
Staphylococcus carnosus	Dry sausage	Ferments the product	Schleifer & Fischer (1982)
Thermococcus celer[a]	Sulfataric, marine water hole	Thermophilic, archaebacterium	Zillig et al. (1983)
Thermodesulfobacterium commune[a]	Volcanic pool	Thermophilic sulphate-reducer	Zeikus et al. (1983)
Vibrio diazotrophicus	Sea urchins, reed surfaces	Marine dinitrogen fixer	Guerinot et al. (1982)
Fungi and yeasts			
Banksiamyces spp.[a]	Dead *Banksia* cones	Discomycete on dead fruits	Beaton & Weste (1982)
Candida succiphila	Peach tree sap	Methanol-assimilating yeast	Lee & Komagata (1980)
Cryptococcus lupi	Antarctic soil	Heterotroph from barren areas	Baharaeen & Vishniac (1982)
Phalangispora constricta[a]	Foam in stream	Hyphomycete with branched conidia	Nawawi & Webster (1982)
Pichia mexicana	Cereoid cacti	Heterothallic yeast	Miranda et al. (1982)
Rhodosporidium paludigenum	Intertidal water	Basidiomycete yeast	Fell & Tallman (1980)
Scytalidium indonesiacum	Soil	Thermophilic hyphomycete	Hedger, Samson & Bazuki (1982)
Trimorphomyces papilionaceus[a]	Host fungi	Parasite of dematiaceous fungi	Oberwinkler & Bandoni (1983)

[a] New genus.

Table 3. *Stages in the isolation of a microbe from its natural habitat*

Stages	Determining factors
1. Selection and sampling of the habitat	The microbe's distribution pattern. Efficiency of the sampling procedure
2. Pretreatment of samples	Concentration or dilution. Enrichment procedures. Selective inhibitory treatments
3. Growth in laboratory media	Growth requirements of the microbe and its competitors. Other selective factors
4. Incubation	Optimal growth conditions. Optimal period
5. Isolate selection	Recognition of the isolates. Establishment of axenic cultures

have emerged (Wiegel & Ljungdahl, 1981; Zeikus, 1983), as they have from studies of alkalophiles in soda lakes and deserts (Grant & Tindall, 1980). There are many other examples of this tendency and some can be gleaned from Table 2.

Most microbial ecologists show an understandable reluctance to discuss the statistical validity of the procedures used to sample natural environments. A few grams of soil or litres of water are unlikely to provide a complete picture of the microbial populations in a hectare field, a lake or the Atlantic Ocean. In contrast, samples taken from more defined habitats, such as the phylloplane, have provided valuable information on the nature of the indigenous microbial community (Ercolani, 1978). Obviously, many sampling procedures are essentially a compromise between feasibility and accuracy (Williams & Gray, 1973). An additional problem is that 'the sample' may consist of a mixture of microhabitats. When these are recognized and analysed separately, the variety of microbes isolated tends to increase. Thus it is well known that the qualitative nature of microbes in soil, the rhizosphere and the rhizoplane is different (Bowen, 1980). However, many other particulate substrates exist in soil and are colonized by different microbes (Williams & Gray, 1973), but these have seldom been considered separately.

Another factor mitigating against the detection of all microbes present in a habitat is the essentially 'blind' nature of the sampling procedure. Relatively few microbes can be seen and isolated directly. Some notable exceptions are the sheathed bacteria, the slimy masses of which can be picked out of water (Mulder & Deinema, 1981), and the myxobacteria (Reichenbach & Dworkin, 1981) and basidiomycetes (Watling, 1971), both of which can be isolated from their fruit bodies. The absence or infrequency of such microbes on isolation plates prepared from water samples or soil suspensions also underlines the fallibility of widely used isolation techniques. Direct isolation of fungal hyphae from soil was achieved by Warcup (1957) and resulted in the detection of one new genus and five new species of the basidiomycetes (Warcup & Talbot, 1962), yet this technique has found few disciples.

The problems of sampling faced by freshwater and marine microbiologists are clearly more extreme than those encountered in studies of terrestial environments. A variety of sampling devices have been constructed to collect uncontaminated samples (Collins et al., 1973; Baumann & Baumann, 1981). The design improvements (e.g. Jannasch, Wirsen & Winget, 1973) together with the wider application of such devices should eventually result in the isolation of many of the unknown microbes which are likely to exist, particularly in marine habitats (see Jannasch, this volume). This has been spectacularly demonstrated by the recent isolation of bacteria from sulphide chimneys or 'black smokers' along the East Pacific Rise (Baross & Deming, 1983). These microbes were isolated from water at 306 °C and were capable of chemolithotrophic growth at a pressure of 265 atm and temperature of at least 250 °C. It was concluded that this indicates that microbial growth is limited not by temperature but by the existence of liquid water, thereby greatly increasing the range of environments, such as within the earth's crust, in which microbes may exist.

Pretreatment of samples

Most samples require some form of treatment before their incorporation into the laboratory media designed to isolate selected microbes. Such pretreatments prolong the period of time between the removal of a microbe from its habitat and its axenic growth in the laboratory, thereby increasing the possibility that it may be inadvertently killed by physical or chemical damage. Unfortunately it is rarely possible to transfer samples directly to growth media, one exception being the isolation of microbes such as the halobacteria which constitute a high proportion of the natural populations and have highly specific growth requirements (Larsen, 1981).

Numbers of the relatively unspecialized but diverse chemo-organotrophs in habitats such as soil and compost are often in the order of several millions per gram. Therefore, detection of individual isolates requires dilution of the samples, which reduces or eliminates the possibility of isolating those micro-organisms which are present in low numbers. Many of the selective enrichment pressures introduced at this and the next stage of isolation are designed to minimize this problem. Standard dilution procedures are not even always efficient in detecting microbes which are known to be prevalent in a habitat. Thus it is very difficult to isolate *Micropolyspora faeni*, a causal agent of 'farmer's lung', from self-heated hay by such procedures, yet it is readily detected in large numbers by sampling the spore clouds released from the deteriorated material (Lacey, 1978; Cross, 1982). This approach, which in essence uses air as the diluent, has been applied to other self-heated materials, resulting in the isolation of some novel actinomycetes (Lacey, 1978).

The isolation of many microbes which exist in relatively low numbers in the habitat or have specific growth requirements is achieved by increasing their proportional numbers in the sample. This is in essence the process of enrichment which is fundamental to isolation. Selective pressures introduced at this point are often reinforced during the subsequent stages of isolation (Table 3). Enrichment procedures often attempt to mimic conditions in the environment and are generally equated with the selective stimulation of growth. However, the proportion of selected microbes in a sample can also be increased by removal of their competitors. The potential of enrichment for the detection of new microbes and producers of metabolites is vast, being limited only by our knowledge of microbial growth requirements and its application. This topic has been reviewed by many workers, including Schlegel & Jannasch (1967), Aaronson (1970), Veldkamp (1970) and Stolp & Starr (1981).

The main variables in enrichment procedures are listed in Table 4, and the approaches will be briefly illustrated. The major factors involved in selective stimulation of growth are nutrition, pH, aeration, temperature and light (Stolp & Starr, 1981). In some cases the rationale is straightforward. Thus denitrifying bacteria are enriched by creating conditions which ensure that nitrate, or a nitrogenous oxide formed from it, serves as the terminal acceptor of the electron transport system that generates the total source of ATP

Table 4. *Some options in enrichment procedures*

Targets	Selected taxa; microbes of particular ecophysiological significance; microbes producing useful or interesting metabolites
Approaches	Selective stimulation of growth; selective inhibition of growth; selective concentration of cells
Systems	Open (batch culture); closed (continuous culture); liquid; solid
Selective pressures	Nutritional; environmental; chemical; physical

(Jeter & Ingraham, 1981). The well-established procedure of incubating samples in nitrogen-free conditions can still initiate the detection of new dinitrogen-fixing microbes such as *Propionispira arboris* (Schink, Thompson & Zeikus, 1982). *Cladosporium resinae*, which can grow in aircraft fuel tanks, was detected in soil by the ingenious technique of baiting soil with creosote-coated matchsticks (Parbery, 1967). In other cases, the procedure may succeed but its rationale is less obvious. The enrichment of Beggiatoaceae is achieved by the incubation of soil samples with extracts of mud, hay or grass (Wiessner, 1981), and that of *Caryophanon* in a slurry prepared from cow dung (Smith & Trentini, 1972). Baits used to isolate aquatic fungi include houseflies, white human hair, hemp seeds and newspaper (Gareth Jones, 1971).

Sometimes it is possible to reduce or eliminate the unwanted microbes in a sample. If the isolates being sought have an extreme or differential tolerance to a particular factor, this can be very effective. Probably the best known example is the use of pasteurization in the isolation of endospore-forming bacteria. Less severe heat treatments have been successfully used to isolate a variety of actinomycete genera (Williams & Wellington, 1982*a, b*; Cross, 1982), but the physiological basis of the differential responses is not clear.

When isolating microbes from habitats such as fresh water and sea water, where the microbial density is low, it is often essential to obtain a non-selective concentration of cells in the sample. This can be done by filtration (e.g. Burman, Oliver & Stevens, 1969) or centrifugation (e.g. Okami & Okazaki, 1972). *Selective* concentration of microbial cells by physical means is also sometimes possible. Thus differential centrifugation has been used to separate *Frankia* from root nodule tissues (Baker, Torrey & Kidd, 1979), flotation to isolate *Actinoplanes* from various habitats (Makkar & Cross, 1982) and migration through agar media to isolate numerous anaerobic *Spirochaeta* strains (Canale-Parola, 1973).

Enrichment procedures are still providing many new and meta-bolically interesting isolates, as is exemplified by the approach of Zeikus and co-workers (Zeikus, 1983). Novel anaerobes isolated by nutritional enrichments from a hot spring included *Clostridium thermosulfurogenes* by using pectin (Schink & Zeikus, 1983) and *Thermodesulfotobacterium commune* by using lactate (Zeikus *et al.*, 1983). *Pectinatus*, a new genus of the Bacteroidaceae, was disco-vered after the unintentional enrichment procedure of storing beer at 30 °C for 30 days (Lee, Mabee & Jangaard, 1978). As Stolp & Starr (1981) emphasized, enrichment procedures may also be directed towards the isolation of hitherto unknown metabolic types, which may lead to the detection of new microbes if such 'con-structed types' exist in nature.

Growth in laboratory media

The growth of microbes in isolation media is also an enrichment process, presenting many of the same opportunities and problems as the preceding stage. The degree of selectivity induced by the composition of the medium is almost infinitely variable. Highly efficient, defined media exist for the isolation of microbes with specialized growth requirements, such as the nitrifying bacteria (Watson, Valois & Waterbury, 1981) and the halobacteria (Larsen, 1981). However, the recent isolation of *Halobacterium sodomense* (Oren, 1983), which differs from other halobacteria in its ion requirements and grows only in the presence of starch or clay minerals, demonstrates that the reservoir of specialized groups is by no means exhausted.

For relatively unspecialized chemo-organotrophs, semi-defined media containing additives such as peptone and yeast extract for soil bacteria (Goodfellow, Hill & Gray, 1968) or beef extract and peptone for *Bacillus* (Norris *et al.*, 1981) are widely used to provide growth factors. Many more ill-defined and exotic additives have been recommended, such as canned tomato juice for lactobacilli (Yoshizumi, 1975), whole chicken eggs for nocardiae (Tsukamura, 1972) and sheep dung extract for thermophilic streptomycetes (Tendler & Burkholder, 1961). These examples give some idea of the 'muck and mystery' approach to which workers studying such organisms are sometimes driven. The rationale behind the formulation of some defined media is also not always apparent. Colloidal chitin agar (Lingappa & Lockwood, 1962; Hsu & Lockwood, 1975) has been widely and successfully used to isolate streptomycetes, yet a recent study of large numbers of species (Williams *et al.*, 1983*a*) revealed that only 25% were strongly chitinolytic. The potential value of complex media should not, however, be underestimated. They have, for example, played a major part in the species 'explosion' of spiroplasmas detected in very diverse habitats since their initial isolation was achieved some 13 years ago (Tully & Whitcomb, 1981).

The pH of the medium is an important selective factor. It is obviously of fundamental relevance to the isolation of obligately acidophilic or neutrophilic microbes such as *Thiobacillus thio-oxidans* and urea-decomposers. However, pH can also have important selective effects on the isolation of microbes with requirements which are less extreme or predictable. Use of media at pH 4.5 revealed, in a variety of environments, streptomycetes which did not appear on the routinely used neutral media (Khan & Williams, 1975). Similarly, *Pseudomonas acidophila* and *P. mesoacidophila*, the producers of powerful new β-lactam antibiotics, were isolated from soil by the simple expedient of using nutrient agar at pH 4.5 (Kintaka *et al.*, 1981*a*, *b*).

A wide range of selective inhibitors has been added to media in attempts to reduce competition from unwanted microbes. Many examples of these were listed by Stolp & Starr (1981), including dyes, metabolic poisons, detergents and antibiotics. The last have often proved to be quite effective due to their greater specificity. Best results are obtained when antifungal antibiotics are used to encourage development of bacteria such as *Streptomyces* (Williams & Wellington, 1982*b*) and *Nocardia* (Orchard, Goodfellow & Williams, 1977). It is difficult to devise combinations of antibiotics which are completely selective for a diversity of bacterial species, even within one genus (Williams & Davies, 1965). However, determination of the antibiotic resistance pattern of a given species often provides the basis for a highly selective isolation medium, such as those for *Nocardia asteroides* (Orchard & Goodfellow, 1974) and *Renibacterium salmoninarum* (Austin, Embley & Goodfellow, 1983).

Incubation

The main incubation variables are temperature, aeration and light, the selection of optimal conditions being assisted by knowledge of a microbe's requirements in the laboratory and often, but not always, by the conditions prevailing in its natural environment.

Samples from high-temperature environments, incubated appropriately, have provided many interesting novel thermophilic or caldoactive isolates such as *Chloroflexus aurantiacus* (Pierson & Castenholz, 1974) and *Caldariella acidophila* (De Rosa, Gambacorta & Bu'lock, 1975). However, the apparently illogical strategy of incubating samples from cold habitats at high temperature has also produced some interesting thermophiles, such as the hydrogen-oxidizing *Bacillus schlegelii* isolated from cold lake sediments (Schenk & Aragno, 1979). Low temperatures predominate in the majority of natural environments and many pyschrophilic isolates have been obtained from habitats such as sea water and Antarctic soils. The growth optima and limits of psychrophiles are by no means consistent though (Baross & Morita, 1978), which must be borne in mind when selecting temperatures for their isolation. Micro-organisms capable of growth at low temperatures also occur in relatively 'mesophilic' habitats; soil isolates, such as certain arthrobacters and other pleomorphic bacteria can grow at temperatures below 5 °C (Baross & Morita, 1978). Conversely, many microbes which are treated as mesophiles in the laboratory, are unlikely to encounter their optimum growth temperature in habitats such as soil and fresh water. Such disparities, and the likelihood that a continuum of microbial temperature requirements exists, suggest that a wider spectrum of isolates may be obtained from many habitats by using a range of incubation temperatures.

The diversity of microbial responses to aeration is well known and accounts for their occurrence and influences in many environments. Conditions for the isolation of obligate or facultative anaerobes are clearly defined. One of the major problems faced in selecting gas mixtures for the isolation of anaerobes is the diversity of their oxygen tolerances; again it is likely that a continuum of responses exists (Schlegel & Jannasch, 1981). The oxygen requirements and tolerances of aerobes also vary, as demonstrated by the use of micro-aerophilic conditions to isolate and cultivate bacteria such as methanotrophs and some dinitrogen-fixers (Pearson, Howsley & Williams, 1982). As with temperature, although obligately anaerobic bacteria are hypersensitive to oxygen, they can be isolated from apparently aerobic habitats (Schlegel & Jannasch, 1981).

While the provision of light, usually from artificial sources, is required for the isolation of phototrophs, inhibition can occur if the intensity is too great. Light is also inhibitory to some non-photo-trophs, but fortuitously they are usually grown in closed incubators (Schlegel & Jannasch, 1981). The chlorophyll absorption maxima of the phototrophic bacteria differ and this can partially determine their distribution patterns within an environment. In aquatic habitats purple and green bacteria often occur beneath layers of cyanobacteria and algae, light harvesting being facilitated by the strong adsorption bands of their chlorophylls in the far-red and infrared regions (Stanier et al., 1981). All anoxygenic phototrophic bacteria which are motile by flagellation show a phobophototactic response that keeps them in an area of favourable light intensity which they have accidentally entered (Schlegel & Jannasch, 1981). There is, however, little information on the significance of this response in vivo. Thus, choice of appropriate light filters and light intensities contributes to selective isolation of the various groups of phototrophs (Biebl & Pfennig, 1981; Pfennig & Trüper, 1981; Rippka, Waterbury & Stanier, 1981a), although other factors, such as aeration and the nitrogen or sulphur content of the medium, are clearly of great relevance. Selection of suitably balanced light–dark cycles is another important variable.

Further opportunities (or problems) for the selective isolation of phototrophic bacteria are provided by the increasing awareness of their metabolic versatility. This is exemplified by the ability of some cyanobacteria to alternate between oxygenic and anoxygenic photosynthesis (Garlick, Oren & Padan, 1977) and of many species of the Rhodospirillaceae to grow chemo-organotrophically in the dark (Stanier et al., 1981). The significance of the presence of bacterio-chlorophyll a in the aerobic bacteria Protaminobacter ruber (Sato, 1978) and Erythrobacter longus (Shiba & Simidu, 1982) remains to be assessed.

The importance of the period of incubation in isolation procedures should not be overlooked; accepted periods vary considerably, ranging, for example, from a few days to isolate thermophilic actinomycetes (Williams & Cross, 1971), to several months to detect nitrifying bacteria (Watson et al., 1981) and novel mycobacteria in soils from Zaïre (Portaels et al., 1982). Extension of the 'normal' incubation period has also contributed to the isolation of microbes such as Frankia, which fixes dinitrogen in the nodules of non-leguminous plants and has eluded isolation until quite recently (Callaham, Tredici & Torrey, 1978). Small colonies are produced on isolation plates after 3 weeks, and in axenic culture colonies of

1–2 mm diameter develop after several months (Becking, 1981). Recently, a novel bacterium with mixed flagellation and both unicellular and multicellular phases (aptly named *Conglomeromonas largomobilis*) was isolated from fresh water on plates incubated for up to 3 weeks (Skerman, Sly & Williamson, 1983).

Isolate selection

The final, but nevertheless often critical stage in isolation of a microbe is its recognition and transfer to axenic culture. Sometimes the colonies on an isolation plate can be distinguished relatively easily by macroscopic or microscopic observation; this is particularly true of fungi. It is well known that some enterobacteria can be recognized by their reactions to pH or redox indicators in the medium and that certain yeasts colonies form a specific colour on media containing ferric ammonium citrate (Beech & Davenport, 1971). Similarly, microbes capable of hydrolysing compounds such as starch, pectin, chitin or cellulose can be detected on isolation plates either directly or after chemical treatments which visualize the hydrolysis zones. Unfortunately it is by no means always feasible to distinguish between the microbes on an isolation plate by superficial examination, and the possibility of overlooking interesting or novel isolates often exists. Many a new microbe has been detected by the painstaking examination of plates by microbiologists with the experience and gift for accurate microscopic observation.

In many cases it is easy to obtain axenic cultures of microbes present in isolation media, but problems are encountered with some groups. For example, the combination of low light intensities to encourage gas-vacuolation followed by differential centrifugation is often required to obtain axenic isolates of planktonic cyanobacteria (Walsby, 1981); such techniques have facilitated the isolation and recognition of species present in aquatic habitats. Some success in isolation of unusual microbes from colonies on plates has also been achieved using micromanipulators (Skerman, 1968; Skerman *et al.*, 1983).

THE SEARCH FOR NEW SECONDARY METABOLITES

The effort devoted by pharmaceutical organizations to the isolation of microbes from natural habitats far exceeds that of any other group of microbiologists. Their aim is obviously to obtain micro-organisms producing novel, potentially useful metabolites. This area is clearly of some relevance to the question we have posed. Although it is well known that microbes produce a wide range of useful secondary metabolites (Nielleman, 1973; Perlman, 1973), the approaches and problems will be illustrated primarily by reference to the search for microbes producing new antibiotics.

The rate of discovery of new microbial antibiotics did not change significantly from the mid-1940s to the mid-1960s (Bérdy, 1974). There was a dramatic increase between 1972 and 1978 (Perlman, 1977), but it has become increasingly difficult to find antibiotics which are both novel and useful (Fleming, Nisbet & Brewer, 1982). This situation will undoubtedly be alleviated to some extent by chemical modification of existing compounds, genetic manipulation of known producers (Hopwood & Chater, 1980; Hopwood, 1981) and improvements in the specificity of screening procedures (Bushell & Nisbet, 1981; Fleming et al., 1982). Nevertheless, new antibiotics are still discovered at the rate of about 50 to 100 per year (Demain, 1981) and many of these originate from new isolates, some recent examples of which are given in Table 5.

Actinomycetes have long been regarded as the principal source of antibiotics; according to Demain (1981) about 2000 antibiotics have been described, of which 1500 originate from actinomycetes. Within the actinomycetes the genus *Streptomyces* excels in its capacity for antibiotic production, possibly because of the extra-large DNA complement of these bacteria (Hopwood & Chater, 1980). There is, however, an increasing awareness of the antibiotic-producing capacity of both other bacterial genera (Table 5) and fungi, such as the basidiomycetes (e.g. Anke et al., 1980), which have been comparatively neglected. A good example of the versatility of streptomycetes is provided by *Streptomyces clavuligerus* (Higgens & Kastner, 1971) which has been reported to produce at least 14 different antibiotics (Cross, 1982). Many streptomycetes producing novel antibiotics are still being isolated and the soil remains a prolific source of all producers. The facility for detecting novel strains and metabolites seems to be a particular attribute of Japanese workers, as might be deduced from Table 5.

Here, we will consider some of the problems faced in devising objective approaches to the search of natural habitats for strains producing new antibiotics *in vitro*, and some possible solutions.

The discovery of isolates producing new antibiotics has usually

Table 5. *Examples of recently isolated bacteria producing novel secondary metabolites*

Isolate	Source	Product	Reference
Actinomadura kijaniata	Soil	Enol antibiotics	Horan & Brodksy (1982)
Actinomadura luzonensis	Soil	Antitumour complex	Tomita *et al.* (1980)
Alteromonas aurantia	Sea water	Polyanionic antibiotics	Gauthier & Breittmayer (1979)
Ampullariella regularis subsp. *mannitophila*	Soil	Candiplanecin	Itoh *et al.* (1981)
Dactylosporangium matsuzakiense	Soil	Dactimicin	Shomura *et al.* (1980)
Micromonospora griseorubida	Soil	Mycinamicins	Satoi *et al.* (1980)
Pseudonocardia azurea	Soil	Azureomycin B	Spiri-Nakagawa *et al.* (1980)
Pseudomonas acidophila	Soil	Sulfazecin	Kintaka *et al.* (1981*b*)
Pseudomonas mesoacidophila	Soil	Isosulfazecin	Kintaka *et al.* (1981*a*)
Saccharopolyspora hirsuta subsp. *kobensis*	Soil	Sporaricin	Iwasaki, Itoh & Mori (1979)
Streptomyces gilvotanareus	Soil	Gilvocarcins	Takahashi *et al.* (1981)
Streptomyces sannanensis	Soil	Sannamycin	Iwasaki, Itoh & Mori (1981)
Streptomyces tenjimariensis	Sea water	Istamycins	Hotta *et al.* (1980)

been achieved by a 'hit and miss' approach, involving the large-scale and indiscriminate sampling of the environment followed by the random isolation and screening of a large number of microbes. While the efficiency of the latter step is being improved by the development of more selective screening procedures, the approach to isolation remains largely subjective. The main reasons for this are:

(i) the lack of knowledge of the ecological significance, if any, of antibiotics;

(ii) inadequate information about the numbers and distribution of antibiotic-producers in natural habitats;

(iii) deficiencies in the classification of some of the major antibiotic-producing microbial taxa.

The ecological significance of antibiotics

The selective inhibitory actions of antibiotics appear to be admirably suited to assist their producers in competitive interactions in habitats such as the soil, from where most originate. Numerous attempts to detect antibiotics in soil were made in the 1940s and 1950s, but they did not reveal the presence of these compounds in natural, unamended samples. Therefore, opposing views as to whether or not antibiotics were natural products were raised (Waksman, 1956; Brian, 1957). This problem remains unresolved and the current opinions and evidence have been reviewed by Gottlieb (1976) and Williams (1982). The likelihood of antibiotics having an ecological significance cannot be discounted, as there are several factors (such as their instability in soil, the inadequacy of detection methods and the ephemeral growth of their potential producers) which may account for the lack of definitive evidence of their occurrence in soil and other habitats (Hill, 1972; Williams & Khan, 1974; Katz & Demain, 1977). It has also been argued that antibiotics must confer a natural selective advantage on their producers, as much chromosomal and extrachromosomal DNA is dedicated to their genetic determination (Hopwood, 1981). However, the possible natural role of all antibiotics is not necessarily confined to competition, as there is some evidence that they may play a part in morphological differentiation, such as the production of spores (Demain, 1981; Hopwood, 1981). Until this problem is resolved, the selection of habitats and microhabitats for obtaining antibiotic-producing isolates will remain largely a matter of chance. The prospects of discerning a natural role for other pharmacologically useful microbial metabolites, such as anti-inflammatory, cardioactive and hypotensive agents (Nielleman, 1973), are even more daunting!

The distribution of antibiotic-producers

Although a host of antibiotic-producing microbes has been isolated from natural habitats, knowledge of their geographical or ecological distribution is very limited. This is due firstly to the general problem of discerning the geographical distribution patterns of most free-living microbes with relatively unspecialized growth requirements. Secondly, the descriptions of the habitats from which antibiotic producers are isolated do not generally provide information of great ecological relevance (Cross, 1982). A typical example is the isolation of a *Pseudomonas* sp., which produces a new antitumour antibiotic, from 'the soil in the Tokai district of the Ibaragi prefecture, Japan' (Ezaki *et al.*, 1980). The reticence to provide more accurate information is no doubt partially explained by commercial considerations. Nevertheless the lack of detailed ecological data impedes the development of more objective approaches to the search for novel antibiotics.

One 'ecological' approach which has met with some success is the screening of isolates from relatively neglected or extreme habitats (Williams & Wellington, 1982a). Novel antibiotics have been detected, for example, from marine isolates of *Alteromonas* (Gauthier & Breittmayer, 1979) and *Streptomyces* (Hotta, Okami & Umezawa, 1980). Although there is no reason to suppose that the reservoir of antibiotic-producers in habitats such as the soil is exhausted, a more extensive application of this approach may be productive.

Taxonomic considerations

As stated previously, the genus *Streptomyces* is the prime source of antibiotics, and the widespread tendency to regard isolates producing novel metabolites as new species has resulted in taxonomic chaos. This situation is detrimental not only to taxonomists but, maybe more importantly, to those studying other aspects of these microbes. A sound classification with a high information content facilitates the selection of species for biochemical, ecological or physiological study, and for genetic manipulation. Recent attempts have been made to construct more objective schemes for the classification (Williams *et al.*, 1983a) and identification (Williams *et al.*, 1983b) of streptomycetes. The data obtained also provide a basis for improving the objectivity of procedures designed to isolate streptomycetes producing antibiotics, and the assessment of their efficiency.

It is clear that the ability to produce a particular type of antibiotic is not necessarily confined to one species. Nevertheless, examples of the antimicrobial activities of species-groups defined by overall phenetic similarity (Table 6) indicate that some broader correlations do exist. Good examples are the high activity of the *Streptomyces*

Table 6. *Antimicrobial activity (percentage positive reactions) within streptomycete species-groups defined by numerical classification*

Species-groups	Streptomyces fulvissimus	S. rochei	S. chromofuscus	S. griseoviridis	S. diastaticus	S. griseoruber	S. lydicus	S. lavendulae	Streptoverticillium griseocarneum
Inhibition of:									
Escherichia coli	0	8	0	0	0	0	0	83	96
Pseudomonas fluorescens	0	8	0	0	0	0	0	50	44
Bacillus subtilis	78	35	11	100	21	0	73	92	78
Micrococcus luteus	89	35	11	100	21	0	100	83	89
Streptomyces murinus	100	39	22	80	5	22	100	100	89
Candida albicans	11	19	0	33	0	0	27	67	100
Saccharomyces cerevisiae	22	15	0	5	0	0	27	67	89
Aspergillus niger	22	27	0	33	11	0	100	75	100
Production of:									
β-Lactamase	40	36	25	33	90	22	40	92	44
β-Lactamase inhibitor	20	8	38	0	0	11	0	0	0

From Williams *et al.* (1983*a*).

lavendulae and *Streptoverticillium griseocarneum* groups against gram-negative bacteria and fungi; the activity of *Streptomyces fulvissimus, S. griseoviridis, S. lydicus* and *S. lavendulae* against gram-positive bacteria; and the general lack of activity of *S. griseoruber*. In the more specific tests, *S. diastaticus* was notable for its ability to produce β-lactamase and *S. chromofuscus* to provide β-lactamase inhibitors.

These data underline the need to maximize the diversity of isolates obtained from a soil sample, but also demonstrate the desirability of developing methods which will facilitate the isolation of selected species or species-groups. To some extent, both of these objectives can be achieved by using the information on the growth requirements and tolerance of species-groups which is contained in numerical classification and identification matrices. Thus the data on streptomycetes provided by Williams *et al.* (1983*a, b*) are being used to devise and assess isolation media which are targeted more specifically than those, such as starch–casein (Küster & Williams,

Table 7. *Streptomycetes isolated from a sample of sand dune soil using three different media*

Species-groups	% total soil isolates on:		
	Starch–casein medium	Starch–casein medium + rifampicin $(50\,\mu\mathrm{g\,ml}^{-1})$	Raffinose–histidine medium
Streptomyces albidoflavus	6.5	13.3	0
S. chromofuscus	2.2	0	0
S. cyaneus	28.3	0	63.6
S. diastaticus	0	80.0	2.3
S. platensis	37.0	0	4.5
S. rochei	8.7	6.6	13.6
Unidentified isolates	17.4	0	15.9

1964), which have been widely used. The formulation of the media is assisted by the application of the DIACHAR program (Sneath, 1980) to the taxonomic data matrix to select the most distinctive growth requirements and tolerance of selected groups (Vickers, Williams & Ross, 1983). The isolates obtained are identified using the probabilistic system of Williams *et al.* (1983*b*).

A comparison of the isolates obtained from one soil sample using two newly devised media with those from starch–casein is given in Table 7. Most of the results could be predicted from the data matrix, such as the inhibition of most groups by rifampicin, allowing detection of *Streptomyces diastaticus*, and the predominance of *S. cyaneus* on raffinose–histidine. These, and other results obtained (Vickers *et al.*, 1983), also demonstrate that it is sometimes impossible to devise an effective 'general' isolation medium even for the species of a single genus. Preliminary results from a screening programme of the isolates obtained, underline the fact that there is often a correlation between objectively defined species-groups and their antimicrobial spectra. Thus improvements in classification and identification have at least some part to play in the development of more effective procedures to detect microbes producing novel antibiotics. There seems to be no reason why the strategy adopted for streptomycetes should not be extended to other antibiotic-producing taxa which are considered to be rare or difficult to isolate.

CREATION OF NEW MICROBES IN THE ENVIRONMENT

The capacity of microbes to transfer DNA *in vitro* under the persuasive guidance of the microbial geneticist requires no emphasis. However, although there is at present little *direct* evidence of the mechanism of genetic transfer between bacteria in the environment it is clear that genetic information is exchanged. In contrast, plasmogamy, karyogamy and meiosis in fungi have long been accepted as natural events, as demonstrated, for example, by the production of the fruit bodies of ascomycetes and basidiomycetes, and the appearance of new races of plant pathogens. The genetic interactions among bacterial communities have been recently reviewed by Reanney, Gowland & Slater (1983), to which the reader is referred for more detailed information.

The best known and most convincing example of genetic exchange by bacteria in the environment is the transfer of drug-resistance plasmids, which sometimes occurs between widely separated populations (Anderson, 1975). It has also been suggested that the genetic pool existing in mixed populations of bacteria in many habitats plays a part in their responses to the introduction of alien chemicals, such as herbicides and pesticides, into their environment (Reanney *et al.*, 1983). The same authors argued that, as in many other instances, the multitude of *in vitro* processes of genetic exchange may not accurately reflect the situation in the natural environment. Indeed, intensive studies of the microbes in a particular habitat often provides evidence of their stability. For example, isolates of a streptomycete strain obtained over a period of 6 years from a forest soil, were consistent in their morphological, physiological and ecological characteristics (Mayfield, 1969). It is possible that bacteria rely primarily on their chromosomal DNA to preserve their individuality, while the transfer of extrachromosomal DNA provides a means of small-scale genetic modifications which may be of ecological significance (Reanney *et al.*, 1983). This would suggest that while it is unrealistic to expect the sudden emergence of an entirely novel microbe, natural environments can generate organisms with relatively few, but possibly important and useful, new properties. Reanney and his colleagues listed the various factors which might promote or retard genetic exchange in natural habitats. A prime requirement is obviously the close proximity of the growth of different strains in their habitat. There is considerable evidence

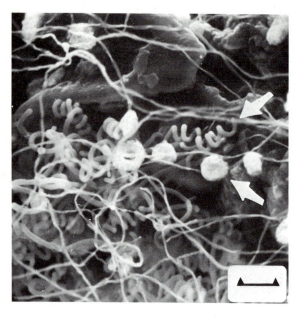

Fig. 1. Scanning electron micrograph of a mixed colony on a root fragment in soil, showing streptomycetes with two types of spore chain (arrowed). Scale bar represents 5 μm.

that this occurs (White, 1983), and it ranges from the development of streptococci in dental plaque (Clewell, 1981) to the growth of mixed streptomycete colonies on root fragments in soil (Mayfield *et al.*, 1972) (Fig. 1).

Phages are well known as agents of genetic transfer *in vitro* and they occur in most, if not all, natural habitats. Their isolation is usually achieved by selective enrichment procedures; these provide valuable tools for laboratory experimentation but give little information on the significance of phages in the environment. Indeed, surprisingly little is known about the ecology of phage. Many of those isolated are polyvalent for species within a genus or, less frequently, for different genera (Jones & Sneath, 1970), although this does not preclude the possibility that they originated from lysogenic associations in the natural habitat, which many can form *in vitro*. It has been suggested that most phages may exist in nature integrated into the DNA of their host's cells (Reanney *et al.*, 1983). However, there are a number of problems faced in accurately enumerating the extracellular phage particles present in habitats such as the soil (Lanning & Williams, 1982), which may explain the low numbers which have been generally reported. Thus, an improved procedure for the enumeration of actinophage in soil (Lanning & Williams, 1982) revealed up to 23 000 per gram dry weight of soil capable of infecting one streptomycete isolate. Similar improvements have been made in the detection of phages in freshwater habitats (Seeley & Primrose, 1982). The numbers of free phage particles in natural environments are, therefore, probably greater than previously assumed, and their concentrations are likely to be even higher in the microsites where growth of their hosts occurs. Overall, there seems no reason to suppose that transduction does not occur *in vivo*. Phages also provide a source of material for genetic manipulation *in vitro* which, for most microbes, has been largely untapped.

Thus it is reasonable to assume that genetic exchange between bacteria in natural habitats, however limited, will contribute to the supply of novel isolates. It is unlikely that all mechanisms of DNA transfer were evolved solely for the convenience of microbial geneticists and biotechnologists. Nevertheless, more direct evidence of genetic transfer *in vivo* is clearly required.

FUTURE PROSPECTS

We hope that this review has indicated some of the reasons for assuming that many more microbes remain to be discovered in the environment. The picture of the natural occurrence and distribution of microbes is far from complete and it is salutary to consider that little is known even about the components of the microflora on our own skins (Noble, 1981)! The continued detection of novel isolates must remain a matter of chance to some extent, but will also be aided and encouraged by:

(i) increased knowledge of microbial ecology and physiology;
(ii) study of neglected habitats and more intensive 'floristic' investigations of the better studied ones;
(iii) improved sampling procedures, particularly for marine and freshwater habitats;
(iv) development of more objective, together with less conservative, isolation procedures;
(v) provision of more efficient identification systems to determine novelty;
(vi) an understanding of the natural roles of secondary metabolites;

(vii) assessment of the frequency and mechanisms of genetic exchange in nature;

(viii) the demands of biotechnology for isolates producing either novel metabolites or known ones more efficiently.

It is apposite to conclude with a remark by the editors of *The Prokaryotes* whose opinions are as weighty as their volumes: 'There are wide areas of the bacteriological landscape in which we have so far detected only some of the highest peaks, while the rest of the beautiful mountain range is still hidden in the clouds and the morning fogs of ignorance. The gold is still on the ground, but we have to bend down to grasp it' (Starr *et al.*, 1981).

We gratefully acknowledge the support of the Science and Engineering Research Council and Glaxo Group Research for some of the work discussed here.

REFERENCES

AARONSON, S. (1970). *Experimental Microbial Ecology*. New York & London: Academic Press.

AINSWORTH, G. C. (1968). The numbers of fungi. In *The Fungi, An Advanced Treatise*, vol. 3, ed. G. C. Ainsworth & A. S. Sussman, p. 505. New York & London: Academic Press.

AINSWORTH, G. C., SPARROW, F. K. & SUSSMAN, A. S. (eds) (1973). *The Fungi, An Advanced Treatise*, vol. 4. New York & London: Academic Press.

ANDERSON, E. S. (1975). Problems and implications of chloramphenicol resistance in typhoid bacillus. *Journal of Hygiene*, **74**, 289–99.

ANKE, T., KUPKA, J., SCHRAMM, G. & STEGLICH, W. (1980). Scorodonin, a new antibacterial and antifungal metabolite from *Marasmius scorodonius* (Fr.) Fr. *Journal of Antibiotics*, **33**, 463–7.

ARAGNO, M. & SCHLEGEL, H. G. (1981). The hydrogen-oxidising bacteria. In *The Prokaryotes. A Handbook on Habitats, Isolation and Identification of Bacteria*, vol. 1, ed. M. P. Starr, H. Stolp, H. G. Trüper, A. Balows & H. G. Schlegel, pp. 865–93. Berlin, Heidelberg & New York: Springer-Verlag.

AUSTIN, B., EMBLEY, T. M. & GOODFELLOW, M. (1983). Selective isolation of *Renibacterium salmoninarum*. *FEMS Microbiology Letters*, **17**, 111–14.

AUSTIN, B. & GOODFELLOW, M. (1979). *Pseudomonas mesophilica*, a new species of pink bacteria isolated from leaf surfaces. *International Journal of Systematic Bacteriology*, **29**, 373–8.

BAHARAEEN, S. & VISHNIAC, H. S. (1982). *Cryptococcus lupi* sp. nov., an Antarctic basidioblastomycete. *International Journal of Systematic Bacteriology*, **32**, 229–32.

BAKER, D., TORREY, J. G. & KIDD, G. H. (1979). Isolation by sucrose-density fractionation and cultivation *in vitro* of actinomycetes from nitrogen-fixing root nodules. *Nature, London*, **281**, 76–8.

BAROSS, J. A. & DEMING, J. W. (1983). Growth of 'black smoker' bacteria at temperature of at least 250 °C. *Nature, London*, **303**, 423–6.

BAROSS, J. A. & MORITA, R. Y. (1978). Life at low temperatures: ecological aspects. In *Microbial Life in Extreme Environments*, ed. D. J. Kushner, pp. 9–72. New York & London: Academic Press.

BAUMANN, P. & BAUMANN, L. (1981). The marine Gram-negative eubacteria:

genera *Photobacterium, Beneckea, Alteromonas, Pseudomonas* and *Alcaligenes*. In *The Prokaryotes. A Handbook on Habitats, Isolation and Identification of Bacteria*, vol. 2, ed. M. P. Starr, H. Stolp, H. G. Trüper, A. Balows & H. G. Schlegel, pp. 1302–31. Berlin, Heidelberg & New York: Springer-Verlag.

BEATON, G. & WESTE, G. (1982). *Banksiamyces* gen. nov. a discomycete on dead *Banksia* cones. *Transactions of the British Mycological Society*, **79**, 271–7.

BECKING, J.-H., (1981). The genus *Frankia*. In *The Prokaryotes. A Handbook on Habitats, Isolation and Identification of Bacteria*, vol. 2, ed. M. P. Starr, H. Stolp, H. G. Trüper, A. Balows & H. G. Schlegel, pp. 1991–2003. Berlin, Heidelberg & New York: Springer-Verlag.

BEECH, F. W. & DAVENPORT, R. R. (1971). Isolation, purification and maintenance of yeasts. In *Methods in Microbiology*, vol. 4, ed. C. Booth, pp. 153–82. New York & London: Academic Press.

BÉRDY, J. (1974). Recent developments of antibiotic research and classification of antibiotics according to chemical structure. *Advances in Applied Microbiology*, **18**, 309–402.

BIEBL, H. & PFENNIG, N. (1981). Isolation of members of the family Rhodospirillaceae. In *The Prokaryotes. A Handbook on Habitats, Isolation and Identification of Bacteria*, vol. 1, ed. M. P. Starr, H. Stolp, H. G. Trüper, A. Balows & H. G. Schlegel, pp. 267–73. Berlin, Heidelberg & New York: Springer-Verlag.

BOOTH, C. (ed.) (1971). *Methods in Microbiology*, vol. 4. New York & London: Academic Press.

BOWEN, G. D. (1980). Misconceptions, concepts and approaches in rhizosphere biology. In *Contemporary Microbial Ecology*, ed. D. C. Ellwood, J. N. Hedges, M. J. Latham, J. M. Lynch & J. H. Slater, pp. 283–304. New York & London: Academic Press.

BRIAN, P. W. (1957). The ecological significance of antibiotic production. In *Microbial Ecology*, ed. R. E. C. Williams & C. C. Spicer, pp. 168–188. Cambridge University Press.

BROCK, T. D., BROCK, K. M., BELLY, R. T. & WEISS, R. L. (1972). *Sulfolobus:* a new genus of sulfur-oxidizing bacteria living at low pH and high temperature. *Archives of Microbiology*, **84**, 54–68.

BULL, A. T. & SLATER, J. H. (ed.) (1982). *Microbial Interactions and Communities*. New York & London: Academic Press.

BURMAN, N. P., OLIVER, C. W. & STEVENS, J. K. (1969). Membrane filtration techniques for the isolation from water of coli-aerogenes, *Escherichia coli*, faecal streptococci, *Clostridium perfringens*, actinomycetes and microfungi. In *Isolation Methods for Microbiologists*, ed. D. A. Shapton & G. W. Gould, pp. 127–34. New York & London: Academic Press.

BUSHELL, M. E. & NISBET, L. J. (1981). A technique for eliminating recurring producers of known metabolites in antibiotic screens. In *Actinomycetes*, ed. K. P. Schaal & G. Pulverer, pp. 507–14. Stuttgart: G. Fischer.

CALLAHAM, D., TREDICI, P. D. & TORREY, J. G. (1978). Isolation and cultivation of the actinomycete causing root nodulation in *Comptonia*. *Science*, **199**, 899–902.

CANALE-PAROLA, E. (1973). Isolation, growth and maintenance of anaerobic free-living spirochetes. In *Methods in Microbiology*, vol. 8, ed. J. R. Norris & D. W. Ribbons, pp. 61–73. New York & London: Academic Press.

CASIDA, E. (1982). *Ensifer adhaerens* gen. nov. sp. nov: a bacterial predator of bacteria in soil. *International Journal of Systematic Bacteriology*, **32**, 339–45.

CLEWELL, D. B. (1981). Plasmids, drug resistance, and gene transfer in the genus *Streptococcus*. *Microbiological Reviews*, **45**, 409–36.

COLLINS, V. G., JONES, J. G., HENDRIE, M. S., SHEWAN, J. M., WYNN-WILLIAMS, D. D. & RHODES, M. E. (1973). Sampling and estimation of bacterial populations in the aquatic environment. In *Sampling: Microbiological Monitoring of Environments*, ed. R. G. Board & D. W. Lovelock, pp. 77–110. New York & London: Academic Press.

COWAN, S. T. (1965). Principles and practice of bacterial taxonomy: a forward look. *Journal of General Microbiology*, **39**, 143–53.

CROMBACH, W. H. J. (1978). *Caseobacter polymorphus* gen. nov. sp. nov. a coryneform bacterium from cheese. *International Journal of Systematic Bacteriology*, **28**, 354–66.

CROSS, T. (1968). Thermophilic actinomycetes. *Journal of Applied Bacteriology*, **31**, 36–53.

CROSS, T. (1981). The monosporic actinomyctes. In *The Prokaryotes. A Handbook on Habitats, Isolation and Identification of Bacteria*, vol. 2, ed. M. P. Starr, H. Stolp, H. G. Trüper, A. Balows & H. G. Schlegel, pp. 2091–102. Berlin, Heidelberg & New York: Springer-Verlag.

CROSS, T. (1982). Actinomycetes: a continuing source of new metabolites. *Developments in Industrial Microbiology*, **23**, 1–18.

DAVIS, R. E., LEE, I.-M. & WORLEY, J. F. (1981). *Spiroplasma floricola*, a new species isolated from surfaces of flowers of the tulip tree, *Liriodendron tulipifera* L. *International Journal of Systematic Bacteriology*, **31**, 456–64.

DEMAIN, A. L. (1981). Applied microbiology: a personal view. In *Essays in Applied Microbiology*, ed. J. R. Norris & M. H. Richmond, pp. 1/1–1/31. New York & Chichester: Wiley.

DE ROSA, M., GAMBACORTA, A. & BU'LOCK, J. D. (1975). Extremely thermophilic acidophilic bacteria convergent with *Sulfolobus acidocaldarius*. *Journal of General Microbiology*, **86**, 156–64.

ERCOLANI, G. L. (1978). *Pseudomonas savastamoi* and other bacteria colonizing the surface of olive leaves in the field. *Journal of General Microbiology*, **109**, 245–57.

EZAKI, N., MIYADOH, S., HISAMATSU, T., KASAI, T. & YAMADA, Y. (1980). BN-183B, a new anti-tumour antibiotic produced by *Pseudomonas*. Taxonomy, isolation, physicochemical and biological properties. *Journal of Antibiotics*, **33**, 213–20.

FELL, J. W. & TALLMAN, A. S. (1980). *Rhodosporidium paludigenum* sp. nov., a basidiomycetous yeast from intertidal waters of South Florida. *International Journal of Systematic Bacteriology*, **30**, 658–9.

FLEMING, I. D., NISBET, L. J. & BREWER, S. J. (1982). Target directed antimicrobial screens. In *Bioactive Microbial Products: Search and Discovery*, ed. J. D. Bu'Lock, L. J. Nisbet & D. J. Winstanley, pp. 107–30. New York & London: Academic Press.

FOSTER, R. C. & ROVIRA, A. D. (1978). The ultra-structure of the rhizosphere of *Trifolium subterraneum* L. In *Microbial Ecology*, ed. M. W. Loutit & J. A. R. Miles, pp. 278–90. Berlin, Heidelberg & New York: Springer-Verlag.

FRANZMANN, P. D. & SKERMAN, V. B. D. (1981). *Agitococcus lubricus* gen. nov. sp. nov., a lipolytic, twitching coccus from freshwater. *International Journal of Systematic Bacteriology*, **31**, 177–83.

GARETH JONES, E. B. (1971). Aquatic fungi. In *Methods in Microbiology*, vol. 4, ed. C. Booth, pp. 335–66. New York & London: Academic Press.

GARLICK, S., OREN, A. & PADAN, E. (1977). Occurrence of facultative anoxygenic photosynthesis among filamentous and unicellular cyanobacteria. *Journal of Bacteriology*, **129**, 623–9.

GAUTHIER, M. J. & BREITTMAYER, V. A. (1979). A new antibiotic-producing bacterium from sea water, *Alteromonas aurantia*, new species. *Journal of Antibiotics*, **29**, 366–72.

GIBBONS, N. E. & MURRAY, R. G. E. (1978). Proposals concerning the higher taxa of bacteria. *International Journal of Systematic Bacteriology*, **28**, 1–6.

GOCHNAUER, M. B., LEPPARD, G. G., KOMARATAT, P., KATES, M., NOVITSKY, T. & KUSHNER, D. J. (1975). Isolation and characterization of *Actinopolyspora halophila*, gen. et sp. nov., an extremely halophilic actinomycete. *Canadian Journal of Microbiology*, **21**, 1500–11.

GOODFELLOW, M., HILL, I. R. & GRAY, T. R. G. (1968). Bacteria in a pine forest soil. In *The Ecology of Soil Bacteria*, ed. T. R. G. Gray & D. Parkinson, pp. 500–15. Liverpool University Press.

GOTTLIEB, D. (1976). The production and role of antibiotics in soil. *Journal of Antibiotics*, **29**, 987–1000.

GRANT, W. D. & TINDALL, B. J. (1980). The isolation of alkophilic bacteria. In *Microbial Growth and Survival in Extremes of Environment*, ed. G. W. Gould & J. E. L. Curry, p. 27. New York & London: Academic Press.

GRAY, T. R. G. & WILLIAMS, S. T. (1971). *Soil Micro-organisms*. Edinburgh: Oliver & Boyd.

GUERINOT, M. L., WEST, P. A., LEE, J. V. & COLWELL, R. R. (1982). *Vibrio diazotrophicus* sp. nov., a marine nitrogen-fixing bacterium. *International Journal of Systematic Bacteriology*, **32**, 350–7.

HARRISON, A. P. (1981). *Acidiphilium cryptum* gen. nov., sp. nov., heterotrophic bacterium from acidic mineral environments. *International Journal of Systematic Bacteriology*, **31**, 327–32.

HEDÉN, C.-G., ILLÉNI, T. & KÜHN, I. (1976). Mechanized identification of micro-organisms. In *Methods in Microbiology*, vol. 9, ed. J. R. Norris, pp. 15–28. New York & London: Academic Press.

HEDGER, J. N., SAMSON, R. A. & BASUKI, T. (1982). *Scytalidium indonesiacum*, a new thermophilous hyphomycete from Indonesia. *Transactions of the British Mycological Society*, **78**, 364–6.

HIGGENS, C. E. & KASTNER, R. E. (1971). *Streptomyces clavuligerus* sp. nov. a β-lactam antibiotic producer. *International Journal of Systematic Bacteriology*, **21**, 326–31.

HILL, L. R. (1974). Theoretical aspects of numerical identification. *International Journal of Systematic Bacteriology*, **24**, 494–9.

HILL, P. (1972). The production of penicillins in soils and seeds by *Penicillium chrysogenum* and the role of penicillin β-lactamase in the ecology of soil bacillus. *Journal of General Microbiology*, **70**, 243–52.

HOPWOOD, D. A. (1981). Genetic studies of antibiotics and other secondary metabolites. In *Genetics as a Tool in Microbiology*, ed. S. W. Glover & D. A. Hopwood, pp. 187–218. Cambridge University Press.

HOPWOOD, D. A. & CHATER, K. F. (1980). Fresh approaches to antibiotic production. *Philosophical Transactions of the Royal Society of London, Series B*, **290**, 313–28.

HORAN, A. C. & BRODSKY, B. C. (1982). A novel antibiotic-producing *Actinomadura, Actinomadura kijaniata* sp. nov. *International Journal of Systematic Bacteriology*, **32**, 195–200.

HOTTA, K., OKAMI, Y. & UMEZAWA, H. (1980). An actinomycete isolated from a marine environment. II. Possible involvement of plasmid in istamycin production. *Journal of Antibiotics*, **33**, 1510–14.

HSU, S. C. & LOCKWOOD, J. L. (1975). Powdered chitin as a selective medium for enumeration of actinomycetes in water and soil. *Applied Microbiology*, **29**, 422–6.

IMHOFF, J. F. & TRÜPER, H. G. (1981). *Ectothiorhodospira abdelmalekii* sp. nov., a new halophilic and alkaliphilic phototrophic bacterium. *Zentralblatt für Bakteri-*

ologie, Parasitenkunde, Infektionskrankheiten und Hygiene, Abteilung 1, Originale, **C2**, 228–34.

ITOH, Y., ENOKITA, R., OSAZAKI, T., IWADO, S., TORIKATA, A., HANEISHI, T. & ARAI, M. (1981). Candiplanecin, a new antibiotic from *Ampullariella regularis* subsp. *mannitophila*, a new sub-species. I. Taxonomy of producing organism and fermentation. *Journal of Antibiotics*, **34**, 929–33.

IWASAKI, A., ITOH, H. & MORI, T. (1979). A new broad-spectrum amino glycoside antibiotic complex sporaricin. II. Taxonomic studies on the sporaricin-producing strain *Saccharopolyspora hirsuta* subsp. *kobensis* new subsp. *Journal of Antibiotics*, **32**, 180–6.

IWASAKI, A., ITOH, H. & MORI, T. (1981). *Streptomyces annanensis* sp. nov. *International Journal of Systematic Bacteriology*, **31**, 280–4.

JANNASCH, H. W., WIRSEN, C. O. & WINGET, C. L. (1973). A bacteriological pressure-retaining deep sea sampler and culture vessel. *Deep Sea Research and Oceanographic Abstracts*, **20**, 661–4.

JETER, R. M. & INGRAHAM, J. L. (1981). The denitrifying prokaryotes. In *The Prokaryotes. A Handbook on Habitats, Isolation and Identification of Bacteria*, vol. 1, ed. M. P. Starr, H. Stolp, H. G. Trüper, A. Balows & H. G. Schlegel, pp. 913–25. Berlin, Heidelberg & New York: Springer-Verlag.

JONES, D. & SNEATH, P. H. A. (1970). Genetic transfer and bacterial taxonomy. *Bacteriological Reviews*, **34**, 40–81.

KATZ, E. & DEMAIN, C. (1977). The peptide antibiotics of *Bacillus:* chemistry, biogenesis and possible functions. *Bacteriological Reviews*, **41**, 449–74.

KAZDA, J. (1980). *Mycobacterium sphagni* sp. nov. *International Journal of Systematic Bacteriology*, **30**, 77–81.

KHAN, M. R. & WILLIAMS, S. T. (1974). Studies on the ecology of actinomycetes in soil. VIII. Distribution and characteristics of acidophilic actinomycetes. *Soil Biology and Biochemistry*, **7**, 345–8.

KINTAKA, K., HAIBARA, K., ASAI, M. & IMADA, A. (1981*a*). Iso-sulfazecin, a new beta-lactam antibiotic produced by an acidophilic pseudomonad, *Pseudomonas mesoacidophila* new species. Fermentation, isolation and characterization. *Journal of Antibiotics*, **34**, 1081–9.

KINTAKA, K., KITANO, K., MOZAKI, Y., KAWASHIMA, F., IMADA, A., NAKAO, Y. & YONEDA, M. (1981*b*). Sulfazecin, a novel beta-lactam antibiotic of bacterial origin. Discovery, fermentation and biological characterization. *Journal of Fermentation Technology*, **59**, 263–8.

KLOOS, W. E. & SCHLEIFER, K. H. (1983). *Staphylococcus auricularis* sp. nov.: an inhabitant of the human external ear. *International Journal of Systematic Bacteriology*, **33**, 9–14.

KRISS, A. E. (1963). *Marine Microbiology: Deep Sea*. Edinburgh: Oliver & Boyd.

KÜSTER, E. & WILLIAMS, S. T. (1964). Selection of media for isolation of streptomycetes. *Nature, London*, **202**, 928–9.

LACEY, J. (1978). Ecology of actinomycetes in fodders and related substrates. In *Nocardia and Streptomyces*, ed. M. Mordarski, W. Kurylowicz & J. Jeljaszewicz, pp. 161–72. Stuttgart & New York: G. Fischer.

LANNING, S. & WILLIAMS, S. T. (1982). Methods for the direct isolation and enumeration of actinophages in soil. *Journal of General Microbiology*, **128**, 2063–71.

LARSEN, H. (1981). The Family Halobacteriaceae. In *The Prokaryotes. A Handbook on Habitats, Isolation and Identification of Bacteria*, vol. 1, ed. M. P. Starr, H. Stolp, H. G. Trüper, A. Balows & H. G. Schlegel, pp. 985–94. Berlin, Heidelberg & New York: Springer-Verlag.

LEE, J. D. & KOMOGATA, K. (1980). *Pichia cellobiosa, Candida cariosilignicola* and

Candida succiphila, new species of methanol-assimilating yeasts. *International Journal of Systematic Bacteriology*, **30**, 514–19.

LEE, S. Y., MABEE, M. S. & JANGAARD, N. O. (1978). *Pectinatus*, a new genus of the family Bacteroidaceae. *International Journal of Systematic Bacteriology*, **28**, 582–94.

LINGAPPA, Y. & LOCKWOOD, J. L. (1962). Chitin media for selective isolation and culture of actinomycetes. *Phytopathology*, **52**, 317–23.

McDADE, J. E., BRENNER, D. E. & BOZEMAN, F. M. (1979). Legionnaires' disease bacterium isolated in 1947. *Annals of Internal Medicine*, **90**, 659–61.

McDADE, J. E., SHEPARD, C. C., FRASER, D. W., TSAI, T. R., REDUS, M. A. & DOWDLE, W. R. (1977). Legionnaires' disease. Isolation of a bacterium and demonstration of its role in other respiratory disease. *New England Journal of Medicine*, **297**, 1197–203.

MADDEN, R. H., BRYDER, M. J. & POOLE, N. J. (1982). Isolation and characterization of an anaerobic, cellulolytic bacterium, *Clostridium papyrosolvens* sp. nov. *International Journal of Systematic Bacteriology*, **32**, 87–91.

MAKKAR, N. S. & CROSS, T. (1982). Actinoplanetes in soil and on plant litter from freshwater habitats. *Journal of Applied Bacteriology*, **52**, 209–18.

MARATEA, D. & BLAKEMORE, R. P. (1981). *Aquaspirillum magnetotacticum* sp. nov., a magnetic spirillum. *International Journal of Systematic Bacteriology*, **31**, 452–5.

MAYFIELD, C. I. (1969). A study of the behaviour of a successful soil streptomycete. PhD thesis, University of Liverpool.

MAYFIELD, C. I., WILLIAMS, S. T., RUDDICK, S. M. & HATFIELD, H. L. (1972). Studies on the ecology of actinomycetes in soil. IV. Observation on the form and growth of streptomycetes in soil. *Soil Biology and Biochemistry*, **4**, 79–91.

MIRANDA, M., HOLZSCHU, D. L., PHAFF, H. J. & STARMER, W. T. (1982). *Pichia mexicana*, a new heterothallic yeast from cereoid cacti in the North American Sonoran Desert. *International Journal of Bacteriology*, **32**, 101–7.

MIYADOH, S., SHOMURA, T., ITO, T. & NIIDA, T. (1983). *Streptomyces sulfonofaciens* sp. nov. *International Journal of Systematic Bacteriology*, **33**, 321–4.

MULDER, E. G. & DEINEMA, M. H. (1981). The sheathed bacteria. In *The Prokaryotes. A Handbook on Habitats, Isolation and Identification of Bacteria*, vol. 1, ed. M. P. Starr, H. Stolp, H. G. Trüper, A. Balows & H. G. Schlegel, pp. 425–40. Berlin, Heidelberg & New York: Springer-Verlag.

NAKAMURA, L. K. (1981). *Lactobacillus amylovorus*, a new starch-hydrolyzing species from cattle waste–corn fermentations. *International Journal of Systematic Bacteriology*, **31**, 56–63.

NAWAWI, A. & WEBSTER, J. (1982). *Phalangispora constricta* gen. et sp. nov., a sporodochial hyphomycete with branched conidia. *Transactions of the British Mycological Society*, **79**, 65–8.

NIELLEMAN, S. L. (1973). Pharmacologically active agents from microbial sources. In *CRC Handbook of Microbiology*, vol. 3, *Microbial Products*, ed. A. I. Laskin & H. A. Lechevalier, pp. 999–1006. Cleveland, Ohio: CRC Press.

NIKITIN, D. I. (1973). Direct electron microscopic techniques for the observation of micro-organisms in soil. In *Modern Methods in the Study of Microbial Ecology*, ed. T. Rosswall, pp. 85–92. Swedish Natural Science Research Council.

NOBLE, W. C. (1981). *Microbiology of Human Skin*. London: Lloyd-Luke.

NONOMURA, H. & OHARA, Y. (1969). Distribution of actinomycetes in soil. VI. A culture method effective for both preferential isolation and enumeration of *Microbispora* and *Streptosporangium* strains in soil: I. *Journal of Fermentation Technology*, **47**, 463–9.

NORRIS, J. R. (1980). Introduction. In *Microbiological Classification and Identifica-*

tion, ed. M. Goodfellow & R. G. Board, pp. 1–10. New York & London: Academic Press.

NORRIS, J. R., BERKELEY, R. C. W., LOGAN, N. & O'DONNELL, A. G. (1981). The genera *Bacillus* and *Sporolactobacillus*. In *The Prokaryotes. A Handbook on Habitats, Isolation and Identification of Bacteria*, vol. 2, ed. M. P. Starr, H. Stolp, H. G. Trüper, A. Balows & H. G. Schlegel, pp. 1711–42. Berlin, Heidelberg & New York: Springer-Verlag.

OBERWINKLER, F. & BANDONI, R. (1983). *Trimorphomyces*: a new genus in the Tremellaceae. *Systematic and Applied Microbiology*, **4**, 105–13.

OKAMI, Y. & OKAZAKI, T. (1972). Studies on marine micro-organisms. I. Isolation from the Japan Sea. *Journal of Antibiotics*, **25**, 456–60.

ORCHARD, V. A. & GOODFELLOW, M. (1974). The selective isolation of nocardia from soil using antibiotics. *Journal of General Microbiology*, **85**, 160–2.

ORCHARD, V. A., GOODFELLOW, M. & WILLIAMS, S. T. (1977). Selective isolation and occurrence of nocardiae in soil. *Soil Biology and Biochemistry*, **9**, 233–8.

OREN, A. (1983). *Halobacterium sodomense* sp. nov., a Dead Sea halobacterium with an extremely high magnesium requirement. *International Journal of Systematic Bacteriology*, **33**, 381–6.

PARBERY, D. G. (1967). Isolation of the kerosene fungus, *Cladosporium resinae*, from Australian soil. *Transactions of the British Mycological Society*, **50**, 682–5.

PEARSON, H. W., HOWSLEY, R. & WILLIAMS, S. T. (1982). A study of nitrogenase activity in *Mycoplana* species and free-living actinomycetes. *Journal of General Microbiology*, **128**, 2073–80.

PERLMAN, D. (1973). Compounds produced by industrial fermentation. In *CRC Handbook of Microbiology*, vol. 3, *Microbial Products*, ed. A. I. Laskin & H. A. Lechevalier, pp. 1007–11. Cleveland, Ohio: CRC Press.

PERLMAN, D. (1977). The roles of the *Journal of Antibiotics* in determining the future of antibiotic research. *Journal of Antibiotics*, **30**, Supplement, S133–S137.

PFENNIG, N. & TRÜPER, H. G. (1981). Isolation of members of the families Chromatiaceae and Chlorobiaceae. In *The Prokaryotes. A Handbook on Habitats, Isolation and Identification of Bacteria*, vol. 1, ed. M. P. Starr, H. Stolp, H. G. Trüper, A. Balows & H. G. Schlegel, pp. 279–89. Berlin, Heidelberg & New York: Springer-Verlag.

PIERSON, B. K. & CASTENHOLZ, R. W. (1974). A phototrophic gliding filamentous bacterium of hot springs, *Chloroflexus aurantiacus* gen. and sp. nov. *Archives of Microbiology*, **100**, 5–24.

PORTAELS, F., GOODFELLOW, M., MINNIKIN, D. E., MINNIKIN, S. M. & HUTCHINSON, I. G. (1982). *Nocardia*-like mycobacteria isolated in natural habitats in Zaïre. *Annales de la Société belge de médecine tropicale*, **61**, 477–87.

PRIDHAM, T. G. & TRESNER, H. D. (1974). *Streptomyces* Waksman and Henrici 1943, 339. In *Bergey's Manual of Determinative Bacteriology*, 8th edn, ed. R. E. Buchanan & N. E. Gibbons, pp. 748–829. Baltimore: Williams & Wilkins.

REANNEY, D. C., GOWLAND, P. C. & SLATER, J. H. (1983). Genetic interactions among microbial communities. In *Microbes in their Natural Environments*, ed. J. H. Slater, R. Whittenbury & J. W. T. Wimpenny, pp. 379–421. Cambridge University Press.

REICHENBACH, H. & DWORKIN, M. (1981). Introduction to the gliding bacteria. In *The Prokaryotes. A Handbook on Habitats, Isolation and Identification of Bacteria*, vol. 1, ed. M. P. Starr, H. Stolp, H. G. Trüper, A. Balows & H. G. Schlegel, pp. 315–27. Berlin, Heidelberg & New York: Springer-Verlag.

RIPPKA, R., DERUELLES, J., WATERBURY, J. B., HERDMAN, M. & STANIER, R. Y. (1979). Generic assignments, strain histories and properties of pure cultures of cyanobacteria. *Journal of General Microbiology*, **111**, 1–61.

RIPPKA, R., WATERBURY, J. B. & STANIER, R. Y. (1981a). Isolation and purification of cyanobacteria: some general principles. In *The Prokaryotes. A Handbook on Habitats, Isolation and Identification of Bacteria*, vol. 1, ed. M. P. Starr, H. Stolp, H. G. Trüper, A. Balows & H. G. Schlegel, pp. 212–20. Berlin, Heidelberg & New York: Springer-Verlag.

RIPPKA, R., WATERBURY, J. B. & STANIER, R. Y. (1981b). Provisional generic assignments for cyanobacteria in pure culture. In *The Prokaryotes. A Handbook on Habitats, Isolation and Identification of Bacteria*, vol. 1, ed. M. P. Starr, H. Stolp, H. G. Trüper, A. Balows & H. G. Schlegel, pp. 247–56. Berlin, Heidelberg & New York: Springer-Verlag.

RIVARD, C. J. & SMITH, P. H. (1982). Isolation and characterization of a thermophilic marine methanogenic bacterium, *Methanogenium thermophilicum* sp. nov. *International Journal of Systematic Bacteriology*, **32**, 430–6.

SANGKHOBOL, V. & SKERMAN, V. B. D. (1981). *Chitinophaga*, a new genus of chitinolytic myxobacteria. *International Journal of Systematic Bacteriology*, **31**, 285–93.

SATO, K. (1978). Bacteriochlorophyll formation by facultative methylotrophs, *Protaminobacter ruber* and *Pseudomonas* AM-1. *FEBS Letters*, **85**, 207–10.

SATOI, S., MUTO, N., HAYASHI, M., FUJII, T. & OTANI, M. (1980). Mycinamicins, new macrolide antibiotics. I. Taxonomy, production, isolation, characterization and properties. *Journal of Antibiotics*, **33**, 364–76.

SCHENK, A. & ARAGNO, M. (1979). *Bacillus schlegelii*, a new species of thermophilic, facultatively chemolithoautotrophic bacterium oxidising molecular hydrogen. *Journal of General Microbiology*, **115**, 333–41.

SCHINK, B., THOMPSON, T. E. & ZEIKUS, J. G. (1982). Characterization of *Propionispira arboris* gen. nov. sp. nov., a nitrogen-fixing anaerobe common to wetwoods of living trees. *Journal of General Microbiology*, **128**, 2771–9.

SCHINK, B. & ZEIKUS, J. G. (1983). *Clostridium thermosulfurogenes* sp. nov., a new thermophile that produces elemental sulphur from thiosulphate. *Journal of General Microbiology*, **129**, 1149–58.

SCHLEGEL, H. G. & JANNASCH, H. W. (1967). Enrichment cultures. *Annual Review of Microbiology*, **21**, 49–70.

SCHLEGEL, H. G. & JANNASCH, H. W. (1981). Prokaryotes and their habitats. In *The Prokaryotes. A Handbook on Habitats, Isolation and Identification of Bacteria*, vol. 1, ed. M. P. Starr, H. Stolp, H. G. Trüper, A. Balows & H. G. Schlegel, pp. 43–82. Berlin, Heidelberg & New York: Springer-Verlag.

SCHLEIFER, K. H. & FISCHER, U. (1982). Description of a new species of the genus *Staphylococcus*: *Staphylococcus carnosus*. *International Journal of Systematic Bacteriology*, **32**, 153–6.

SEELEY, N. D. & PRIMROSE, S. B. (1982). The isolation of bacteriophages from the environment. *Journal of Applied Bacteriology*, **53**, 1–17.

SHIBA, T. & SIMIDU, U. (1982). *Erythrobacter longus* gen. nov. sp. nov., an aerobic bacterium which contains bacteriochlorophyll *a*. *International Journal of Systematic Bacteriology*, **32**, 211–17.

SHOMURA, T., KOJIMA, M., YOSHIDA, J., ITO, M., AMANO, S., TOTSUGAWA, K., NIWA, T., INOUYE, S., ITO, T. & NIIDA, T. (1980). Studies on a new amino glycoside antibiotic, dactimicin 1. Producing organism and fermentation. *Journal of Antibiotics*, **33**, 924–30.

SKERMAN, V. B. D. (1968). A new type of micromanipulator. *Journal of General Microbiology*, **54**, 287–98.

SKERMAN, V. B. D., McGOWAN, V. & SNEATH, P. H. A. (1980). Approved lists of bacterial names. *International Journal of Systematic Bacteriology*, **30**, 225–420.

SKERMAN, V. B. D., SLY, L. I. & WILLIAMSON, M. L. (1983). *Conglomeromonas largomobilis* gen. nov. sp. nov., a sodium-sensitive, mixed-flagellated organism from fresh waters. *International Journal of Systematic Bacteriology*, **33**, 300–8.

SLOTS, J. & POTTS, T. V. (1982). *Fusobacterium simiae*, a new species from monkey dental plaque. *International Journal of Systematic Bacteriology*, **32**, 191–4.

SMITH, D. L. & TRENTINI, W. C. (1972). Enrichment and selective isolation of *Caryophanon latum*. *Canadian Journal of Microbiology*, **18**, 1197–200.

SNEATH, P. H. A. (1977). The maintenance of large numbers of strains of micro-organisms, and the implications for culture collections. *FEMS Microbiology Letters*, **1**, 333–4.

SNEATH, P. H. A. (1978). Identification of micro-organisms. In *Essays in Microbiology*, ed. J. R. Norris & M. H. Richmond, pp. 10/1–10/32. Chichester & New York: Wiley.

SNEATH, P. H. A. (1980). BASIC program for the most diagnostic properties of groups from an identification matrix of percent positive characters. *Computers and Geosciences*, **6**, 21–6.

SNEATH, P. H. A. & SOKAL, R. R. (1973). *Numerical Taxonomy. The Principles and Practice of Numerical Classification*. San Francisco: W. H. Freeman.

SNIPES, K. P. & BIBERSTEIN, E. L. (1982). *Pasteurella testudinis* sp. nov., a parasite of desert tortoises (*Gophorus agassizi*). *International Journal of Systematic Bacteriology*, **32**, 201–10.

SPIRI-NAKAGAWA, P., OIWA, R., TANAKA, Y. & OMURA, S. (1980). The site of inhibition of bacterial cell wall peptidoglycan synthesis by azureomycin B, a new antibiotic. *Journal of Biochemistry (Tokyo)*, **88**, 565–70.

STANIER, R. Y., PFENNIG, N. & TRÜPER, H. G. (1981). Introduction to the phototrophic prokaryotes. In *The Prokaryotes. A Handbook on Habitats, Isolation and Identification of Bacteria*, vol. 1, ed. M. P. Starr, H. Stolp, H. G. Trüper, A. Balows & H. G. Schlegel, pp. 197–211. Berlin, Heidelberg & New York: Springer-Verlag.

STARR, M. P., STOLP, H., TRÜPER, H. G., BALOWS, A. & SCHLEGEL, H. G. (1981). *The Prokaryotes. A Handbook on Habitats, Isolation and Identification of Bacteria*, vols. 1 and 2. Berlin, Heidelberg & New York: Springer-Verlag.

STETTER, K. O., THOMM, M., WINTER, G., WILDGRUBER, G., HUBER, H., ZILLIG, W., JANEKOVIC, D., KONIG, H., PALM, P. & WUNDERL, S. (1981). *Methanothermus fervidus*, sp. nov., a novel extremely thermophilic methanogen isolated from an Icelandic hot spring. *Zentralblatt für Bakteriologie, Parasitenkunde, Infektionskrankheiten und Hygiene, Abteilung 1, Originale*, **C2**, 166–78.

STOLP, H. & GADKARI, D. (1981). Nonpathogenic members of the genus *Pseudomonas*. In *The Prokaryotes. A Handbook on Habitats, Isolation and Identification of Bacteria*, vol. 1, ed. M. P. Starr, H. Stolp, H. G. Trüper, A. Balows & H. G. Schlegel, pp. 719–41. Berlin. Heidelberg & New York: Springer-Verlag.

STOLP, H. & STARR, M. P. (1981). Principles of isolation, cultivation and conservation of Bacteria, vol. 1, ed. M. P. Starr, H. Stolp, H. G. Trüper, A. Balows & *Identification of Bacteria*, vol. 1, ed. M. P. Starr, H. Stolp, H. G. Trüper, A. Balows & H. G. Schlegel, pp. 135–75. Berlin, Heidelberg & New York. Springer-Verlag.

TAKAHASHI, K. YOSHIDA, M., TOMITA, F. & SHIRAHATA, K. (1981). Gilvocarcins, new anti-tumour antibiotics. II. Structural elucidation. *Journal of Antibiotics*, **34**, 271–5.

TENDLER, M. D. & BURKHOLDER, P. R. (1961). Studies on the thermophilic actinomycetes. I. Methods of cultivation. *Applied Microbiology*, **9**, 394–9.

TOMITA, K., HOSHINO, Y., SASAHIRA, T. & KAWAGUCHI, H. (1980). BBM-928, a new anti-tumour antibiotic complex. II. Taxonomic studies on the producing organism. *Journal of Antibiotics*, **33**, 1098–102.

TREJO, W. H. (1970). An evaluation of some concepts and criteria used in the speciation of streptomycetes. *Transactions of the New York Academy of Sciences*, **32**, 989–97.

TRÜPER, H. G. & KRÄMER, J. (1981). Principles of characterization and identification of prokaryotes. In *The Prokaryotes. A Handbook on Habitats, Isolation and Identification of Bacteria*, vol. 1, ed. M. P. Starr, H. Stolp, H. G. Trüper, A. Balows & H. G. Schlegel, pp. 176–93. Berlin, Heidelberg & New York: Springer-Verlag.

TSUKAMURA, M. (1972). An improved selective medium for atypical mycobacteria. *Japanese Journal of Microbiology*, **16**, 243–6.

TULLY, J. G. & WHITCOMB, R. F. (1981). The genus *Spiroplasma*. In *Prokaryotes. A Handbook on Habitats, Isolation and Identification of Bacteria*, vol. 2, ed. M. P. Starr, H. Stolp, H. G. Trüper, A. Balows & H. G. Schlegel, pp. 2271–84. Berlin, Heidelberg & New York: Springer-Verlag.

VELDKAMP, H. (1970). Enrichment cultures of prokaryotic organisms. In *Methods in Microbiology*, vol. 3A, ed. J. R. Norris & D. W. Ribbons, pp. 305–61. New York & London: Academic Press.

VICKERS, J. C., WILLIAMS, S. T. & ROSS, G. W. (1983). A taxonomic approach to selective isolation of streptomycetes from soil (in press).

VISHNIAC, H. S. & BAHARAEEN, S. (1982). Five new basidioblastomycetous yeast species segregated from *Cryptococcus vishniacii* emend. auct., an Antarctic yeast species comprising four new varieties. *International Journal of Systematic Bacteriology*, **32**, 437–45.

VISHNIAC, H. S. & HEMPFLING, W. P. (1979). *Cryptococcus vishniacii* sp. nov., an Antarctic yeast. *International Journal of Systematic Bacteriology*, **29**, 153–8.

WAKSMAN, S. A. (1956). The role of antibiotics in natural processes. *Giornale di microbiologia*, **2**, 1–14.

WALSBY, A. E. (1981). Cyanobacteria: planktonic gas-vacuole forms. In *The Prokaryotes. A Handbook on Habitats, Isolation and Identification of Bacteria*, vol. 1, ed. M. P. Starr, H. Stolp, H. G. Trüper, A. Balows & H. G. Schlegel, pp. 224–35. Berlin, Heidelberg & New York: Springer-Verlag.

WARCUP, J. H. (1957). Studies on the occurrence and activity of fungi in a wheat-field soil. *Transactions of the British Mycological Society*, **40**, 237–62.

WARCUP, J. H. & TALBOT, P. H. B. (1962). Ecology and identity of mycelia isolated from soil. *Transactions of the British Mycological Society*, **45**, 495–518.

WATLING, R. (1971). Basidiomycetes: Homobasidiomycetidae. In *Methods in Microbiology*, vol. 4. ed. C. Booth, pp. 219–36. New York & London: Academic Press.

WATSON, S. W., VALOIS, F. W. & WATERBURY, J. B. (1981). The family Nitrobacteracea. In *The Prokaryotes. A Handbook on Habitats, Isolation and Identification of Bacteria*, vol. 1, ed. M. P. Starr, H. Stolp, H. G. Trüper, A. Balows & H. G. Schlegel, pp. 1005–22. Berlin, Heidelberg & New York: Springer-Verlag.

WAYNE, L. G., KRICHEVSKY, E. J., LOVE, L. L., JOHNSON, R. & KRICHEVSKY, M. I. (1980). Taxonomic probability matrix for use with slowly growing mycobacteria. *International Journal of Systematic Bacteriology*, **30**, 528–38.

WHITE, D. C. (1983). Analysis of micro-organisms in terms of quantity and activity in natural environments. In *Microbes in their Natural Environments*, ed. J. H. Slater, R. Whittenbury & J. W. T. Wimpenny, pp. 37–66. Cambridge University Press.

WIEGEL, J. & LJUNGDAHL, L. G. (1981). *Thermoaerobacter ethanolicus* gen. nov. sp. nov., a new extreme thermophilic anaerobic bacterium. *Archives of Microbiology*, **128**, 343–51.

WIESSNER, W. (1981). The family Beggiatoaceae. In *The Prokaryotes. A Handbook* on *Habitats, Isolation and Identification of Bacteria*, vol. 1, ed. M. P. Starr, H. Stolp, H. G. Trüper, A. Balows & H. G. Schlegel, pp. 380–9. Berlin, Heidelberg & New York: Springer-Verlag.

WILLIAMS, S. T. (1982). Are antibiotics produced in soil? *Pedobiologia*, **23**, 427–35.

WILLIAMS, S. T. & CROSS, T. (1971). Actinomycetes. In *Methods in Microbiology*, vol. 4, ed. C. Booth, pp. 295–334. New York & London: Academic Press.

WILLIAMS, S. T. & DAVIES, F. L. (1965). Use of antibiotics for selective isolation and enumeration of actinomycetes in soil. *Journal of General Microbiology*, **38**, 251–61.

WILLIAMS, S. T. & GRAY, T. R. G. (1973). General principles and problems of soil sampling. In *Sampling: Microbiological Monitoring of Environments*, ed. R. G. Board & D. W. Lovelock, pp. 111–22. New York & London: Academic Press.

WILLIAMS, S. T. & KHAN, M. R. (1974). Antibiotics: a soil microbiologist's viewpoint. *Postepy Higieny i Medycyny Doswiadczalnej*, **28**, 395–408.

WILLIAMS, S. T. & WELLINGTON, E. M. H. (1982*a*). Principles and problems of selective isolation of microbes. In *Bioactive Products: Search and Discovery*, ed. J. D. Bu'Lock, L. J. Nisbet & D. J. Winstanley, pp. 9–26. New York & London: Academic Press.

WILLIAMS, S. T. & WELLINGTON, E. M. H. (1982*b*). Actinomycetes. In *Methods of Soil Analysis*, part 2, *Chemical and Microbiological Properties*, 2nd edn, ed. A. L. Page, R. H. Miller & D. R. Keeney, pp. 969–87. Madison, Wisconsin: American Society of Agronomy/Soil Science Society of America.

WILLIAMS, S. T., GOODFELLOW, M., ALDERSON, G., WELLINGTON, E. M. H., SNEATH, P. H. A. & SACKIN, M. J. (1983*a*). Numerical classification of *Streptomyces* and related genera. *Journal of General Microbiology*, **129**, 1743–813.

WILLIAMS, S. T., GOODFELLOW, M., WELLINGTON, E. M. H., VICKERS, J. C., ALDERSON, G., SNEATH, P. H. A., SACKIN, M. J. & MORTIMER, A. M. (1983*b*). A probability matrix for identification of some streptomycetes. *Journal of General Microbiology*, **129**, 1815–30.

YOSHIZUMI, H. (1975). A malo-lactic bacterium and its growth factor. In *Lactic Acid Bacteria in Beverages and Food*, ed. J. G. Carr, C. V. Cutting & G. C. Whiting, pp. 87–102. New York & London: Academic Press.

ZEIKUS, J. G. (1983). Metabolic communication between biodegradative populations in nature. In *Microbes in their Natural Environments*, ed. J. H. Slater, R. Whittenbury & J. W. T. Wimpenny, pp. 67–118. Cambridge University Press.

ZEIKUS, J. G., DAWSON, M. A., THOMPSON, T. E., INGVORSEN, K. & HATCHIKIAN, E. C. (1983). Microbial ecology of volcanic sulphidogenesis: isolation and characterization of *Thermodesulfobacterium commune* gen. nov. and sp. nov. *Journal of General Microbiology*, **129**, 1159–69.

ZILLIG, W., HOLZ, I., JANEKOVIC, D., SCHÄFER, W. & REITER, W. D. (1983). The archaebacterium *Thermococcus celer* represents a novel genus within the thermophilic branch of the archaebacteria. *Systematic and Applied Microbiology*, **4**, 88–94.

ZILLIG, W., STETTER, K. O., SCHÄFER, W., JANEKOVIC, D., WUNDERL, S., HOLZ, I. & PALM, P. (1981). Thermoproteales: a novel type of extremely thermoacidophilic anaerobic archaebacteria isolated from Icelandic sulfataras. *Zentralblatt für Bakteriologie, Parasitenkunde, Infektionskrankheiten und Hygiene, Abteilung 1, Originale*, **C2**, 205–27.

MICROBE CREATION BY GENETIC ENGINEERING

DAVID A. HOPWOOD AND ANDREW W. B. JOHNSTON

John Innes Institute, Colney Lane, Norwich NR4 7UH, UK

The Society for General Microbiology, whose hundredth meeting we are celebrating, was founded in 1945. Soon afterwards, in the first volume of the Society's new journal, Gordon *et al.* (1947) described an improved fermentation vessel in which a selected strain of *Penicillium chrysogenum* yielded about 0.5 g of penicillin per litre. Such yields were already dramatically higher than those obtained with the original strains and growth conditions used for the production of the antibiotic (Abraham, 1974), but now seem paltry compared with titres of at least 50 g per litre produced by current commercial strains of *Penicillium* grown under a modern fermentation regime. The contribution of genetic engineering (in the broad sense of making planned changes in the genotype of an organism by any technique of *in vivo* or *in vitro* recombination) to this dramatic improvement can be summarized in a single word – nil! Strain improvement in *Penicillium* (and indeed in all antibiotic-producing bacteria and fungi) has been achieved essentially by the principle of brute force and ignorance: brute force in the sense that strains have been subjected to successive rounds of intense mutagenesis to generate large populations of genetically altered organisms to be laboriously screened for the enhanced production of antibiotics; and ignorance because the biochemical and genetic changes required to produce such improved strains have been, at best, poorly understood.

In conventional breeding regimes, whether for higher yielding lines of wheat or for faster racehorses, selection is made for the best offspring of sexual crosses between preselected parents. In industrial microbiology the use of sexual recombination to re-assort favourable characters from different parents has been used only to a small extent. Amongst fungi, notable examples are found in the breeding of improved yeast strains for alcohol production, or strains which combine the ability to utilize particular polysaccharides, such as dextrins, derived from wild yeasts with the favourable attributes of flavour found in established beer yeasts (Johnston & Oberman,

1979; Johnston, 1981). Amongst bacteria, the few examples of the use of *in vivo* recombination to develop economically superior strains are in situations in which useful traits are carried naturally by plasmids, so that transfer of whole plasmids from one strain to another, rather than the re-assortment of chromosomal genes, suffices to generate new genotypes. One example, the introduction of the uptake hydrogenase genes into strains of the root nodule bacterium *Rhizobium*, will be considered in more detail later in this paper. Another case concerns the construction of *Pseudomonas* organisms that catabolize diverse components of petroleum from strains each capable of using a different class of hydrocarbon (octane and related paraffins, toluene and xylene, camphor, or naphthalene) by virtue of the possession of a specific type of plasmid (OCT, TOL, CAM or NAH) (Friello, Mylroie & Chakrabarty, 1976).

Why were the enormous possibilities for genetic analysis and strain construction in bacteria and fungi which had been revealed by fundamental research on a few selected laboratory strains in the 1950s and 1960s not seized upon by industrial microbiologists and applied to micro-organisms that make useful products? The answer is probably to be found in the very considerable research effort that would have been required to develop a new organism into a subject for sophisticated *in vivo* genetic manipulation. By 1970, some 310 genes had been located on the genetic map of *E. coli* K-12, as a result of the development of versatile conjugation and transduction systems (Taylor, 1970). Such systems, and such knowledge, could be used to tailor the genotype of the organism for specific purposes to a very considerable degree and one could even isolate pure genes by exploiting these systems in particularly elegant ways (Shapiro *et al.*, 1969), but it had taken many man-millennia of sustained high-quality research to reach this capability. Amongst fungi, the basic genetic architecture of the most-studied organisms – *Neurospora crassa*, *Aspergillus nidulans* and *Saccharomyces cerevisiae* – had been revealed (Fincham & Day, 1971). These examples of 'academic' bacteria and fungi suggested, then, that the tools of *in vivo* genetic analysis and manipulation could be developed for virtually any microbial strain of interest (Hopwood, 1977). However, the large investment of time and money that would have been needed could not, apparently, be justified, even in such 'golden' microbes as *Penicillium chrysogenum* or *Streptomyces aureofaciens*.

The advent of *in vitro* genetic manipulation some ten years ago completely changed the prospects for the rational application of genetics to industrial microbiology, and the first commercial products of this revolution are already with us. Moreover, it is clear that such materials as human insulin and improved vaccines are but the first products of a potentially vast new industry.

The reasons why the *in vitro* genetics has caught on in industry, and in the space of only a few years, while the *in vivo* genetics did not, even over several decades, are not hard to see. Whereas in a purely *in vivo* study every microbial strain must be approached anew, by the isolation of mutations, the development of a system of genetic exchange, and the laborious charting of a linkage map, *in vitro* methods allow breathtaking short-cuts to be taken by such devices as the use of the genetics of a developed strain to isolate and analyse the genes of another. Thus the researcher can realistically start work with hitherto unstudied strains, rather than remaining locked into a single organism. The key is the power of genetic engineering to cross the normal barriers to genetic exchange between unrelated organisms. More than that: it leads directly to the analysis of the structure and control of expression of natural genes at the level of the base sequences (see Reznikoff, this volume); to the possibility of creating artificial, eventually tailor-made genes by chemical synthesis; and so to the possibility of altering the quantity and quality of gene products in precise and potentially predictable ways. In other words it leads to microbe creation.

THE CHOICE OF HOST ORGANISM

The basic requirement for the creation of a new microbe by insertion of artificially manipulated DNA into an existing organism is a suitable host, to be used as recipient for the DNA. Apart from all other considerations, a potential host is only suitable if we have available vectors into which DNA can be inserted *in vitro*, which can be introduced into and replicated within the host, and which will result in expression of the introduced genetic information, either from signals on the foreign DNA compatible with the transcription and translation apparatus of the host, or from signals on the vector already matched to that apparatus. It was therefore inevitable, as soon as the feasibility of *in vitro* joining of unrelated DNA segments was realized – by homopolymer tailing (Jackson, Symons & Berg, 1972; Lobban & Kaiser, 1973) or by the use of type II restriction enzymes (Cohen *et al.*, 1973) – that *E. coli* K-12 would be the first host. Only in *E. coli* was there the right combination of potential plasmid and/or phage vectors with an adequate transformation/transfection system to make the experiments feasible. Once the first successful experiments were made, the system became autocatalytic, with knowledge gained in each round of clonings being related to the enormous store of genetic experience of *E. coli*, thus leading to improvements in the technology. These in turn produced more theoretical insights, generating further technical advances, to produce an unstoppable bandwaggon. Thus, although *E. coli* K-12 had not been chosen rationally as an ideal host for foreign DNA – everything could be traced back to the serendipitous choice of this strain to isolate nutritional mutants in the Stanford Biological Laboratories in the early 1940s (Gray & Tatum, 1944) – there would need to be cogent reasons for choosing alternative hosts for recombinant DNA once the *E. coli* show was on the road. There *are* such reasons and several other organisms have recently been developed as cloning hosts, notably *Bacillus subtilis*, *Streptomyces lividans*, *Saccharomyces cerevisiae*, *Methylophilus methylotrophus* (AS1) and various *Rhizobium* species. Cloning into many other organisms has been reported, or can be expected over the next few years.

The motives for studying 'alternatives to *E. coli*' are varied. One aim has been to compare the suitability of other hosts with that of *E. coli* for the production of foreign proteins such as insulin, interferon or growth hormone. In such a venture one is using the host as a 'test-tube' to express specific genetic information coming exclusively from the introduced DNA; the genotype of the host is relevant only insofar as it bears on the expression, yield, stability, post-translational modification or export of a foreign protein. This being the case, it seems inconceivable that *E. coli* K-12, chosen on completely different criteria, could be *the* 'ideal' host for the purpose – unless, as is not impossible, the power of genetic engineering is so great that any intrinsic shortcomings of the natural host can be put to rights. Much of the thrust to the development of *Bacillus subtilis* and *Saccharomyces cerevisiae* as sophisticated cloning hosts has come from such considerations: further hosts, such as *Streptomyces* and AS1, studied previously for other reasons, are also being evaluated as 'test-tube' hosts, or 'biophores' (a term coined by J. Lederberg to describe a generalized cloning host).

Different considerations apply when the objective is to improve a microbe which is already of importance for commercial reasons and whose desirable attributes are determined by many genes. A good example is provided by the methylotroph AS1 whose metabolism is quite different from that of *E. coli*, being able to grow well on ammonia and methanol. The metabolism of this organism (which has been approved as a suitable protein-rich biomass for animal feed) could be modified in a specific way – by the rational replacement of a gene for glutamate synthase by one for glutamate dehydrogenase – to make ammonia assimilation more efficient (Windass *et al.*, 1980). Clearly, in this case, one needed to clone *into* AS1 rather than *out* into *E. coli*. Another example of a multigenic situation is provided by symbiotic nitrogen-fixation by *Rhizobium*; it would (presumably) be folly to try to create a better bacterial partner for a legume crop by completely re-modelling *E. coli* when one can re-tune *Rhizobium* instead. We shall use a detailed consideration of the *Rhizobium* case, in a later section, as an example of microbe creation by a combination of genetic techniques.

Antibiotic production at first sight appears to represent an intermediate case between the single gene product such as interferon or growth hormone, and the complex multigenic situation of *Rhizobium* or AS1. Antibiotics are metabolites, not direct gene products. A simple molecule such as chloramphenicol, which consists of a single substituted aromatic nucleus, probably requires about seven specific enzyme-catalysed steps to be derived from a generally available metabolite, in this case chorismic acid (Malik, 1983). For an aminoglycoside such as streptomycin, however, the number of steps is probably nearer 30 to convert glucose into the finished antibiotic, which consists of two specialized sugar residues and one substituted cyclohexane moiety (Davies & Yagisawa, 1983). These 'oligogene' products might therefore conceivably be suitable subjects for transplantation into a biophore, especially because the genes for antibiotic biosynthesis in actinomycetes (but not fungi) are often clustered together in presumed operons, making cloning much easier (Hopwood, 1983). The situation is, however, much more complex. The efficient synthesis of antibiotics requires an abundant supply of metabolic precursors, their exact nature depending on the chemistry of the antibiotic: sugars, acetylcoenzyme A, amino acids, etc. These are generally available in micro-organisms, but in the comparatively small amounts required for normal balanced growth. Efficient producers of a particular class of antibiotics must have their primary metabolism specialized and channelled in specific ways to provide the building blocks for the antibiotic in disproportionate amounts (Demain, Aharonowitz & Martin, 1983). Moreover, they usually need specific resistance mechanisms to prevent self-destruction of the host. These considerations make efficient antibiotic production a phenotype dependent on many tens of genes, effectively precluding the creation of useful antibiotic-producing strains of *E. coli, Bacillus subtilis* or yeast. The goal of understanding, and thence manipulating, the genes for antibiotic production provided one of the main spurs to the recent development of endogenous gene cloning systems in *Streptomyces*, making use of protoplast transformation and plasmid and phage vectors engineered from *Streptomyces* replicons and selectable marker genes (Chater *et al.*, 1982; Hopwood & Chater, 1982; Bibb, Chater & Hopwood, 1983). This cloning capability offers considerable prospects for the creation of streptomycetes able to synthesize unnatural antibiotics, as well as organisms producing, for defined reasons, a higher titre of a known compound (Hopwood *et al.*, 1983).

In the rest of this paper we first consider the features of a cloning host to be used to create microbes capable of producing foreign gene products, and then examine in detail a selected example of a complex genetic engineering task: the development of bacterial strains capable of more efficient symbiotic nitrogen fixation. Although the main emphasis is to be *in vitro* genetic manipulation (genetic engineering in the strict sense) in microbe creation, two further considerations are of great importance. One is that genetic engineering is most powerful as a synthetic tool (for microbe creation) when allied to other genetic methods – such as mutant isolation, genetic transposition, and recombination mediated by a variety of means including plasmid transfer, conjugation, transduction or protoplast fusion – rather than being used in isolation. For this reason alone, the 'ideal' biophore will not be a genetically uncharacterized wild-type strain but will be the subject of considerable genetic knowledge and experimentation. The other is that genetic engineering has enormous analytical power and often its immediate benefits will be knowledge about the control of a relevant aspect of the phenotype, rather than a directly useful new microbe.

POTENTIAL BIOPHORES

From amongst the select group of microbial strains with a good background of genetic knowledge, and adequate vectors and delivery systems for the introduction of DNA into the cells, the choice of a suitable host for the production of a particular foreign gene product depends on both technical (engineering) and biological considerations, few of which are understood well enough at the present time for a completely informed choice to be made. Many pointers are discernible, though, and here we review some of these factors; Table 1 analyses some of the most obvious potential hosts in respect of these factors.

Technical factors

Most industrial-scale fermentations have to be cooled to remove excess heat generated both metabolically and, especially, from the mechanical energy of the stirring equipment required to disperse and aerate the culture; this constitutes a considerable cost, which might be reduced if a thermophilic host were used (or perhaps a facultative anaerobe which would not require aeration). There is little published information on attempts to use such hosts. Alternatively, a change in fermenter design can be made, matched to the characteristics of a particular organism, as in the air-lift fermenter used in the continuous culture of the AS1 organism on methanol and ammonia (Smith, 1980).

In a relatively small-scale process generating a high-value product such as human growth hormone (Goeddel et al., 1979b), attempts to reduce the cost of the fermentation medium by choice of a microbial host capable of using an inexpensive medium may be relatively unimportant, whereas for a large-volume, lower cost commodity such as albumin or rennin (Lawn et al., 1981; Emtage et al., 1983; Mellor et al., 1983), a host able to grow on cheaper substrates may be crucial. Moreover, in a large-scale process, steps to increase the efficiency of substrate utilization, even by comparatively small amounts, may be well worth while. Thus, replacement of the glutamine synthetase + glutamate synthase pathway of ammonia assimilation (which uses ATP) in the AS1 organism by the energy-saving glutamate dehydrogenase pathway was a useful application of genetic engineering (Windass et al., 1980).

Much argument surrounds the choice between unicellular and mycelial hosts, and between cloning strategies and hosts that lead to export of the product into the culture medium compared with those that result in retention of the product in the cytoplasm. Cultures of unicellular organisms such as eubacteria and yeasts can in general be grown to high cell density with less input of energy than mycelial

Table 1. *Attributes of some potential hosts for recombinant DNA products*

Organism	Growth form	DNA delivery system	Vectors	RNA intron splicing	Glycosylation	Export to medium	Special features
Escherichia coli	Prokaryotic single cell	Efficient transformation of shocked cells: *in vitro* phage packaging followed by infection	Numerous high- and low-copy plasmids and lambda phage derivatives	No	No	No, but potential for accumulation in periplasmic space	Outer membrane pyrogens need to be rigorously removed from pharmaceutical products
Bacillus subtilis	Prokaryotic single cell	Efficient transformation of competent cells or protoplasts; transfection of protoplasts	Numerous plasmids and several phages	No	No	Yes	Possibility of developmentally regulated expression
Saccharomyces cerevisiae	Eukaryotic single cell	Relatively inefficient transformation of protoplasts (sphaeroplasts); shuttle transfer from *E. coli* normally needed	Integrating (Yip), autonomous (Yep) or minichromosomal (Ycp) vectors, all combined with *E. coli* replicon	Yes, but not faithful to higher eukaryote signals	Yes, but may not conform to higher eukaryotic pattern	Yes	
Methylophilus methylotrophus (ASI)	Prokaryotic single cell	No transformation; conjugal transfer from *E. coli* used	P1 and Q group plasmids	Presumably not	Presumably not	?	Large-scale inexpensive culture on methanol and ammonia
Streptomyces lividans	Prokaryotic mycelium	Efficient transformation or transfection of protoplasts	High- and low-copy plasmids and phage derivatives	Presumably not	Presumably not	Probably	Possibility of developmentally regulated expression
Penicillium chrysogenum	Eukaryotic mycelium	Under development	Under development	?	?	Probably	
Bacillus stearothermophilus	Prokaryotic single cell	?	?	Presumably not	Presumably not	Probably	Thermophilic growth should reduce cooling costs

organisms such as streptomycetes and filamentous fungi, because of the greater resistance to stirring presented by mycelial cultures (this is especially true for fungi with their larger mycelial diameter). On the other hand, mycelial organisms can be harvested more easily – by simple filtration rather than by the more costly centrifugation.

When a desired protein is exported it is automatically separated from most of the proteins of the host, providing a significant purification step compared with a protein retained by the cells; on the other hand an exported protein usually ends up at a lower concentration in a large volume of fluid from which it must be recovered. Whether the product is more or less vulnerable to degradation by host proteases if exported will depend on the abundance and specificity of exported or intracellular proteases.

Biological factors

The literature on the molecular biology of gene cloning for the production of foreign proteins, especially in *E. coli*, is copious and detailed and it would be inappropriate to attempt to review it here. The reader is referred to the excellent recent review by Harris (1983).

It goes without saying that any serious candidate for a biophore must be a host for good cloning vectors. Many vectors can be suitable for laboratory-scale gene cloning, when high product yield and good stability of structure and inheritance of the vector and cloned DNA are relatively unimportant. At the production stage such criteria are crucial and are a major preoccupation of industrial cloners. So many factors bear on stability that few satisfying generalizations are discernible. In principle, a similar yield could be produced from a low-copy gene driven by maximally expressed transcription and translation signals, or from multiple copies less efficiently expressed. Perhaps the most elegant solution to many problems will be a single copy inserted stably into the chromosome via a phage vector capable of lysogenic integration or a 'scaffold' type of delivery system (Young, 1980).

The main expression signals for transcription and translation vary between eukaryotes and prokaryotes, and between members of each class. Thus there are many barriers to the natural expression of genes in foreign hosts. They need not bear strongly on the choice of cloning host because expression vectors can presumably be developed for any host which is a serious candidate as a biophore.

Such vectors may supply transcription signals only, signals for transcription and ribosome binding, or complete signals for transcription and translation, leading to the production of a foreign protein fused to the amino terminus of a host protein (Itakura *et al.*, 1978; Goeddel *et al.*, 1979*a*, *b*). However, the achievement of good expression, by the use of efficient promoters and ribosome binding sites, does involve considerable vector development and much is still to be done in this area, even in *E. coli*. For example the precise spacing between ribosome binding site and the start of translation, and the base composition of the spacer, may have a considerable bearing on expression (Gheysen *et al.*, 1982).

An important and interesting aspect of vector development for fermentation is the use of regulated expression vectors so that product formation can be arranged to occur after the main period of vegetative growth is over, thus avoiding the possible poisoning of the host by toxic products, as well as perhaps minimizing product degradation. Several possibilities are being investigated, such as the use of a sudden temperature shift late in the fermentation in order to activate expression (this presupposes that the associated engineering problems could be solved); a combination of a temperature-sensitive repressor and a temperature-induced amplification of plasmid copy number has been used on a laboratory-scale (Remaut, Tsao & Fiers, 1983). Other possibilities could include a metabolic induction occasioned by a change in nutrient supply.

Such strategies as these are needed in a non-differentiating host such as *E. coli*. Interesting alternative procedures may apply in organisms such as *Bacillus subtilis*, whose normal cycle of morphological differentiation may involve activation of special classes of promoters by a change in RNA polymerase specificity at a late stage in the growth cycle (Losick & Pero, 1981) (the same may be true in *Streptomyces*: J. Westpheling & R. Losick, personal communication). Expression vectors might, in principle, be designed to contain such promoters so that the culture would progress naturally to a production phase as the fermentation proceeded or could be made to respond to an external stimulus to sporulation.

Two other fundamental aspects of gene structure – the presence and nature of introns, and codon usage – have to be approached differently since they are features of the cloned DNA itself rather than of the switching signals. Because the introns of those natural eukaryotic genes which contain them are not removed by RNA processing in any known prokaryote, interest turned to *Saccharo-*

myces in those cases where the hope was to express higher eukaryotic genomic DNA rather than synthetic genes or cDNA clones. The result has been disappointing; it appears that, even though some yeast chromosomal genes (such as that coding for actin: Gallwitz & Sures, 1980) do contain introns, accurate processing of the introns of higher eukaryotic genes cloned in yeast does not occur (Beggs *et al.*, 1980; Langford *et al.*, 1983).

Codon usage was a preoccupation in the first successful attempt to express an artificial gene for a mammalian protein in *E. coli* (Itakura *et al.*, 1978), since large numbers of alternative sequences can code for the same polypeptide, and attempts were made to choose a better than random sequence. There is little published evidence to evaluate such choices for practical cloning purposes in *E. coli*, but the natural level of gene expression does appear to correlate strongly with codon usage in both *E. coli* and yeast (Grosjean & Fiers, 1982; Bennetzen & Hall, 1982). *Streptomyces* could represent an extreme case since the DNA of members of this genus has an unusually high guanine plus cytosine (G + C) content (73% instead of the 50% in *E. coli*). This might be expected to affect codon usage. Thus the first *Streptomyces* gene to be sequenced – the *S. fradiae* neomycin phosphotransferase – uses only 38 out of the 61 possible sense codons (with 97% G + C in the third codon position) (Thompson & Gray, 1983), whereas the corresponding *E. coli* gene of Tn5 uses 56 sense codons, with 63% G + C in the third position (Beck *et al.*, 1982). Even though *Streptomyces lividans* can evidently translate all 61 sense codons (Bibb *et al.*, 1983), the relative abundance of different tRNA species may possibly be a very significant factor in the choice of codons for good expression of a synthetic gene in *Streptomyces*, but this is yet to be determined.

Post-translational events, also, have to be considered in the choice of a suitable biophore. These include possible glycosylation of the protein as well as its export and degradation. No prokaryote proteins are known to be glycosylated and so, presumably, prokaryotes cannot add carbohydrate residues to the products of foreign genes, whereas many eukaryotic proteins carry such residues, often in quite complicated patterns. Presumably, at least in some cases, specific glycosylation is important for the biological properties of the protein, even though fibroblast interferon, normally glycosylated when made by human cells, was biologically active when synthesized in *E. coli* (Taniguchi *et al.*, 1980; Goeddel *et al.*, 1980). *Saccharomyces cerevisiae* is capable of carrying out at least some glycosylation reactions (e.g. Taussig & Carlson, 1983) but the pattern of glycosylation may differ from that of mammalian proteins.

In a gram-negative bacterium such as *E. coli*, proteins may reside in one of several compartments: they may be retained in the cytoplasm or in the cytoplasmic membrane; they may be exported through the cytoplasmic membrane into the periplasmic space; they may become inserted into the outer membrane; or they may cross the outer membrane and appear free in the culture medium. In contrast, in a gram-positive bacterium such as *Bacillus subtilis*, or in yeast, the choice is essentially between retention in the cytoplasm or its membrane, or export to the medium. In most well-studied examples in *E. coli*, *B. subtilis* and yeast, export through the cytoplasmic membrane occurs by virtue of a 'leader' sequence of predominantly hydrophobic amino acid residues which constitutes the amino terminus of the protein as it is synthesized; once the protein has crossed the cytoplasmic membrane the leader sequence is removed by proteolytic cleavage by specific processing enzymes (Davis & Tai, 1980). The leader sequence does not allow export across the outer membrane of *E. coli* (in some of the very few examples of truly exported proteins in *E. coli*, the haemolysins, transport across the outer membrane does not involve protein processing but the activity of specific outer membrane proteins which export the periplasmic form of the protein unchanged: Wagner, Vogel & Goebel, 1983). It follows from these considerations that the choice of host (as well as the cloning strategy to place a suitable leader sequence on the protein) is important when export is to be achieved.

Many foreign proteins are degraded by intracellular or extracellular proteases and various devices have been adopted to decrease degradation, for example by fusion of the desired polypeptide to the amino terminus (often a considerable segment) of a host protein (Goeddel *et al.*, 1979*a*, *b*) or by the use of host strains lacking certain proteases (Hardy, Stahl & Küpper, 1983). Several foreign proteins, when produced in *E. coli* to a level approaching 20% of the total protein content, accumulate as insoluble cytoplasmic inclusion bodies. This may result in reduced proteolytic degradation, but has the disadvantage that it can be very difficult to recover the material as biologically active protein.

We see from this brief survey of the considerations that apply to the choice of host and cloning strategy for creating a microbe which efficiently produces a genetically simple product that, in spite of the very considerable effort (and huge financial support) that has already been invested, a satisfactory understanding of the various factors at work is still some way off. Many more genes need to be cloned into many more hosts before the subject progresses from its present rather empirical state to one of rational choice. In the following section we move to a multifactorial phenotype, very much more complex than that of a single gene product. Here the main emphasis is on the use of the various techniques of genetic analysis and manipulation for understanding the biological situation in order to begin to change it.

SYMBIOTIC NITROGEN FIXATION: A MULTIDIMENSIONAL PROBLEM IN MICROBE CREATION

This topic of biological nitrogen fixation offers two quite different approaches to the speculating genetic engineer. One possibility would be the introduction of nitrogen-fixing ability into species which do not normally possess it; depending on the scale of one's ambition the new host could be a green plant (a problem outside the scope of this volume) or a bacterium that inhabits the roots of a crop plant. The second approach would involve the introduction of favourable traits into bacteria which already have the capacity to fix nitrogen and which may also have evolved the ability to interact in a close symbiotic relationship with plants. We shall consider the ways in which such goals might be reached and the problems likely to be encountered. Much of the impetus for such schemes stems from a study of the genetics and molecular biology of the nitrogen fixation (*nif*) genes in the bacterium *Klebsiella pneumoniae*, which fixes nitrogen in the free-living state. Although it seems unlikely that *K. pneumoniae* itself will ever be used to increase the levels of biological nitrogen fixation in agriculture, the academic work done on this organism has generated detailed information on the organization, control and function of the *nif* genes; this information has already been exploited directly in analysing the analogous genes of other nitrogen-fixing organisms such as *Azotobacter*, *Anabaena* and *Rhizobium*. It therefore provides a fine example of the use of *in vivo* and *in vitro* genetics in the analysis of a relatively complex biological phenomenon.

The nif *region of* K. pneumoniae

From mapping studies using P1 transduction it was shown that the *nif* genes of *K. pneumoniae* are in a cluster close to the *hisD* gene (Streicher, Gurney & Valentine, 1971). Since then a whole battery of techniques of molecular genetics has been applied to this system. Such techniques include the isolation of F-primes (Dixon & Postgate, 1972) to allow the complementation analyses to be done and the use not only of chemical mutagens but also of transposons and bacteriophage Mu as insertion mutagens which have facilitated the physical mapping of mutant alleles. More recently, the region has been cloned and mapped with restriction enzymes; gradually a matching of the physical and genetic maps has been achieved (Reidel, Ausubel & Cannon, 1979; Kennedy *et al.*, 1981).

At present, particular interest is focused on the means by which the *nif* genes are regulated. A very powerful tool in such studies has been the use of *lac* fusions generated *in vitro* in which the state of expression of the gene is determined by measuring β-galactosidase activity rather than by attempting to estimate the concentration of the native gene product. Such *lac* fusions with the different *nif* genes have been obtained either through the use of Mu [(lac)amp] (Casadaban & Cohen, 1979) insertions (Dixon *et al.*, 1980) or by the *in vitro* construction of *lac* fusions (e.g. Buchanan-Wollaston *et al.*, 1981).

From Table 2 it can be seen that 17 *nif* genes have been identified, arranged in seven separate transcriptional units. The functions of several but not all of these genes have been identified. The nitrogenase genes are subject to a precise regulation. In the presence of ammonia or oxygen (which inactivates nitrogenase irreversibly) none of the seven *nif* operons is transcribed; in the absence of ammonia and oxygen all seven are actively expressed. This control is determined at two levels. Within the *nif* gene complex two genes in the same operon – *nifA* and *nifL* – code for regulatory proteins; the *nifA* gene product acts as a positive regulator required for the induction of all the *nif* promoters (Buchanan-Wollaston *et al.*, 1981; Hill *et al.*, 1981), while the *nifL* product, in the presence of ammonia or oxygen, prevents transcription of the *nif* genes. The *nif* genes are also subject to another level of control, mediated by *ntrC*, which acts on several complexes of genes (e.g. for glutamine synthetase, proline and histidine utilization) that are involved in other aspects of nitrogen metabolism. The

Table 2. *Identification and functions of the* K. pneumoniae nif *genes*

Gene	Function
J	Electron transport to nitrogenase
H	Structural gene for nitrogenase reduction
D	Structural gene for nitrogenase α subunit
K	Structural gene for nitrogenase β subunit
Y	Unknown
E	Required for iron–molybdenum cofactor activity
N	Required for iron–molybdenum cofactor activity
X	Unknown
U	Unknown
S	Unknown
V	Affects substrate specificity of nitrogenase
M	Activation of nitrogenase reductase
F	Electron transport to nitrogenase
L	Negative regulator of other *nif* genes
A	Positive regulator of other *nif* genes
B	Required for iron–molybdenum cofactor activity
Q	Unknown

From Kennedy *et al.* (1981).
Genes in groups are in the same transcriptional units. The genes are presented in sequence, *nifJ* being to the right of the *nif* cluster as it is normally represented.

presence of high levels of ammonia prevents the expression of *ntrC*, whose product is a positively acting regulatory protein required for transcription of the *nifLA* operon: without the *nifA* product, no other *nif* genes are transcribed (Drummond *et al.*, 1983).

Transfer of the nif *genes to other bacteria*

This fine-scale analysis of the *Klebsiella nif* region has led to the possibility that the *nif* genes might usefully be transferred, as a regulated unit, into non-fixing organisms. Potential hosts for such genes would be bacteria, such as species of *Pseudomonas* and *Bacillus*, that colonize the roots of crop plants. For both of these genera plasmid cloning vectors exist. Plasmids of the P1 and Q incompatibility groups can replicate in and be transferred between *E. coli* or *Pseudomonas*. Several suitable plasmids and cosmids have been derived, by *in vitro* manipulation, from the Q-group plasmid RSF1010 (Bagdasarian *et al.*, 1981) and from the P1-group plasmid RK2 (Ditta *et al.*, 1980; Friedman *et al.*, 1982). Using these vectors, the primary genetic manipulations can be done in *E. coli* and the resultant recombinant plasmids transferred by conjugation into *Pseudomonas* or other gram-negative species (no good DNA uptake systems are available for many such strains). Shuttle vectors for gene transfer between *E. coli* and *Bacillus* are also available (Ehrlich, Niaudet & Michel, 1982), but many gram-positive bacteria can be transformed directly, making shuttle vectors less crucial.

There would appear, then, to be no great difficulties in transferring the cloned *K. pneumoniae nif* genes into novel hosts. Will they be expressed? When a plasmid containing the whole *nif* region of *Klebsiella* was transferred into Nif⁻ mutants of *Azotobacter vinelandii*, a Nif⁺ phenotype resulted, showing that the *K. pneumoniae nif* genes could function in this background (Cannon & Postgate, 1976). However, when the same plasmid was transferred to *Rhizobium* or to *Agrobacterium*, no nitrogenase activity was detected (Dixon, Cannon & Kondorosi, 1976).

Even if the *K. pneumoniae nif* genes are found not to be naturally expressed in a particular bacterial host the detailed analysis of the promoters of the different *nif* operons and their regulation has opened the possibility of putting the *nif* genes under the control of promoters which function in *Pseudomonas*, *Bacillus* or any other chosen bacterial host, though the engineering for expression of seven independent operons would be quite a *tour de force*. Such an approach would also allow the construction of strains capable of expressing the *nif* genes even in the presence of ammonia, an attribute that might be desirable under field conditions. Unfortunately, other problems must be overcome before any constructed strains will be valuable to the farmer.

Oxygen protection

Nitrogenase is irreversibly inactivated by oxygen and many naturally occurring nitrogen-fixing microbes have evolved complex mechanisms that protect nitrogenase from this attack. These range from the development of relatively oxygen-free heterocysts in some cyanobacteria to the synthesis of leghaemoglobin in legume root nodules, which buffers the oxygen tension surrounding the nitrogen-fixing *Rhizobium* bacteroids. *Azotobacter* synthesizes a protein which protects nitrogenase from the irreversible effects of oxygen by causing the enzyme molecules to form a polymeric complex (Haaker & Veeger, 1977); however, this complex cannot fix nitrogen, so this system is likely to be of limited value in protecting nitrogenase when it has been introduced into a new host.

Energy cost

A further problem concerns the energy cost of nitrogen fixation. About 20 moles of ATP are believed to be required for the reduction of 1 mole of nitrogen (Daesch & Mortenson, 1968). Thus, the survival of an engineered bacterium in competition with members of the rhizosphere microflora depends on satisfaction of this extra drain on its resources. Since the immediate rhizosphere of plants is not normally deficient in fixed nitrogen, severe problems may impede establishment of the novel bacterial strain around plant roots if its competitors can obtain organic nitrogen by leakage from the roots. Success may therefore depend less on one's ability to clone and achieve expression of the *nif* genes in a novel host than on management of the constructed organism when it is liberated into the field. The limiting factor is a need for much greater understanding of the ecology of soil and rhizosphere bacteria so that we may in turn be able to construct strains that are more competitive or which are more efficient in energy consumption.

In the hypothetical case described above there is also the question of how best to make available to the plant the ammonia that is fixed. Normally, nitrogen-fixing bacteria assimilate all the nitrogen that they fix and no ammonia is released, but certain mutants of *K. pneumoniae* export ammonia. Such a trait is, however, likely to deal another severe blow to fitness; moreover, it is hard to see how the ammonia could be channelled to the plant rather than being assimilated by other members of the microflora.

The Rhizobium symbiosis

Fortunately, nature has provided us with several cases which avoid these problems associated with competition with other soil bacteria. These are the symbioses in which a bacterium fixes nitrogen within a plant organ. By such means the plant sequesters the nitrogen-fixing bacterium and thus has sole access to the ammonia that is synthesized; in return the bacterium finds a niche in which it is free from competition from other species. Agronomically, the most important nitrogen-fixing symbiosis is that between *Rhizobium* and legumes; this system has been the subject of the most detailed genetic and molecular analysis, although our understanding of the *nif* genes of *Rhizobium* is still rudimentary compared with our knowledge of the *Klebsiella nif* region.

At least in the fast-growing strains of *Rhizobium*, which nodulate such crops as peas, clover, *Phaseolus* beans or alfalfa, the genes for nodulation are located on large indigenous plasmids; in the examples analysed so far, nodulation genes are relatively close (within 20 kb) to the *nif* genes (Long, Buikema & Ausubel, 1981; Downie *et al.*, 1983*a*). Some of the *nif* genes are so highly conserved amongst bacteria that the cloned *K. pneumoniae nif* genes have been used successfully as hybridization probes to identify and locate the corresponding genes in *Rhizobium* (Nuti *et al.*, 1979; Ruvkun & Ausubel, 1980), providing a good example of the use of gene cloning to speed up genetic analysis of a hitherto unexplored organism.

The requirement to analyse the function of cloned *Rhizobium* genes in *Rhizobium* provided the impetus for the construction of cloning vectors that would replicate in members of this genus. Thus Ditta *et al.* (1980) constructed pRK290, a 20 kb derivative of the wide-host-range plasmid RK2, and Friedman *et al.* (1982) made pLAFR1, a cosmid derivative of pRK290, specifically for the purpose of making clone libraries of *R. meliloti* DNA.

Such vectors have been used to demonstrate that host range and nodulation ability are apparently determined by fewer than ten genes (Downie *et al.*, 1983*b*). Soon we should have a very detailed knowledge of these genes and of the biochemical functions required for *Rhizobium* to recognize its correct host plant, to enter the plant, and to induce the plant to form the root nodule. Only then may it be possible to construct, in a rational way, strains of *Rhizobium* that have improved symbiotic properties. Meanwhile, we can only speculate on the outcome of schemes in which (for example) the nodulation genes of *Rhizobium* are put under the control of abnormally strong promoters. Would such strains be more aggressive in their ability to nodulate in the face of competition from resident strains?

Breeding by plasmid transfer

One phenotype which apparently bears on the efficiency of nitrogen fixation is the possession of the enzyme uptake hydrogenase (Hup).

The nitrogenase from all nitrogen-fixing bacteria liberates hydrogen, which represents a substantial energy drain – 25% of the flux of electrons through nitrogenase is devoted to the reduction of protons (Evans *et al.*, 1981). Many nitrogen-fixing bacteria, but only a minority of *Rhizobium* strains, possess an uptake hydrogenase, thought to be potentially beneficial for three reasons: it recoups some of the energy that would otherwise have been lost; it removes gaseous hydrogen, an inhibitor of nitrogenase; and, acting as a hydrogen oxidase, it may reduce the oxygen tension in the vicinity of nitrogenase.

Brewin *et al.* (1980) showed that the *hup* genes reside on the symbiotic plasmid of a particular strain of *Rhizobium leguminosarum*. A transmissible derivative (called pIJ1008) of this plasmid was transferred to another, Hup⁻, field isolate, and the resulting transconjugants were more effective as inoculants than were either of the parental strains, as judged by pea plant growth (DeJong *et al.*, 1982). It remains to be seen whether the improvement was due to the hydrogenase or to some other gene(s) present on pIJ1008; it is known, for example, that the nitrogenase enzyme specified by pIJ1008 is different from that of the strain used as the recipient in this cross (N. J. Brewin, personal communication).

The fact that many (if not all) of the symbiotic genes are located on single, often transmissible, plasmids offers the opportunity of transferring such a plasmid from one strain to another in the hope that this will lead empirically to an improved hybrid strain. For example Brewin *et al.* (1983) found that the competitive ability of *Rhizobium* was markedly influenced by the identity of the symbiotic plasmid. This type of approach, though not intellectually very stimulating, may provide the best short-term opportunity for improved strain construction, until such time as the techniques of molecular biology will give us the knowledge for the planned construction of strains which are resistant to environmental stress (e.g. drought, salinity); which are highly competitive; and which fix nitrogen efficiently at high levels. If we can understand the *Rhizobium*–legume symbiosis we might conceivably gain enough insight to obtain infections on non-leguminous crop plants. Perhaps some genetic information might usefully be transplanted from strains of the actinomycete genus *Frankia*, which can form nitrogen-fixing root nodules on the roots of woody plants belonging to several different families.

CONCLUSION

We have intended that an air of guarded optimism should pervade this article. Given the ubiquity of viruses and plasmids in microbes and the fact that systems have been developed for delivering foreign DNA into a range of organisms that includes *E. coli*, yeast, tobacco and mouse, there seems no insurmountable barrier to prevent any gene from being cloned and introduced into any microbe. Thus, using current techniques or extensions of such techniques, the prospects for microbial improvement seem bright indeed for those cases in which the microbe is simply to be used as a factory for the production of a foreign gene product. Moreover, it will doubtless be possible to clone whole sets of genes involved in the production of relatively complex metabolites, even though the prospect of inducing *E. coli* to make (for example) a compound such as quinine, the product of several biochemical steps determined by unlinked genes in a higher plant, is at present daunting (for reasons touched upon earlier, the resulting microbe may not, at least at first, be a very good factory for such compounds).

The prospects for using engineered microbes as agents for ecological or agricultural improvement are fraught with rather more problems. Some of these were discussed in relation to the construction of a hypothetical rhizosphere bacterium that was capable of fixing nitrogen. The problems that must be solved before we can routinely improve crop yield or deal with pollution by the widespread dissemination of a strain that on paper should be purpose-built have more to do with the challenges of microbial ecology than of molecular biology.

For all the elegance with which a gene can be cloned, its product obtained and its DNA sequence established, we are still far from understanding the precise function of a protein as determined by its primary amino acid sequence. If (perhaps at the 200th meeting of the Society?) we, or our computers, could scan an open reading frame and deduce that the resultant polypeptide would catalyse the decarboxylation of substance X and would have a K_m of Y and a turnover rate of Z it would mean that molecular biology had overcome the next great hurdle and the repercussions would be enormous. It would mean that genes to degrade or synthesize novel compounds could be made *de novo* and that logical strategies could be used for improving natural genes by mutagenesis (e.g. an RuBP carboxylase with a higher affinity for carbon dioxide, or a nitrogenase that did not liberate hydrogen or was resistant to oxygen damage). Considering the giant strides that have been taken in genetics since the first meeting of the Society, such feats are probably much nearer than we might imagine.

We are very grateful to Dr Gwyn Humphreys for discussion of many of the ideas in this article.

REFERENCES

ABRAHAM, E. P. (1974). Some aspects of the development of the penicillins and cephalosporins. *Developments in Industrial Microbiology*, **15**, 3–15.

BAGDASARIAN, M., LURZ, R., RUCKERT, B., FRANKLIN, F. C. H., BAGDASARIAN, M. M., FREY, J. & TIMMIS, K. N. (1981). Specific purpose cloning vectors. II. Broad host range high copy number RSF1010-derived vectors and a host vector system for gene cloning in *Pseudomonas*. *Gene*, **16**, 237–47.

BECK, E., LUDWIG, G., AUERSWALT, E. A., REISS, B. & SCHOLLER, H. (1982). Nucleotide sequence and exact localization of the neomycin phosphotransferase gene from transposon Tn5. *Gene*, **19**, 327–36.

BEGGS, J. D., VAN DEN BERG, J., VAN OOYEN, A. & WEISSMANN, C. (1980). Abnormal expression of chromosomal rabbit β-globin gene in *Saccharomyces cerevisiae*. *Nature, London*, **283**, 835–40.

BENNETZEN, J. L. & HALL, B. D. (1982). Codon selection in yeast. *Journal of Biological Chemistry*, **257**, 3026–31.

BIBB, M. J., CHATER, K. F. & HOPWOOD, D. A. (1983). Developments in *Streptomyces* cloning. In *Experimental Manipulation of Gene Expression*, ed. M. Inouye, pp. 53–82. New York & London: Academic Press.

BREWIN, N. J., DEJONG, T. M., PHILLIPS, D. A. & JOHNSTON, A. W. B. (1980). Co-transfer of determinants for hydrogenase activity and nodulation ability in *Rhizobium leguminosarum*. *Nature, London*, **288**, 77–9.

BREWIN, N. J., WOOD, E. A. & YOUNG, P. W. (1983). Contribution of the symbiotic plasmid to the competitiveness of *Rhizobium*. *Journal of General Microbiology* (in press).

BUCHANAN-WOOLASTON, A., CANNON, M. C., BEYNON, J. L. & CANNON, F. C. (1981). Role of the *nifA* gene product in the regulation of *nif* expression in *Klebsiella pneumoniae*. *Nature, London*, **294**, 776–8.

CANNON, F. C. & POSTGATE, J. R. (1976). Expressing *Klebsiella* nitrogen fixation genes (*nif*) in *Azotobacter*. *Nature, London*, **260**, 271–2.

CASADABAN, M. J. & COHEN, S. N. (1979). Lactose genes fused to exogenous promoters in one step using a Mu–lac bacteriophage *in vivo* probe for transcriptional control sequences. *Proceedings of the National Academy of Sciences, USA*, **76**, 4530–3.

CHATER, K. F., HOPWOOD, D. A., KIESER, T. & THOMPSON, C. J. (1982). Gene cloning in *Streptomyces*. *Current Topics in Microbiology and Immunology*, **96**, 69–95.

COHEN, S. N., CHANG, A. C. Y., BOYER, H. W. & HELLING, R. B. (1973). Construction of biologically functional bacterial plasmids *in vitro*. *Proceedings of the National Academy of Sciences, USA*, **70**, 3240–4.

DAESCH, G. & MORTENSON, L. E. (1968). Sucrose metabolism in *Clostridium pasteurianum* and its relation to N_2 fixation. *Journal of Bacteriology*, **96**, 346–51.

DAVIES, J. E. & YAGISAWA, M. (1983). The aminocyclitol glycosides (aminoglyco-sides). In *Biochemistry and Genetic Regulation of Commercially Important Antibiotics*, ed. L. C. Vining, pp. 329–54. Reading, Mass.: Addison-Wesley.

DAVIS, B. D. & TAI, P.-C. (1980). The mechanism of protein secretion across membranes. *Nature, London*, **283**, 433–8.

DEJONG, T. M., BREWIN, N. J., JOHNSTON, A. W. B. & PHILLIPS, D. A. (1982). Improvement of symbiotic properties in *Rhizobium leguminosarum* by plasmid transfer. *Journal of General Microbiology*, **128**, 1829–38.

DEMAIN, A. L., AHARONOWITZ, Y. & MARTIN, J.-F. (1983). Metabolic control of secondary biosynthetic pathways. In *Biochemistry and Genetic Regulation of Commercially Important Antibiotics*, ed. L. C. Vining, pp. 49–72. Reading, Mass.: Addison-Wesley.

DITTA, G., STANFIELD, S., CORBIN, D. & HELINSKI, D. R. (1980). Broad host range DNA cloning system for Gram-negative bacteria: construction of a gene bank of *Rhizobium meliloti*. *Proceedings of the National Academy of Sciences, USA*, **77**, 7347–51.

DIXON, R. A., CANNON, F. C. & KONDOROSI, A. (1976). Construction of a P plasmid carrying nitrogen fixation genes from *Klebsiella pneumoniae*. *Nature, London*, **260**, 268–71.

DIXON, R., EADY, R. R., ESPIN, G., HILL, S., IACCARINO, M., KAHN, D. & MERRICK, M. (1980). Analysis of regulation of *Klebsiella pneumoniae* nitrogen fixation (*nif*) gene cluster with gene fusions. *Nature, London*, **286**, 128–32.

DIXON, R. A. & POSTGATE, J. R. (1972). Genetic transfer of nitrogen fixation from *Klebsiella pneumoniae* to *Escherichia coli*. *Nature, London*, **237**, 102–3.

DOWNIE, J. A., MA, Q.-S., KNIGHT, C. D., HOMBRECHER, G. & JOHNSTON, A. W. B. (1983a). Cloning of the symbiotic region of *Rhizobium leguminosarum*; the nodulation genes are between the nitrogenase genes and a *nifA*-like gene. *EMBO Journal*, **2**, 947–52.

DOWNIE, J. A., HOMBRECHER, G., MA, Q.-S., KNIGHT, C. D., WELLS, B. & JOHNSTON, A. W. B. (1983b). Cloned nodulation genes of *Rhizobium leguminosarum* determine host range specificity. *Molecular and General Genetics*, **190**, 359–65.

DRUMMOND, M., CLEMENTS, J., MERRICK, M. & DIXON, R. (1983). Positive control and autogenous regulation of the *nifLA* (1983). Positive control and autogenous regulation of the *nifLA* promoter in *Klebsiella pneumoniae*. *Nature, London*, **301**, 302–7.

EHRLICH, S. D., NIAUDET, B. & MICHEL, B. (1982). Use of plasmids from *Staphylococcus aureus* for cloning of DNA in *Bacillus subtilis*. *Current Topics in Microbiology and Immunology*, **96**, 19–30.

EMTAGE, J. S., ANGAL, S., DOEL, M. T., HARRIS, T. J. R., JENKINS, B., LILLEY, G. & LOWE, P. A. (1983). Synthesis of calf prochymosin (prorennin) in *Escherichia coli*. *Proceedings of the National Academy of Sciences, USA*, **80**, 3671–5.

EVANS, H. J., PURPHIT, K., CANTRELL, M. A., EISBRENNER, G., RUSSELL, S. A., HANUS, F. J. & LEPO, J. E. (1981). Hydrogen losses and hydrogenases in nitrogen-fixing organisms. In *Current Perspectives in Nitrogen Fixation*, ed. A. H. Gibson & W. E. Newton, pp. 84–96. Canberra: Australian Academy of Sciences.

FINCHAM, J. R. S. & DAY, P. R. (1971). *Fungal Genetics*, 3rd edn. Oxford: Blackwell Scientific.

FRIEDMAN, A. M., LONG, S. R., BROWN, S. E., BUIKEMA, W. J. & AUSUBEL, F. M. (1982). Construction of a broad host range cosmid cloning vector and its use in the analysis of *Rhizobium* mutants. *Gene*, **18**, 289–96.

FRIELLO, D. A., MYLROIE, J. R. & CHAKRABARTY, A. M. (1976). Use of genetically-engineered multi-plasmid microorganisms for rapid degradation of fuel hydrocarbons. In *Proceedings of the Third Biodegradation Symposium*, ed. J. M. Sharpley, pp. 205–14. Essex: Applied Science Publications.

GALLWITZ, D. & SURES, I. (1980). Structure of a split yeast gene: complete nucleotide sequence of the actin gene in *Saccharomyces cerevisiae*. *Proceedings of the National Academy of Sciences, USA*, **77**, 2546–50.

GHEYSEN, D., ISERENTANT, D., DEROM, C. & FIERS, W. (1982). Systematic alteration of the nucleotide sequence preceding the translation initiation codon and the effects on bacterial expression of the cloned SV40 small-t antigen gene. *Gene*, **17**, 55–63.

GOEDDEL, D. V., KLEID, D. G., BOLIVAR, F., HEYNECKER, H. C., YANSURA, D. G., CREA, R., HIROSE, T., KRASZEUSKI, A., ITAKURA, K. & RIGGS, A. D. (1979a). Expression in *E. coli* of chemically synthesised genes for human insulin. *Proceedings of the National Academy of Sciences, USA*, **76**, 106–10.

GOEDDEL, D. V., HEYNECKER, H. L., HOZUMI, T., ARENTZEN, R., ITAKURA, K., YANSURA, D. G., ROSS, M. J., MIOZZARI, G., CREA, R. & SEEBURG, P. H. (1979b). Direct expression in *Escherichia coli* of a DNA sequence coding for human growth hormone. *Nature, London*, **281**, 544–8.

GOEDDEL, D. V., SHEPARD, H. M., YELVERTON, E., LEUNG, D. & CREA, R. (1980). Synthesis of human fibroblast interferon by *E. coli*. *Nucleic Acids Research*, **8**, 4057–74.

GORDON, J. J., GRENFELL, E., KNOWLES, E., LEGGE, B. J., McALLISTER, R. C. A. & WHITE, T. (1947). Methods of penicillin production in submerged culture on a pilot-plant scale. *Journal of General Microbiology*, **1**, 187–202.

GRAY, G. H. & TATUM, E. L. (1944). X-ray induced growth factor requirements in bacteria. *Proceedings of the National Academy of Sciences, USA*, **30**, 404–10.

GROSJEAN, H. & FIERS, W. (1982). Preferential codon usage in prokaryotic genes: the optimal codon–anticodon interaction energy and the selective codon usage in efficiently expressed genes. *Gene*, **18**, 199–209.

HAAKER, H. & VEEGER, C. (1977). Involvement of the cytoplasmic membrane in nitrogen fixation by *Azotobacter vinelandii*. *European Journal of Biochemistry*, **77**, 1–10.

HARDY, K. G., STAHL, S. & KÜPPER, H. (1983). Expression of eukaryotic and viral genes in *Bacillus*. In *Genetics of Industrial Microorganisms*, ed. Y. Ikeda & T. Beppu, pp. 136–40. Tokyo: Kodansha.

HARRIS, T. J. R. (1983). Expression of eukaryotic genes in *Escherichia coli*. In *Genetic Engineering*, vol. 4, ed. R. Williamson, pp. 127–85. New York & London: Academic Press.

HILL, S., KENNEDY, C., KAVANAGH, E., GOLDBERG, R. B. & HANAU, R. (1981). Nitrogen fixation gene (*nifL*) involved in oxygen regulation of nitrogenase synthesis in *K. pneumoniae*. *Nature, London*, **290**, 424–6.

HOPWOOD, D. A. (1977). Genetic recombination and strain improvement. *Developments in Industrial Microbiology*, **18**, 9–21.

HOPWOOD, D. A. (1983). Actinomycete genetics and antibiotic production. In *Biochemistry and Genetic Regulation of Commercially Important Antibiotics*, ed. L. C. Vining, pp. 1–23. Reading, Mass.: Addison-Wesley.

HOPWOOD, D. A., BIBB, M. J., BRUTON, C. J., CHATER, K. R., FEITELSON, J. S. & GIL, J. A. (1983). Cloning of *Streptomyces* genes for antibiotic production. *Trends in Biotechnology*, **1**, 42–8.

HOPWOOD, D. A. & CHATER, K. F. (1982). Cloning in *Streptomyces*: systems and strategies. In *Genetic Engineering*, vol. 4, ed. J. K. Setlow & A. Hollaender, pp. 119–45. New York: Plenum Press.

ITAKURA, K., HIROSE, T., CREA, R., RIGGS, A. D., HEYNECKER, H. L., BOLIVAR, F. & BOYER, H. W. (1978). Expression in *Escherichia coli* of a chemically synthesized gene for the hormone somatostatin. *Science*, **198**, 1056–63.

JACKSON, D. A., SYMONS, R. H. & BERG, P. (1972). Biochemical method for inserting new genetic information into DNA of Simian virus 40: circular SV40 DNA molecules containing lambda phage genes and the galactose operon of *Escherichia coli*. *Proceedings of the National Academy of Sciences, USA*, **69**, 2904–9.

JOHNSTON, J. R. (1981). Recent developments in the application of genetics to brewing and wine strains of yeast. In *Fermentation Alcolique*, ed. P. Barre & G. Durand, pp. 125–46. Reims: Société Française de Microbiologie.

JOHNSTON, J. R. & OBERMAN, H. (1979). Yeast genetics in industry. *Progress in Industrial Microbiology*, **15**, 151–205.

KENNEDY, C., CANNON, F., CANNON, M., DIXON, R., HILL, S., JENSEN, J., KUMAR, S., MCLEAN, P., MERRICK, M., ROBSON, R. & POSTGATE, J. (1981). Recent advances in the genetics and regulation of nitrogen fixation. In *Current Perspectives in Nitrogen Fixation*, ed. A. H. Gibson & W. E. Newton, pp. 146–56. Canberra: Australian Academy of Sciences.

LANGFORD, C., NELLEN, W., NIESSING, J. & GALLWITZ, D. (1983). Yeast is unable to excise foreign intervening sequences from hybrid gene transcripts. *Proceedings of the National Academy of Sciences, USA*, **80**, 1496–500.

LAWN, R. M., ADELMAN, J., BOCK, S. C., FRANKE, A. E., HOUCK, C. M., NAJARIAN, R. K., SEEBURG, P. H. & WION, K. L. (1981). The sequence of human serum albumin cDNA and its expression in *E. coli*. *Nucleic Acids Research*, **9**, 6103–14.

LOBBAN, P. E. & KAISER, A. D. (1973). Enzymatic end to end joining of DNA molecules. *Journal of Molecular Biology*, **78**, 453–71.

LONG, S. R., BUIKEMA, W. J. & AUSUBEL, F. M. (1982). Cloning of *Rhizobium meliloti* nodulation genes by direct complementation of Nod⁻ mutants. *Nature, London*, **298**, 485–8.

LOSICK, R. & PERO, J. (1981). Cascades of sigma factors. *Cell*, **25**, 582–4.

MALIK, V. (1983). Chloramphenicol. In *Biochemistry and Genetic Regulation of Commercially Important Antibiotics*, ed. L. C. Vining, pp. 293–309. Reading, Mass.: Addison-Wesley.

MELLOR, J., DOBSON, M. J., ROBERTS, N. A., TUITE, N. S., EMTAGE, J. S., WHITE, S., LOWE, P. A., PATEL, T., KINGSMAN, A. J. & KINGSMAN, S. M. (1983). Efficient synthesis of enzymatically active calf chymosin in *Saccharomyces cerevisiae*. *Gene*, **24**, 1–14.

NUTI, M. P., LEPÍDI, A. A., PRAKASH, R. K., SCHILPEROORT, R. A. & CANNON, F. C. (1979). Evidence for nitrogen fixation (*nif*) genes on indigenous *Rhizobium* plasmids. *Nature, London*, **282**, 533–5.

REIDEL, G. E., AUSUBEL, F. M. & CANNON, F. C. (1979). Physical map of chromosomal nitrogen fixation (*nif*) genes of *Klebsiella pneumoniae*. *Proceedings of the National Academy of Sciences, USA*, **76**, 2866–70.

REMAUT, E., TSAO, H. & FIERS, W. (1983). Improved plasmid vectors with a thermoinducible expression and temperature-regulated runaway replication. *Gene*, **22**, 103–13.

RUVKUN, G. B. & AUSUBEL, F. M. (1980). Interspecies homology of nitrogenase genes. *Proceedings of the National Academy of Sciences, USA*, **77**, 191–5.

SHAPIRO, J., MACHATTIL, L., FROM, L., IHLER, G., IPPEN, K. & BECKWITH, J. (1969). Isolation of pure *lac* operon DNA. *Nature, London*, **224**, 768–74.

SMITH, S. R. L. (1980). Single cell protein. *Philosophical Transactions of the Royal Society of London, Series B*, **290**, 341–54.

STREICHER, S., GURNEY, E. & VALENTINE, R. C. (1971). Transduction of the nitrogen-fixation genes in *Klebsiella pneumoniae*. *Proceedings of the National Academy of Sciences, USA*, **68**, 1174–7.

TANIGUCHI, T., GUARENTE, L., ROBERTS, T. M., KINELMAN, D., DOUHAN, J. & PTASHNE, M. (1980). Expression of the human fibroblast interferon gene in *Escherichia coli*. *Proceedings of the National Academy of Sciences, USA*, **77**, 5230–3.

TAUSSIG, R. & CARLSON, M. (1983). Nucleotide sequence of the yeast SUC2 gene for invertase. *Nucleic Acids Research*, **11**, 1943–54.

TAYLOR, A. L. (1970). Current linkage map of *Escherichia coli*. *Bacteriological Reviews*, **34**, 155–75.

THOMPSON, C. J. & GRAY, G. S. (1983). The nucleotide sequence of a streptomycete aminoglycoside phosphotransferase gene and its relationship to phosphotranferase encoded by resistance plasmids. *Proceedings of the National Academy of Sciences, USA*, **80**, 5190–4.

WAGNER, W., VOGEL, M. & GOEBEL, W. (1983). Transport of hemolysin across the outer membrane of *Escherichia coli* requires two functions. *Journal of Bacteriology*, **154**, 200–10.

WINDASS, J. D., WORSEY, M. J., PIOLI, E. M., PIOLI, D., BARTH, P. T., ATHERTON, K. T., DART, E. C., BYROM, D., POWELL, K. & SENIOR, P. J. (1980). Improved conversion of methanol to single-cell protein by *Methylophilus methylotrophus*. *Nature, London*, **287**, 396–401.

YOUNG, F. E. (1980). Impact of cloning in *Bacillus subtilis* on fundamental and industrial microbiology. *Journal of General Microbiology*, **119**, 1–15.

THE IMPACT OF THE MICROBE IN MEDICINE

J. P. ARBUTHNOTT

Department of Microbiology, Moyne Institute, Trinity College, Dublin, Ireland

Advances in clinical medicine depend heavily on the application of new principles gleaned from the basic sciences. It is fair to say that the emergence of the discipline of microbiology, with the discovery of micro-organisms as the causes of infection, was one of the most significant events of modern medicine. The early discoveries of pioneers such as Jenner, Pasteur, Lister, Koch, Behring, Ehrlich, Fleming and Domagk provided the framework for the development of strategies for the diagnosis, treatment and prevention of a major scourge of mankind – infectious disease. They also laid the basis for the development of safe modern aseptic surgical procedures. No one would doubt the successes of antimicrobial agents and vaccines in reducing the incidence of infectious disease. To be convinced of this one need only examine the statistics of the incidence of fatal bacterial meningitis (Smith, 1982) (Table 1) or consider the success of the WHO campaign which resulted in the global eradication of smallpox (Fenner, 1982).

Immunization and antimicrobial therapy are not, however, the only factors that have contributed to the reduced incidence of infectious disease. For example, the downward trend in incidence of tuberculosis began between the First and Second World Wars before the introduction of streptomycin, and deaths from scarlet fever had already markedly declined before the introduction of penicillin (Smith, 1982). Improvements in adverse social factors such as overcrowding, poor nutrition and poor hygiene have doubtless played a role. So too, almost certainly, have changes in the pathogenicity of the causative organisms.

In the developed world a combination of these factors has led to reduced mortality from infection. At the same time there is no room for complacency. Morbidity resulting from infectious illnesses is still common and comprises much of the work of family doctors and hospital medical teams. It has been estimated that about 25% of illness in the community is due to infection, and infection in

Table 1. *Meningococcal meningitis in England and Wales*

Year	Notification	Deaths	Mortality rate (%)
1920	559	553	95
1930	661	631	95
1940	12771	2449	19
1950	1150	283	24
1960	632	95	15
1970	525	143	25
1980	509	61	12

From Smith (1982).

hospitals runs at about 5–10% (Report, 1981). The microbe continues to be highly adaptable, as evidenced by the emergence of resistance to antimicrobial agents. Multiple resistance to antibiotics sets a formidable test of the management skills of clinical microbiologists, physicians and surgeons. This holds especially for new approaches to the treatment of heart disease, transplantation surgery and cancer. Also the immunocompromised patient is a prime target for infection and rapid diagnosis followed by careful management of antibiotic therapy are crucial to the patient's survival.

New diseases such as non-gonococcal urethritis, diarrhoeal disease caused by *Campylobacter jejuni*, rotavirus or Norwalk virus agent, pseudomembranous colitis due to *Clostridium difficile*, neonatal meningitis due to Group B streptococci, neonatal botulism, Legionnaires' disease, Toxic Shock Syndrome and Acquired Immune Deficiency Syndrome continue to emerge as a result of improved procedures for isolation and diagnosis, or from changing patterns of social behaviour. For instance the apparent decline of sexually transmitted diseases in the early 1950s was followed by a rapid increase in the 1960s and 1970s; fortunately the Public Health teams responsible for monitoring these diseases had been maintained and were available to cope with the increased workload and changing pattern of infection.

Epidemics of acute respiratory infection, meningitis and gastrointestinal infection still occur in less developed countries, especially following famine or the outbreak of war, and many chronic bacterial or parasitic diseases remain endemic in these regions. Costly antimicrobial agents often cannot be provided and mass programmes of immunization run into organizational difficulties. In these countries the provision of an adequate health care system remains a formidable challenge.

For these reasons the microbe continues to have a major impact on medicine. It is impossible to deal with this vast subject in detail in a single paper. Of necessity the coverage will be rather general; in many instances it is possible only to refer to major review articles rather than to original references. Of the many ways of tackling this subject I have chosen to illustrate the impact of the microbe on medicine by examining four aspects: changes in diagnostic methods, trends in antimicrobial therapy, existing and new possibilities for vaccines, and lastly the changing pattern of infection and the emergence of 'new' infectious diseases. Other papers in this volume emphasize advances in our knowledge of the microbe's physiology, genetics and ecology; I have attempted to illustrate where these advances are relevant to medical microbiology.

ADVANCES IN DIAGNOSTIC METHODS

The study of pure cultures of micro-organisms began in 1881 with the introduction of solid media by Robert Koch. This, together with Gram's differential staining reaction, formed the basis of identification of micro-organisms. Refinements were added at the turn of the century when assays for fermentation products and serological tests gave greater discrimination. Little change occurred in the next 60 years. However, increased interest in changing patterns of hospital infection and antibiotic resistance, and increased awareness of the need for rapid diagnosis of microbial infection in compromised patients, provided the impetus for changing attitudes to diagnosis and identification. Increasingly the emphasis is on standardization, speed, reproducibility, mechanization and automation. These developments have been admirably reviewed by D'Amato, Holmes & Bottone (1981). It should be noted in passing, that advances in this area require important input by 'academic' microbiologists; in many cases new techniques exploit fundamental knowledge of microbial genetics, structure and physiology. For instance it is now possible to examine clinical specimens with a DNA hybridization technique. This approach is illustrated by Moseley *et al.* (1982) who used radiolabelled fragments of DNA encoding *E. coli* ST or LT entero-toxin to detect enterotoxigenic strains of *E. coli* in stool samples from patients with diarrhoea.

Miniaturization of test systems for identification (involving single or multiple compartments, with substrates being supplied in various forms including impregnated discs or filter strips) are becoming commonplace and provide for high reproducibility. More recently these systems have been coupled to computerized data bases. End-points can be measured after incubation times of a few hours; automated inoculation, test reading and identification are possible. Test results can be represented in numerical code suitable for computerization. Also, automated systems are being developed to circumvent the time delay involved in working with pure cultures, such that organisms can be grouped or identified directly from clinical specimens even in the presence of mixed populations (Aldridge *et al.*, 1977).

The incidence of bacteraemia has increased in the past 20 years and considerable priority has been placed on devising rapid methods for its detection (Tilton, 1982). Non-traditional methods involve systems for measuring microbial synthesis of an end-product such as $^{14}CO_2$. Radiometric detection of $^{14}CO_2$ produced from $[^{14}C]$glucose forms the basis of the Bactec method, which is rapidly gaining acceptance (Randall, 1975). This has been progressively modified by an extension of the range of labelled substrates, provision of a means of agitation and inclusion of anaerobic pre-reduced samples for detection of anaerobes.

Changes in electrical impedance resulting from bacterial growth can be used to detect the presence of micro-organisms. Hadley & Senyk (1975) were able to detect bacteria at levels of 10^6–10^7 per millilitre. Recently lysis-filtration has been combined with impedance monitoring; blood is lysed and filtered through a membrane which is then placed in a nutrient medium containing stainless steel electrodes. Monitoring of changes in electrical potential associated with metabolic activity is also a promising new method of detection.

Alternative approaches which avoid the need to culture the infecting agent have also been actively pursued (see Yolken, 1982). This is particularly relevant to the diagnosis of viral, bacterial or fungal agents which cannot be easily cultivated. Another advantage is the potential for rapid diagnosis of infection directly in clinical specimens, allowing rapid institution of appropriate therapy. For instance microbial metabolites can be detected by gas–liquid chromatography, and the method is particularly applicable to anaerobes which form characteristic fermentation products. Most such methods, however, depend on immunoassays such as radioimmuno-assay (RIA) or enzyme-linked immunoabsorbent assay (ELISA). ELISA methods are particularly useful because they avoid the requirement for costly unstable radioactive reagents, they are extremely sensitive and they are easy to perform. For these reasons they have been developed for measurement of viral and bacterial antigens such as hepatitis B surface antigen, rotavirus, gastrointes-tinal adenovirus, *E. coli* LT toxin, *Clostridium difficile* toxin and the capsular polysaccharides of *Haemophilus influenzae* type b, *Neisseria meningitidis* and *Streptococcus pneumoniae* (see Yolken, 1982, for references). It must be pointed out, however, that so far ELISA methods have been unsuccessful for some viruses, for *Chlamydia trachomatis* and for *Mycoplasma pneumoniae*.

Other immunological methods such as counter-immunoelec-trophoresis (Alter, Holland & Purcell, 1971), agglutination proce-dures that make use of *Staphylococcus aureus* protein A or latex particles (Thirumoorthi & Dajani, 1979) and fluorescent antibody tests (Fulton & Middleton, 1974) have also proved extremely useful for detecting antigen in clinical specimens.

TREATMENT AND PREVENTION OF INFECTIOUS DISEASE BY ANTIMICROBIAL AGENTS

The era of antimicrobial agents was heralded by the discovery of penicillin by Fleming (1929), and by Domagk's work (1935) in the laboratories of I. G. Farben in Germany with the synthetic sulpha compound prontosil.

The reappraisal of Fleming's contribution to the development of penicillin as a therapeutic antibiotic by Abraham (1980) correctly credits Fleming with the flair for recognizing the unusual in exploit-ing his chance observation of the antibacterial activity of *Penicillium notatum*. It equally emphasizes the fact that Florey's group in Oxford turned the basic observation into a practical reality by isolating and purifying penicillin and showing in a clinical trial that the antibiotic was potentially of outstanding value as a therapeutic agent.

The continuing search for antibiotics has yielded about 5000 substances with antimicrobial activity (Aharonowitz & Cohen,

Table 2. *Introduction of antibiotics of clinical value*

Group of antibiotics	Mode of action	Examples	Introduction to clinical use	Antibacterial activity
β-Lactams	Binding to penicillin-binding proteins: inhibition of the enzymes of wall synthesis	Standard penicillins e.g. Benzylpenicillin	1940s	Active against gram-positive bacteria and gram-negative cocci
		Semi-synthetic penicillins e.g. Methicillin Cloxacillin	Early 1960s	Active against penicillinase-producing staphylococci
		e.g. Ampicillin Amoxycillin Carbenicillin		Active against gram-negative bacilli
		Cephalosporins Cephaloridine Cefuroxime Cefotaxime Moxalactam	mid 1960s–1980s	Broad-spectrum antibiotics resistant to β-lactamase enzymes and active against gram-positive and gram-negative organisms
	Potent inhibitor of β-lactamase	Clavulanic acid		Used in conjunction with amoxycillin against β-lactamase-producing organisms
Aminoglycosides	Binding to ribosomes and inhibition of translation	Streptomycin Neomycin Kanamycin Gentamicin	1944 1950 1958 1969	Active mainly against gram-negative bacilli and *S. aureus*

Class	Mode of action	Examples	Date	Uses/activity
Tetracyclines	Prevents binding of aminoacyl transfer RNA to ribosome	Tetracycline Semi-synthetic tetracyclines Doxycycline minocycline	1953 1966 1977	Broad-spectrum antibiotics
Chloramphenicol	Reacts with 50S ribosome subunit to inhibit peptidyltransferase step in translation	Chloramphenicol	1949	Broad-spectrum antibiotic
Macrolide–lincosamide–streptogramin B group	Bind to 50S ribosome subunit and inhibit protein synthesis	Erythromycin Lincomycin Clindamycin	1950 Early 1960s Early 1960s	Active against gram-positive bacteria and gram-negative cocci Active against anaerobic organisms
Vancomycin	Inhibition of biosynthesis of peptidoglycan wall polymers	Vancomycin	1956	Active against gram-positive bacteria and gram-negative cocci
Rifampicin	Inhibition of DNA-dependent RNA polymerase	Rifampicin	1967	Useful in treatment of tuberculosis
Polyenes	Membrane-damaging agents reacting with sterols	Nystatin Amphotericin B	1958 1958	Antifungal agents

1981). About 1000 antibiotics have come from six genera of filamentous fungi including *Penicillium* and *Cephalosporium*, and about 3000 from three genera of actinomycetes of which *Streptomyces* is the richest source. Of this myriad of substances only about 100 have been marketed. Table 2 summarizes the discovery and mode of action of the major groups of antibiotics and indicates their current use as first-choice drugs. In commercial as well as clinical terms antibiotics represent an extremely important group of pharmaceutical agents. The estimated world-wide bulk sales in 1978 of the four most important groups of antibiotics (the penicillins, the cephalosporins, the tetracyclines and the erythromycins) amounted to \$4.2 billion (Aharonowitz & Cohen, 1981). In the UK, expenditure on antibiotics amounts to about £100 million per annum, which is approximately 1.5% of the total cost of the National Health Service. The cost of developing a new antibiotic is currently estimated at around \$70 million.

The importance of antibiotics as therapeutic drugs is self-evident. To quote from a recent Editorial (1982) in the *Journal of Infection*:

It is hard to believe that within the memory of many practising doctors there was a time when antibiotics and most other antimicrobial agents we take for granted did not exist. The impact such drugs have made in almost every aspect of medicine, the illnesses prevented or curtailed and the lives saved are incalculable.

It is important to remember that behind the dramatic clinical success obtained with antibiotics lie decades of patient research in the folllowing areas: (*a*) enhancement of production by improved cultural methods and strain improvement; (*b*) the molecular analysis of mode of action; and (*c*) methods of assay and quality control.

The basic objective in developing an antibiotic for clinical use is of course to obtain an agent that is selectively toxic for the pathogen without harming the host. Although there have been notable successes it has to be recognized that all antibiotics are to some extent toxic for man. These toxicities limit the use of antibiotics in certain patients and represent one of the many factors that have to be taken into account in the management of antibiotic therapy.

Antibiotic resistance

By far the most important limiting factor in the effectiveness of antibiotics has been the emergence of resistant strains. Resistant strains were known even in the early days of antibiotic research but the dramatic increase, especially in multiply-resistant isolates, gives cause for great concern.

A single example serves to emphasize this point for gram-positive organisms. Since 1975 strains of *Staphylococcus aureus* resistant to methicillin and one or more aminoglycoside antibiotics have become prominent as causes of life-threatening infection in compromised patients (Soussy *et al.*, 1976; Crossley, Landesman & Zaska, 1979; Hone *et al.*, 1981). These organisms were known in the early 1960s but were rare; the reason for their recent appearance in several centres (France, USA and Ireland) is unknown. Experience in Dublin (Cafferkey, Hone & Keane, 1981) has shown 50% mortality from bacteraemia arising from such organisms and vancomycin has proved to be the only clinically effective antimicrobial agent.

Resistance among the gram-negative bacilli has also proved a major problem. These organisms can acquire multi-resistance plasmids and therefore cause rapid changes in patterns of sensitivity which require new treatment regimens. The acquisition among organisms in the community of resistance to a commonly used effective antibiotic can lead to major epidemics. This has been the case with chloramphenicol-resistant *Salmonella typhi* and *Shigella sonnei* (Anderson & Smith, 1972; WHO, 1978).

A disturbing feature has been the recent spread of resistance from the enterobacteria to other organisms (Williams, 1978). For instance plasmid transfer has led to the acquisition of β-lactamase by *Haemophilus influenzae* (Williams, Katlan & Cavanagh, 1974) and *Neisseria gonorrhoeae* (Percival *et al.*, 1976), with serious consequences in terms of therapy. Another alarming event has been the development of penicillin resistance in *Streptococcus pneumoniae* (Jacobs *et al.*, 1978). That *Streptococcus pyogenes* has remained sensitive to small amounts of penicillin seems little short of a miracle.

As pointed out by Williams (1981) there is little hope of finding a single agent that will act uniformly against antibiotic-resistant gram-negative bacilli. The cephalosporin cefotaxime, for instance, is effective against resistant *Citrobacter*, *Serratia* and *Alcaligenes* but species of *Pseudomonas* and *Acinetobacter* are resistant.

Superimposed on the problems of resistance are several practical difficulties that limit the use of effective antibiotics. An antibiotic that is effective *in vitro* may be of limited value because of: (*a*) toxicity (e.g. chloramphenicol causes aplastic anaemia in about 1 in

25 000 patients); (*b*) poor absorption by the oral route or limited penetration to the site of infection (e.g. eye or cerebrospinal fluid); (*c*) delay in detection of infection (e.g. Group B streptococci are sensitive to penicillin yet there is a high failure rate of penicillin treatment in infections of neonates caused by this organism: Parker, 1979).

Mechanisms of antibiotic resistance

Most antibiotic resistance of clinical significance is due to plasmids (Falkow, 1975; Broda, 1979; Hardy, 1981). Also, resistance genes are sometimes found in discrete genetic units called transposons (Kleckner, 1981) which have the ability to 'hop' from one DNA molecule to another. Transposition can involve mobilization of genes from a plasmid location to the chromosome and vice versa. There is little doubt that transposons contribute to the rapid dissemination of antibiotic resistance. Although less common, resistance can arise from mutations in chromosomal genes.

The molecular mechanisms of antibiotic resistance have been under intensive study since the 1950s when the clinical significance of the problem first became apparent. Recent useful reviews of antimicrobial resistance mechanisms include those of Neu (1982) and Foster (1983). There are basically four categories of mechanisms: (1) alteration in transport system or permeability, (2) enzymatic inactivation, (3) alteration in target site by synthesis of a resistant or alternate target, (4) loss of the sensitive metabolic reaction. Examples of the occurrence of these mechanisms are summarized in Table 3.

Countering the problem of resistance

The pharmaceutical industry has reacted with considerable initiative and chemical ingenuity to the problem of resistance. This is illustrated by the development over the past 20 years of successive generations of β-lactam antibiotics designed to counter resistance to the penicillins and cephalosporins (Waldvogel, 1982). The factors that determine the susceptibility of bacteria to β-lactams are: (1) the rate of penetration of the compound through the outer wall to reach a target penicillin binding protein, (2) presence, type and specific activity of β-lactamase, and (3) effectiveness of the β-lactam in inhibition of peptidoglycan synthesis.

Table 3. *Mechanisms of bacterial resistance to antimicrobial agents*

Agent(s)	Mechanism	Examples
β-Lactams	Plasmid-encoded enzymatic inactivation by β-lactamases	Gram-positive and gram-negative bacteria
	Mutations causing altered affinity of penicillin binding proteins	*S. pneumoniae, N. gonorrhoeae, P. aeruginosa,* methicillin-resistant *S. aureus*
Aminoglycosides	Plasmid-encoded production of inactivating enzymes (phosphotransferase, acetyl transferase, adenyl transferase)	Staphylococci, streptococci, Enterobacteriaceae, pseudomonads
	Mutational change in target site (ribosome) or in transport across cytoplasmic membrane	*N. gonorrhoeae, S. aureus, P. aeruginosa*
Tetracyclines	Plasmid-encoded reduction in accumulation of drug: rapid efflux of drug occurs in resistant cells	Widely encountered in gram-positive and gram-negative bacteria
Chloramphenicol	Plasmid-encoded enzymatic inactivation by chloramphenicol acetyl transferase	Widely distributed in gram-negative and gram-positive bacteria
	Plasmid-encoded reduction in drug accumulation	Gram-negative bacteria
Macrolide–lincosamide–streptogramin B group	Plasmid-encoded alteration of target: methylation of 23S RNA	*S. aureus*, streptococci, *Bacteroides*
Sulphonamide	Chromosomal determinant similar to plasmid-specified mechanism	*S. pneumoniae*
	Plasmid-encoded sulphonamide-resistant dihydropteroate synthase. No longer sensitive to competitive inhibition by sulphonamide	Enterobacteriaceae
Trimethoprim	Plasmid-encoded trimethoprim-resistant dihydrofolate reductase	Enterobacteriaceae

Compiled from Foster (1983).

The chemical modifications used to design semi-synthetic β-lactam compounds having different spectra of activity became possible after the isolation by Batchelor *et al.* (1959) of the 6-amino-penicillanic acid nucleus. Similarly the production of 7-amino-cephalosporanic acid allowed the synthesis of extremely active cephalosporin derivatives. Fig. 1 summarizes some of these modifications.

An alternative strategy to the synthesis of antibiotics that are resistant to inactivating enzymes such as β-lactamase, is to find agents which possess little or no antibiotic activity on their own but which are potent inhibitors of the inactivating enzyme (see Bush & Sykes, 1983). Clavulanic and olivanic acids (Howarth, Brown & King, 1976; Brown *et al.*, 1976) were identified as inhibitors of β-lactamase which block the active site of the enzyme (Charnas, Fischer & Knowles, 1978). Clavulanic acid when combined with amoxycillin (a β-lactamase-sensitive antibiotic) in the combination known as augmentin has proved clinically effective.

Space does not permit detailed consideration of the numerous antimicrobial chemotherapeutic agents made entirely by chemical synthesis. These include: (1) the folic acid antagonists, sulphonamide and trimethoprim, which when used in combination as co-trimoxazole have proved so effective especially in the treatment of urinary tract and respiratory infection (Salter, 1982); (2) the nitro-imidazoles which are effective against anaerobes; (3) the anti-tuberculosis agents, such as isoniazid and ethambutol which have been used in combination therapy along with streptomycin or rifampicin; (4) the antileprosy drug diaminophenyl sulphone (Dapsone); (5) antifungal agents such as the imidazole derivatives, for example ketoconazole (Borgers, 1980), which inhibit biosynthesis of ergosterol, the main sterol in fungal membranes; (6) the new antiviral agent, acyclovir(9-(2-hydroxyethoxymethyl)-guanine), active against herpes viruses (see Timbury, 1982). The phosphorylated active form of the latter drug, which is formed by viral thymidine kinase in infected cells, acts as a potent inhibitor of viral DNA polymerase.

This section would be incomplete without mention of interferon. The interest created in commercial production of interferon from leukocytes or by gene cloning methods for possible use as an antiviral and antitumour agent has been remarkable. It was estimated that in 1982 some 30 different companies were involved in production of natural or 'recombinant' interferon (see Merigan,

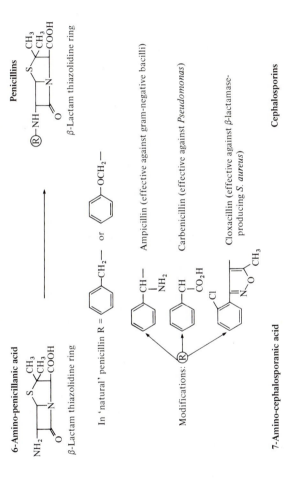

Fig. 1. Modifications leading to development of semi-synthetic β-lactam antibiotics. (Modified from Waldvogel, 1982.)

6-Amino-penicillanic acid

β-Lactam thiazolidine ring

In 'natural' penicillin R =

Modifications:

Ampicillin (effective against gram-negative bacilli)

Carbenicillin (effective against *Pseudomonas*)

Cloxacillin (effective against β-lactamase-producing *S. aureus*)

Penicillins

β-Lactam thiazolidine ring

7-Amino-cephalosporanic acid

β-Lactam dihydrothiazine ring

Modifications in R₁ and R₂ → first-generation cephalosporins (e.g. cephalothin)
second-generation cephalosporins (e.g. cefazolin) } increased potency, broader spectrum
third-generation cephalosporins (e.g. cefotaxime)

Modifications in R₁ and R₂ + modifications in dihydrothiazine ring → moxalactam: active in low concentration against gram-negative bacilli

Cephalosporins

β-Lactam dihydrothiazine ring

1982). Experience with interferon has done much to spearhead the application of recombinant DNA techniques to problems of human medicine. Unfortunately public expectation was raised to unreasonable levels. Experimental findings in animals when interferon was given early in the disease process (either viral infection or experimentally induced tumour) showed striking protective effects. Some promising results have also been obtained in clinical trials (see Merigan, 1982, for references). But many questions remain to be answered before interferon produced by recombinant DNA can be fully evaluated. Are the side-effects (fever, malaise, marrow suppression) acceptable? Gene cloning has revealed numerous interferon genes; so which recombinant interferon should be selected for treatment of particular diseases? Beta interferon and fibroblast-interferons are normally glycosylated but the cloned product from bacterial cells is not glycosylated; how can this problem be overcome? Will interferon emerge as an effective agent after critical evaluation of clinical trials?

The therapeutic challenge from serious viral infection and from neoplasia remains daunting. It is to be hoped that the immense financial and technological effort expended in interferon research will be rewarded by the addition to our armament of a powerful therapeutic agent.

VACCINES

Knowledge that exposure to an attenuated infectious agent could protect against subsequent infection with the virulent form of a pathogen has been the spur that has motivated the search for effective, yet safe vaccines. There have been a limited number of striking successes with both viral and bacterial vaccines in man and animals (Table 4). There is increasing hope that increased knowledge of pathogenic mechanisms coupled with monoclonal antibody techniques and gene cloning technology will lead to improvements in existing vaccines and to the production of an extended range of vaccines effective against protozoan pathogens and parasitic worms as well as bacteria and viruses. The search for effective vaccines is a necessary complement to the use of antimicrobial agents for several reasons. Where possible, prophylactic immunization is a more efficient way of protecting susceptible populations than the use of antimicrobial agents. Also, as indicated previously, resistance to antimicrobial agents is an increasing problem; and in some parts of the world socioeconomic factors exclude the use of these costly agents.

Eradication and control of viral diseases

One of the most outstanding achievements of prophylactic vaccination has been the global eradication of smallpox. Several factors contributed favourably to the eradication of the disease (Fenner, 1982). The most important of these were probably the absence of recurrent infectivity, absence of an animal reservoir, availability of a stable effective vaccine, seasonal fluctuation which permitted interruption of transmission, and the determination and leadership of those responsible for the WHO Smallpox Eradication Programme.

A recent gathering of world experts analysed the lessons of the smallpox campaign and considered other potential candidates for eradication (Stuart-Harris, 1982). Considerable attention focused on measles, especially in the light of the success achieved in the USA (Hinman, 1982). Before the introduction of measles vaccine in 1963, the disease was almost universal in the USA. By 1979 the annual total of reported cases had dropped to less than 5% of that in the pre-vaccine period and the number of deaths had declined similarly. This success was achieved by an intensive immunization campaign; all 50 states now have laws requiring measles vaccination before entry to primary school. There are high hopes that the disease will soon be eliminated from the USA. This contrasts markedly with the situation in the UK, where uptake of the vaccine is only about 48%. A sustained aggressive approach on an international level would be necessary to achieve global eradication.

Great success has also been achieved in elimination of poliomyelitis (Fig. 2) in several developed countries, following introduction of the inactivated poliovirus vaccines in 1955 and oral poliovirus vaccine in 1963 (Nathanson, 1982). However, the picture is very different in less developed countries, where oral poliovirus vaccine has often been found to be ineffective. It is not clear to what extent this is due to inadequate coverage with the vaccine, interference from other endemic enteroviruses or immunosuppression due to schistosomiasis or malaria.

Recent advances in the study of the genome of the polio virus hold out the possibility of improved live vaccine viruses. For instance the total genome of the Sabin type 1 attenuated vaccine

Table 4. *Some current and experimental bacterial and viral vaccines*

Vaccine	Nature of immunogen	Comment
Anthrax	Protein fractions from *Bacillus anthracis*	High efficacy of short duration. Used on persons at risk only
Cholera	Whole phenol-killed *Vibrio cholerae* of Ogawa and Inaba serotypes	Poor efficacy of short duration. Attenuated or genetically engineered vaccine strains under development. Toxoid also a possibility
Dental caries	(i) Cell-associated proteins	Experimental vaccines
	(ii) Whole killed *Streptococcus mutans*	
Diphtheria	Toxoid	Almost completely effective for 10 years. Given alone or in combination with tetanus and pertussis
Enterotoxigenic *E. coli*	(i) Fimbrial antigens	Effective vaccines for animals; experimental for man
	(ii) Enterotoxin	
Gonorrhoea	(i) Fimbrial antigens	Experimental vaccines
	(ii) Outer membrane proteins	
	(iii) IgA protease	
Haemophilus influenzae b	(i) High-molecular-weight polysaccharide from capsule	Experimental vaccines
	(ii) Protein–polysaccharide complexes	
	(iii) Live *E. coli* 100	
Influenza	Killed virus or inactivated protein subunit vaccine	Used in persons at risk from respiratory disease. Effectiveness depends on serotype of prevalent strain
Measles	Live attenuated virus	Highly effective where vaccine uptake is high
Meningococcal Group A	Pure capsular polysaccharide from Group A organisms	Highly effective and long duration
Meningococcal Group B	(i) Serotype-specific outer membrane protein	Experimental vaccine; Group B polysaccharide alone is not immunogenic
	(ii) *O*-acetylated **K** polysaccharide	
Meningococcal Group C	Pure capsular polysaccharide of Group C organisms	Effective in persons over 2 years old. Long duration

Disease	Vaccine	Comments
Meningococcal Group Y, 29E and W135	Pure capsular polysaccharide	Experimental vaccine
Mumps	Live attenuated virus	Use in adolescents and adults, especially males. Probably long duration
Pertussis	Whole heat or chemically killed phase I pertussis organisms	Usually given in combination with diphtheria and tetanus toxoid. 90% effective. Small risk of neurological damage
Pneumococcus	Pure capsular polysaccharide from 14 serotypes	High efficacy; long-lasting
Polio	(i) Attenuated live virus (3 serotypes) given orally (Sabin)	High efficacy; long-lasting
	(ii) Inactivated virus (3 serotypes) given parenterally (Salk)	High efficacy; short duration; boosters required
Rubella	Attenuated live virus	Effective protection against rubella congenital deformity. Children or girls at puberty
Shigella	Live attenuated type-specific *Shigella* organisms given orally	Experimental vaccine
Smallpox	Live attenuated virus	Disease eradicated globally
Streptococcal Group B	Type III$_a$ capsular polysaccharide	Experimental vaccine
Tetanus	Toxoid	Usually given in combination with diphtheria toxoid and pertussis. Highly effective for 10 years
Tuberculosis	Live attenuated *Mycobacterium tuberculosis* (BCG strain)	Efficacy seems variable in different geographical areas. Immunity probably long duration when present
Typhoid	(i) Whole phenol-killed *Salmonella typhi* organisms	Efficacy poor; short duration
	(ii) Whole acetone-killed *Salmonella typhi* organisms	Vi antigen preserved; efficacy high
	(iii) Live attenuated Ty21a gal E mutant (oral)	Experimental; efficacy 85–95% over 3 years at least
Yellow fever	Live attenuated virus	Control of disease depends on vaccination *and* elimination of vector

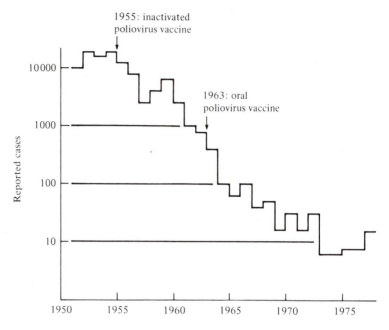

Fig. 2. Eradication of poliomyelitis in the USA. (From Yorke *et al.*, 1979.)

strain has been sequenced (see review by Minor, Kew & Schild, 1982). This revealed 57 base differences (in a total of 7441 nucleotides) from the parent Mahoney strain and suggested that attenuation might involve modification in the virion surface. It is to be hoped that in future gene sequencing will aid development of improved viral vaccines (see below).

The importance of maintaining a high level of immunity in a population ('herd immunity') in order to avoid re-emergence of a disease brought under control by vaccination has been highlighted recently by Anderson & May (1982). The influence of national vaccination policy on the approach to disease control can be seen by comparing the UK and American policies on vaccination against rubella. In the USA, where the objective is to achieve high levels of 'herd immunity' (90% of children vaccinated), and subsequent eradication of rubella, children are immunized at 2 years of age. By contrast in the UK, where immunization is not enforceable by law, the objective is to allow girls to become naturally infected before they reach childbearing age. Remaining susceptible adolescent girls aged 12–13 are immunized to ensure prevention of congenital rubella syndrome.

The structure and antigenic determinants of influenza virus have been thoroughly studied; so too has the immune response to this virus. However, the virus's characteristic ability to change its antigenic composition from year to year continues to frustrate efforts to achieve prophylactic protection by immunization. Currently available vaccines (inactivated intact virus, disrupted virus, purified subunit virus) are not satisfactory (McMichael *et al.*, 1982). Stable live attenuated virus vaccine seems to offer the best hope, though the problem of genetic change remains.

Vaccines are available or are becoming available against several other viral infections (Table 4) and the new approach of developing synthetic vaccines (see below) holds considerable promise. It is appropriate at this point to pay tribute to the early researchers whose work at the turn of the century laid the groundwork for the discovery of the viruses and the development of virology. For instance in 1900 Walter Reed and his colleagues of the Yellow Fever Commission proved that the mosquito was the vector of yellow fever and that the causative agent was ultramicroscopic and filterable. Destruction of the vector and subsequent development of an effective vaccine has done much to bring under control a disease which plagued the modern world for 200 years (see Hruska, 1979).

Bacterial vaccines

Bacterial vaccines have had a major impact on the incidence of certain life-threatening infections, especially where a single major virulence determinant such as a potent exotoxin (diphtheria, tetanus, botulism) or capsular polysaccharide (meningococcal meningitis) has been identified see (Table 4). Immunization with diphtheria and tetanus toxoids has been outstandingly successful, to the extent that cases of diphtheria and tetanus in developed countries are now rare. The number of bacterial vaccines of proven effectiveness in man is, however, still disturbingly low for a number of reasons: in many instances bacterial pathogenesis is multifactorial (e.g. *Staphylococcus aureus*, *Neisseria gonorrhoeae*, *Shigella dysenteriae*); in the case of several important bacterial pathogens the major virulence determinant exists in a large number of serological types (e.g. *Streptococcus pneumoniae*, *Streptococcus pyogenes*, *Neisseria meningitidis*); sometimes 'toxic' reactions have limited public acceptance of the vaccine (e.g. *Bordetella pertussis*). For these reasons and because of the striking success of antibiotics and

Table 5. *Examples of determinants of bacterial pathogenesis*

Virulence determinant	Host defence system affected	Examples
Exotoxins	Altered secretion of and/or damage to mucosal epithelial surfaces (e.g. intestine, respiratory tract)	Cholera enterotoxin; E. coli LT and ST; shigella enterotoxin; salmonella enterotoxin; C. perfringens enterotoxin; B. cereus enterotoxin; S. aureus enterotoxins; B. pertussis toxin
	Damage to host defence cells (inhibition of chemotaxis, killing of phagocytes), local blood vessels, connective tissue	Cytolytic membrane-damaging toxins of S. aureus; C. perfringens; E. coli; and Pasteurella haemolytica
	Damage to tissues distant from focus of infection	Diphtheria toxin; tetanus toxin; botulinum toxins; C. perfringens β- and ε-toxins
Exoenzymes	Damage to connective tissue, possibly acting synergistically with exotoxin	C. perfringens collagenase, hyaluronidase; S. aureus proteases, lipase, DNAase; S. pyogenes hyaluronidase, proteases; DNAase, NADase; P. aeruginosa elastase
Filamentous protein surface antigens	Adherence to mucosal epithelial surfaces	E. coli K88, K99, 987, F41, CFAI, CFAII, P-fimbriae; N. gonorrhoeae fimbriae; B. pertussis filamentous haemagglutinin
Lipoteichoic acid	Adherence to mucosal epithelial surfaces	S. pyogenes
Other surface proteins	Prevention of phagocytosis	S. pyogenes M protein; S. aureus protein A (binds to Fc region of immunoglobulins and prevents opsonization)
Capsular polysaccharides	Prevention of phagocytosis	S. pneumoniae; H. influenzae b; N. meningitidis A. B, C; E. coli K1; Group B streptococci; S. aureus (some strains)
Lipopolysaccharide	Many systems affected. Effects include: lethality, complement activation, disseminated intravascular coagulation, circulatory hypotension	Gram-negative bacteria
Outer membrane proteins A	Resistance to serum killing	Protein I class of N. gonorrhoeae; Tra T protein of E. coli
		Protein II class of N. gonorrhoeae
	Adhesion to epithelial cells and leukocytes	
	Iron availability	Outer membrane protein receptors for iron uptake systems in E. coli and Vibrio anguillarium

chemotherapeutic agents, interest in developing bacterial vaccines declined sharply in the 1950s. Emergence of drug resistance as a major chemical problem, together with the recent rapid improvement in knowledge of bacterial pathogenesis and host defence systems, has stimulated renewed interest. Some of the main bacterial virulence determinants are listed in Table 5. The rapidly advancing developments in bacterial pathogenesis have recently been reviewed in some detail (O'Grady & Smith, 1981; Smith, Arbuthnott & Mims, 1983).

Progress is being made in developing effective vaccines against encapsulated organisms (see Arbuthnott, Owen & Russell, 1983). An encouraging feature of such vaccines is their lack of reactogenicity in man. It has been estimated that about 130 million people have been injected with these antigens with little evidence of adverse side-effects (Robbins, 1978). The efforts of Austrian and others (see Austrian, 1981) have led to the development of a polyvalent vaccine containing purified capsular polysaccharide from 14 serotypes responsible for most of the serious pneumococcal infections. Successful prevention of meningococcal meningitis caused by Groups A, C and Y has been achieved with capsular polysaccharide vaccines (WHO, 1976; Peltola *et al.*, 1977) and these are now being manufactured in five countries. Protection against Group B meningococci presents a problem because Group B polysaccharide is a poor immunogen. In this case outer membrane protein antigens have been identified as possible immunogens (Frasch & Robbins, 1978). Also the type b polysaccharide of *Haemophilus influenzae* is non-toxic and immunogenic in adults and in children older than 14 months (Makela *et al.*, 1977), but the protective effect appears to be short-lived.

The role of the capsular polysaccharides in *E. coli* K1 strains and Group B streptococci, both important causes of neonatal meningitis, are well established as determinants of virulence and it is to be hoped that progress will be made in achieving anticapsular immunity in neonates.

The remarkable progress in analysis of the molecular mechanisms of pathogenesis of enterotoxigenic strains of *E. coli* (ETEC) has provided useful strategies for prevention of diarrhoeal diseases by immunization, especially in domestic animals. Vaccines against ETEC strains that possess the adhesive fimbrial antigens K88, K99 and 987 are now commercially available. An alternative approach is to immunize against the heat-labile enterotoxin (LT toxin) produced by many of these strains. Attention is now focused on making effective human *E. coli* vaccines based on the fimbrial antigens CFA I and II or LT enterotoxin.

Recently Holmgren (1981) and Greenberg & Guerrant (1981) pointed to yet other possibilities. Binding of adhesive fimbrial antigens to their receptors on mucosal surfaces might be prevented by administration of agents that block the receptor. Also the loss of fluid from the intestinal mucosa as a result of enterotoxin action might be countered by using anti-secretory drugs that prevent activation of adenylate or guanylate cyclase.

Cholera remains a common and serious infection in many less developed countries. Cholera enterotoxin, the cause of the massive outpouring of fluid in the intestine, has been thoroughly characterized in molecular terms. It is an example of a bipartite toxin in which the B, or binding, region attaches to a specific receptor, GM1 ganglioside, allowing subsequent penetration of the toxic A portion of the toxin molecule. The B region of cholera enterotoxin induces antibodies that neutralize the intact toxin. However, vaccines designed to confer antitoxic immunity in man have as yet given disappointing results. Recently an alternative approach was followed by Honda & Finkelstein (1979) who isolated a variant of *Vibrio cholerae* with a mutation in the gene for toxin production. This mutant produced only choleragenoid (i.e. the B subunit of cholera toxin) and no A subunit. Such a variant or a suitable genetically engineered train with similar properties could be the forerunner of a live vaccine that would stimulate a specific IgA response in the gut.

A potentially valuable type of vaccine has recently been shown to give good protection against typhoid fever (Wahdan *et al.*, 1982). The mutant strain Ty21a is a gal E mutant of *Salmonella typhi* which is attenuated and can be used as a live oral vaccine. Gal E mutants form smooth-type lipopolysaccharide only in the presence of galactose; however, they lack the ability to metabolize galactose, which accumulates intracellularly as galactose-1-phosphate or UDP galactose. This in turn leads to lysis and death of the cells. Such self-destroying strains are therefore attractive candidates as vaccine strains. In experimental challenge studies in volunteers the vaccine gave a high protection against a challenge dose that caused typhoid fever in 53% of unimmunized subjects. In a controlled field trial in Egypt carried out between 1978 and 1981 in children aged 6–7, protection was demonstrated in approximately 95% of vaccinated children, with minimal side-effects.

Considerable controversy surrounds the use of pertussis vaccine against whooping cough. In the UK confidence in the pertussis vaccine collapsed in the early 1970s as a result of public concern about its safety. This led to a marked decrease in uptake of the vaccine and was followed in 1977–9 by an epidemic which was responsible for 36 deaths and 17 cases of brain damage (see Anderson & May, 1982). Also, the course of the illness in non-fatal cases that occurred during the epidemic was protracted (10–12 weeks) and there were an estimated 5000 hospital admissions. A similar pattern of events occurred in Japan following reduced vaccination as a result of public misgivings in 1974. Careful studies of neurological disorders among children vaccinated against pertussis in the UK between 1976 and 1979 revealed an estimated risk of 1 in 310000 of persisting neurological damage one year after vaccination (see Editorial, 1981). Calculations based on birth rate and number of immunizing doses indicate that the consequences of non-vaccination are distinctly worse than those of vaccination and major efforts are being made to increase the uptake of the vaccine.

In the meantime research is being actively directed towards identification of the principal virulence determinants of *Bordetella pertussis* with a view to developing an improved vaccine. So far gene cloning techniques have not been successful but the genetic approach of Weiss & Falkow (1983), in which transposon mutagenesis has been employed to create derivatives defective in individual virulence factors, appears promising. Also progress is being made in purifying possible virulence components such as filamentous haemagglutinin by conventional methods.

The possibility of a malaria vaccine

Intensive efforts by the WHO directed towards eradicating malaria by the use of insecticides to kill mosquitos and chloroquine to treat affected individuals have not been successful. Resistance to the insecticides and to chloroquine has been the main problem. Accordingly considerable attention is now focused on exploiting recombinant DNA and monoclonal techniques to identify protection parasite antigens (see Newmark, 1983). Three different vaccines directed against the sporozoite, merozoite and gametocyte stages of the life cycle of *Plasmodium* are being pursued and promising results have been obtained for a sporozoite vaccine prepared by gene cloning techniques (Ellis *et al.*, 1983). With an estimated 200 million people infected with malaria at any one time and 2 million deaths from malaria per year, there is understandable urgency to determine the value of a prophylactic approach based on vaccination. Other parastic diseases caused by protozoans (e.g. trypanasomiasis and amoebiasis) or parasitic worms (e.g. schistosomiasis) also affect many millions of people and are the targets for possible vaccine development.

Synthetic vaccines

Problems associated with the instability of some killed virus vaccines and the possibility of mutation of live attenuated bacterial or viral vaccines have led to interest in the possibility of devising synthetic peptide vaccines that contain the amino acid sequence present in the antigenic determinant of a protective antigen (see Lerner, 1983). The technical problems of synthesizing peptides corresponding to short sequences of individual proteins have been overcome; amino acid sequences can be determined by traditional sequencing methods or from the nucleotide sequences of cloned genes, and peptides can be generated by Merrifield synthesis. Computer modelling has also been employed to identify peptides that are surface antigenic determinants. The potential for synthetic vaccines is illustrated by recent advances with foot-and-mouth disease virus. A peptide corresponding to positions 141–160 of virus protein 1 (VP1) of type O virus was synthesized which induced protective antibodies in guinea-pigs (Bittle *et al.*, 1982). Progress has also been made with synthetic peptides corresponding to regions of the major polypeptide portion of hepatitis B surface antigen (Lerner *et al.*, 1981; Dreesman *et al.*, 1982). Thus the groundwork has been laid for a rational approach to the design of synthetic viral vaccines.

To elicit an adequate level and duration of immunity to synthetic peptides, safe effective carriers and adjuvants will be required. Encouraging results have been reported by Audibert *et al.* (1982) who have produced a totally synthetic diphtheria vaccine comprising an octadecapeptide linked along with adjuvant to a synthetic carrier.

THE CHANGING PATTERN OF INFECTION: EXAMPLES OF 'NEW' DISEASES

Childhood immunization, improved hygiene, sanitation and housing, and the availability of a powerful range of antimicrobial agents

Table 6. *Decline in incidence of major notifiable infectious diseases in England and Wales*

	1960	1970	1979
Diphtheria	49	22	0
Poliomyelitis	378	6	8
Measles	159 364	307 408	77 386
Bacillary dysentery	43 285	10 775	2 786
Infectious hepatitis	Not available	23 569	3 203
Pulmonary tuberculosis	24 000	10 000	9 688[a]

From Havard (1982).

[a] 1978 figure.

have been responsible for the decline in notifiable infectious diseases in the past 60 years. The trends over the past two decades can be seen in the incidence of major notifiable infectious diseases for England and Wales (Havard, 1982) (Table 6).

This decrease has been counterbalanced by the parallel increase in 'new' diseases. In part these result from new forms of treatment: the development of organ transplantation, cardiac surgery and aggressive cytotoxic drug treatment of leukaemia and other neoplasms, for example. Changes in eating habits such as greater use of processed food and the reheating of food in canteen and 'fast-food' restaurants has led to an increase in food poisoning. The greater availability of air travel has made the holiday-maker more adventurous and has brought a large increase in visitors from distant parts; both have brought problems of infection. For example, diarrhoea from enterotoxigenic *E. coli* affects many travellers and there has been a large increase in notifications of cases of imported malaria in the UK (over 2000 cases in 1979).

Other changes in the modern way of life have resulted in an altered pattern of infection. This can be seen in the incidence and causes of sexually transmitted diseases.

Sexually transmitted diseases

After the Second World War the incidence of syphilis and gonorrhoea in the UK dropped from a peak in 1946 to low levels in 1955 and 1956. Although the incidence of syphilis has remained steady at around 4500 new cases per year, the incidence of gonorrhoea increased rapidly during the 1960s and 1970s before levelling off at a peak of 66 000 new cases in 1977 (Thin, 1982). In 1976 β-lactamase-producing penicillin-resistant gonococci first appeared. The number of resistant isolates has continued to rise and has reached the alarming level of 19.2% in Singapore (Rajan, Thirumoorthy & Tan, 1981). Non-gonococcal urethritis in men was first reported in 1951. The numbers of cases in England and Wales rose dramatically from about 11 000 in 1951 to about 80 000 in 1979 (Thin, 1982). *Chlamydia trachomatis* is the causative organism in 30–50% of non-gonococcal urethritis. The development in some cases of serious consequences of chlamydial infection (Dunlop, 1981) emphasizes the need for careful clinical management of the disease. Recurrent infection due to herpes simplex virus 2 is another form of sexually transmitted disease that has increased recently.

Currently intensive research is being directed towards attempting to establish whether an infectious agent is responsible for the recently described Acquired Immune Deficiency Syndrome (AIDS) which first appeared in 1979 in the USA among male homosexuals, who account for 72% of all cases so far. An as yet unrecognized bacterial or viral agent might be the cause of what could be a new form of sexually transmitted disease. The high mortality rate associated with AIDS, together with the fact that the condition has been recognized in other groups (drug addicts, Haitian immigrants and haemophiliacs) and is no longer restricted to the USA is reason for concern (see Waterson, 1983). The degree of immune deficiency in AIDS patients is greater than that seen in patients receiving immunosuppressive therapy. The main features are: (i) deficiency in T-helper and inducer subsets such that the helper/suppressor cell ratio among T-lymphocytes is reversed; (ii) lowering of natural killer cell activity; (iii) development of autoimmune phenomena. Those affected develop infections usually seen in immunocompromised individuals; the organisms responsible include *Pneumocystis carnii*, *Candida albicans*, *Cryptococcus neoformans* and cytomegalovirus. It is notable, however, that the humoral immune system remains unaffected.

Epidemiological investigations strongly suggest that the cause of immune deficiency is communicable and the effects on the cellular immune system resemble those of cytomegalovirus. Intensive efforts are being made to identify the cause of AIDS – without success as yet. An unusual ultrastructural feature (vesicular rosette) has been reported in an electron microscopic study of the lymph nodes taken at autopsy (Ewing *et al.*, 1983), but the significance of this finding is not certain.

Legionellosis

The isolation of the causative organism responsible for Legionnaires' disease illustrates how carefully conducted epidemiological investigation coupled with close attention to growth requirements of a possible infectious agent can solve the challenging problems posed by 'new' infectious diseases (Fraser & McDade, 1979). The commendable intelligence gathering of the Centre for Disease Control of the US Public Health Service rapidly established that the disease did not spread from person to person but that the most likely mode of transmission was airborne. The previously unrecognized bacterium responsible, *Legionella pneumophila*, was isolated not during routine bacteriological investigations but in tests in guinea-pigs and chick embryos designed to detect rickettsias. The fastidious nature of the organism led to a delay in obtaining suitable *in vitro* growth conditions, though once perfected, the medium gave good recovery of the organism from fresh pathological specimens. It soon became evident that legionellosis was not a rare disease; retrospective studies revealed several earlier outbreaks. The past 5 years has seen rapid development of knowledge in the following areas: characterization of *L. pneumophila*; the aetiology, pathogenesis and epidemiology of infection; role of aerosols in transmission; antimicrobial therapy; rapid diagnosis (Meyer, 1983). Also six other species of *Legionella* have been identified as being associated with pneumonia in man.

Gastrointestinal infections

In global terms acute gastroenteritis is a common and devastating disease. In the tropics it is estimated to cause 3–5 million deaths in infants and young children each year (see O'Brien, 1982). There is a link with malnutrition, and malnourished children are more vulnerable to attack. In the past 20 years the causes of gastrointestinal infection have been studied intensively. To some extent interest in this area has been stimulated by the successful analysis in man and animals of the pathogenic mechanisms of cholera and diarrhoeal disease caused by enterotoxigenic *E. coli*.

Contrary to previous views, viral diarrhoea has proved to be of great importance. Rotavirus (Bishop, Davidson & Holmes, 1973) is now recognized as being responsible (alone or together with bacterial pathogens) for most episodes of infantile diarrhoea in Europe, Australia, North America and the tropics (O'Brien, 1982). Studies in Bangladesh have established that this virus (alone or together with enterotoxigenic *E. coli*) accounts for 60–80% of diarrhoea and life-threatening dehydration in young children (Sack *et al.*, 1978; Black *et al.*, 1981*a*). Infection is by the faecal–oral route and the disease affects children aged from 6 months to 3 years. The virus can be demonstrated in stool filtrates by electron microscopy or by radioimmune or enzyme-linked immunoabsorbent assay; diagnosis can be confirmed by a rising titre of antibody in serum.

Entertoxigenic *E. coli* have now been firmly implicated as a cause of diarrhoea in patients of all ages in Bangladesh (Black *et al.*, 1981*b*). There is little doubt that these organisms are responsible for much diarrhoeal disease in children in other areas, especially in less developed countries, and that they cause traveller's diarrhoea (see Greenberg & Guerrant, 1981). The role of the fimbrial antigens CFA I and CFA II and the LT and ST enterotoxins in the pathogenesis of enterotoxigenic *E. coli* has been clearly established.

Another important 'new' cause of diarrhoea in Europe, Asia and Africa is the minute slender gram-negative vibrio *Campylobacter jejuni*, which was first recognized in the early 1970s (see review by O'Brien, 1982). Polluted water and unpasteurized milk appear to be the main sources of infection. Diarrhoea may be accompanied by the presence of blood and mucus and attacks can last for 2 weeks or more. Care is required in the culture of the organism, which is rapidly overgrown by other faecal organisms; isolation can be achieved in selective media containing added antibiotics, incubated at 42 °C (Billingham, 1981).

The final example of a recently recognized infectious disease of the intestinal tract reveals yet another complexity of antibiotic therapy. In pseudomembranous colitis (PMC) the presence of certain antibiotics encourages growth and toxin production in the colon by the organism *Clostridium difficile*; two toxins have been implicated (a potent cytotoxin and an enterotoxin). This results in the development of diarrhoea and the formation of a necrotic pseudomembrane in the colon (Chang, Bartlett & Taylor, 1981). In recent years clindamycin has been associated with PMC with incidences ranging from 0.01% to 10% in treated patients. An understanding of the pathogenesis of this disease arose from studies in an animal model of PMC in hamsters (Bartlett, 1979) which implicated the involvement of a clostridial toxin. Also, *C. difficile* was the only organism isolated from affected animals that caused the symptoms of caecitis when injected into healthy animals.

These findings in animals complemented studies in patients. A potent cytotoxin was recognized in the stools of patients with PMC by tests in tissue cultures (Larson & Price, 1977) and cultures of stools from such patients usually yielded *C. difficile*. It is now generally accepted that PMC results from the action of *C. difficile* toxins on the colon following outgrowth of the organism as a result of oral administration of antibiotics.

Staphylococcal Toxic Shock Syndrome

Staphylococcal Toxic Shock Syndrome (TSS) leapt to prominence as a matter of public concern in the USA in the spring and summer of 1980. Through the media it became known as a potentially fatal illness suddenly striking young women, especially adolescents, during menstruation. The term TSS had previously been coined by Todd *et al.* (1978) to describe an illness that had symptoms of high fever, rash, lowered blood pressure and, in survivors, subsequent desquamation of the palms and soles. Epidemiological studies carried out by the Centre for Disease Control in the USA at the peak of the disease, when over 100 cases per month were being reported, indicated an association with the use of vaginal tampons and with carriage of certain phage Group I strains of *Staphylococcus aureus*. The notable absence of bacteraemia in most cases suggested the involvement of diffusible toxic factors produced by *S. aureus*, and the search for the toxin(s) responsible began.

There have been three independent reports of the isolation of a 'toxic' protein, termed staphylococcal enterotoxin F (Bergdoll *et al.*, 1981), pyrogenic exotoxin C (Schlievert *et al.*, 1981) and toxic shock factor (de Azavedo, Hartigan & Arbuthnott, 1983); all three groups appear to have isolated the same protein. Although the role of this factor in the pathogenesis of TSS remains to be firmly established, its ability (either alone or in combination with endotoxin) to induce pyrogenicity and a lethal shock-like syndrome in rabbits makes it worthy of further study. There is no doubt, however, that the disease is multi-factorial and depends on the susceptibility of the host, the virulence of the *S. aureus* strain and precipitating factors such as use of tampons.

These examples of newly recognized infectious disease serve to underline the continuing importance of the microbe as a cause of disease. They also illustrate the continuing challenge that faces the microbiologist and his clinical colleagues.

CONCLUDING REMARKS

I have attempted to highlight the dramatic progress that has been made in conquering infectious disease and to demonstrate the impact that microbiology has had, and is having, on the development of modern medicine. The basic strategies for attacking the microbe (diagnostic methods, selective antimicrobial action, protective immunity) were developed with startling insight by the pioneers of the subject. Modern advances in microbial physiology, structure and genetics together with the harnessing of the most powerful techniques of biochemistry, immunology and molecular biology have seen a remarkable refinement of these basic strategies and a new understanding of the complexities of host/pathogen interactions.

In developed countries these advances will be used to secure our advantage over major infections that affect mankind and to evolve methods of combating the 'new' pathogens that have emerged to take advantage of our altered social behaviour.

It is to be hoped that the less developed countries will benefit from a new determination to tackle the enormous microbiological and public health problems associated with protecting or treating the infections that are endemic or epidemic in large, and often disadvantaged populations. The next decade could be one of the most exciting yet in man's constant attempts to conquer the microbe.

I am grateful to Drs F. Falkiner, T. J. Foster and R. Russell for their help in preparing this paper, and to Gillian Johnston for typing the manuscript.

REFERENCES

ABRAHAM, E. P. (1980). Fleming's discovery. *Reviews of Infectious Diseases*, **2,** 140–1.

AHARONOWITZ, Y. & COHEN, G. (1981). The microbiological production of pharmaceuticals. *Scientific American*, **245**, 106–18.

ALDRIDGE, C., JONES, P. W., GIBSON, S., LANHAM, J., MEYER, M., VANNEST, R. & CHARLES, R. (1977). Automated microbiological detection/identification system. *Journal of Clinical Microbiology*, **6**, 406–13.

ALTER, H. J., HOLLAND, P. V. & PURCELL, R. H. (1971). Counter-immunoelectrophoresis for detection of hepatitis-associated antigen: methodology and comparison with gel diffusion and complement fixation. *Journal of Laboratory and Clinical Medicine*, **77**, 1000–10.

ANDERSON, E. S. & SMITH, H. R. (1972). Chloramphenicol resistance in the typhoid bacillus. *British Medical Journal*, **iii**, 329–31.

ANDERSON, R. & MAY, R. (1982). The logic of vaccination. *New Scientist*, **96**, 410–15.

ARBUTHNOTT, J. P., OWEN, P. & RUSSELL, R. J. (1983). Bacterial antigens. In *Topley and Wilson's Principles of Bacteriology, Virology and Immunity*, 7th edn, vol. 1., ed. A. Miles & G. Wilson. London: Edward Arnold (in press).

AUDIBERT, F., JOLIVET, M., CHEDID, L., ARNON, R. & SELA, M. (1982). Successful immunization with a totally synthetic diphtheria vaccine. *Proceedings of the National Academy of Sciences, USA*, **79**, 5042–6.

AUSTRIAN, R. (1981). Some observations on the pneumococcus and on the current status of pneumococcal disease and its prevention. *Reviews of Infectious Diseases*, **2**, *Supplement*, S1–S18.

BARTLETT, J. G. (1979). Antibiotic-associated pseudomembranous colitis. *Reviews of Infectious Diseases*, **1**, 530–9.

BATCHELOR, F. R., DOYLE, F. P., NAYLER, J. H. C. & ROLINSON, G. N. (1959). Synthesis of penicillin: 6-amino-penicillanic acid in penicillin fermentations. *Nature, London*, **183**, 257–8.

BERGDOLL, M. S., CROSS, B. A., REISSER, R. F., ROBBINS, R. N. & DAVIS, J. P. (1981). A new staphylococcal enterotoxin, enterotoxin F, associated with toxic-shock-syndrome *Staphylococcus aureus* isolates. *Lancet*, **II**, 1017–21.

BILLINGHAM, J. D. (1981). A comparison of two media for the isolation of *Campylobacter* in the tropics. *Transactions of the Royal Society of Tropical Medicine and Hygiene*, **75**, 645–6.

BISHOP, R. F., DAVIDSON, G. P. & HOLMES, I. H. (1973). Virus particles in epithelial cells of duodenal mucosa from children with acute non-bacterial gastroenteritis. *Lancet*, **II**, 1281–3.

BITTLE, J. L., HOUGHTEN, R. A., ALEXANDER, H., SHINNICK, T. M., SUTCLIFFE, J. G., LERNER, R. A., ROWLANDS, D. J. & BROWN, F. (1982). Protection against foot-and-mouth disease by immunization with a chemically synthesized peptide predicted from the viral nucleotide sequence. *Nature, London*, **298**, 30–3.

BLACK, R. E., MERSON, M. H., HUG, I., ALIM, A. R. & YUMUS, M. (1981*a*). Incidence and severity of rotavirus and *Escherichia coli* diarrhoea in rural Bangladesh: implications for vaccine development. *Lancet*, **I**, 141–3.

BLACK, R. E., MERSON, M. H., ROWE, B., TAYLOR, P. R., ABDUL ALIM, A. R., GROSS, R. J. & SACK, D. A. (1981*b*). Enterotoxigenic *Escherichia coli* diarrhoea, acquired immunity and transmission in an endemic area. *Bulletin of the World Health Organization*, **59**, 263–8.

BORGERS, M. (1980). Mechanism of action of antifungal drugs, with special reference to the imidazole derivatives. *Reviews of Infectious Diseases*, **2**, 520–34.

BRODA, P. (1979). *Plasmids*. London: W. H. Freeman.

BROWN, A. G., BUTTERWORTH, D., COLE, M., HANSCOMB, G., HOOD, J. D. & READING, C. (1976). Naturally-occurring β-lactamase inhibitors with antibacterial activity. *Journal of Antibiotics*, **29**, 668–9.

BUSH, K. & SYKES, R. B. (1983). β-Lactamase inhibitors in perspective. *Journal of Antimicrobial Chemotherapy*, **11**, 97–107.

CAFFERKEY, M. T., HONE, R. & KEANE, C. T. (1982). Severe staphylococcal infections treated with vancomycin. *Journal of Antimicrobial Chemotherapy*, **9**, 69–74.

CHANG, T.-W., BARTLETT, J. G. & TAYLOR, N. S. (1981). *Clostridium difficile* toxin. *Pharmacology and Therapeutics*, **13**, 441–52.

CHARNAS, R. L., FISCHER, J. & KNOWLES, J. R. (1978). Chemical studies on the inactivation of *Escherichia coli* RTEM β-lactamase by clavulanic acid. *Biochemistry*, **17**, 2185–9.

CROSSLEY, K., LANDESMAN, B. & ZASKA, D. (1979). An outbreak of infections caused by strain of *Staphylococcus aureus* resistant to methicillin and aminoglycosides. II. Epidemiological studies. *Journal of Infectious Diseases*, **139**, 280–7.

D'AMATO, R. F., HOLMES, B. & BOTTONE, E. J. (1981). The systems approach to diagnostic microbiology. *CRC Critical Reviews in Microbiology*, **9**, 1–44.

DE AZAVEDO, J. C. S., HARTIGAN, P. J. & ARBUTHNOTT, J. P. (1983). Toxins in relation to Toxic Shock Syndrome. In *Proceedings of the European Workshop on Bacterial Toxins*, ed. J. H. Freer, F. Fehrenbach & J. Jeljaszewicz. New York & London: Academic Press (in press).

DOMAGK, G. (1935). Chemotherapy of bacterial infections. *Deutsche medizinische Wochenschrift*, **61**, 250–3.

DREESMAN, G. R., SANCHEZ, Y., IONESCU-MATIU, I., SPARROW, J. T., SIX, H. R., PETERSON, D. C., HOLLINGER, F. B. & MELNICK, J. L. (1982). Antibody to hepatitis B surface antigen after a single inoculation of uncoupled synthetic HBsAg peptides. *Nature, London*, **295**, 158–60.

DUNLOP, E. M. C. (1981). Chlamydial infection: terminology, disease and treatment. In *Recent Advances in Sexually Transmitted Diseases*, vol. 2, ed. J. R. W. Harris, pp. 101–19. Edinburgh: Churchill Livingstone.

EDITORIAL (1981). Pertussis vaccine. *British Medical Journal*, **282**, 1563–4.

EDITORIAL (1982). How long can we go on winning? *Journal of Infection*, **4**, 1–3.

ELLIS, J., OZAKI, L. S., GWADZ, R. W., COCHRANE, A. H., NUSSENZWEIG, V., NUSSENZWEIG, R. S. & GODSON, G. N. (1983). Cloning and expression in *E. coli* of the malarial sporozoite surface antigen gene from *Plasmodium knowlesi*. *Nature, London*, **302**, 536–8.

EWING, E. P., SPIRA, T. J., CHANDLER, F. W., CALLAWAY, C. S., BRYNES, R. K. & CHAN, W. C. (1983). Unusual cytoplasmic body in lymphoid cells of homosexual men with unexplained lymphadenopathy: a preliminary report. *New England Journal of Medicine*, **308**, 819–22.

FALKOW, S. (1975). *Infectious Multiple Drug Resistance*. London: Pion Ltd.

FENNER, R. (1982). A successful eradication campaign. The global eradication of smallpox. *Reviews of Infectious Diseases*, **4**, 916–22.

FLEMING, A. (1929). On the antibacterial action of cultures of a *Penicillium* with special reference to their use in the isolation of *B. influenzae*. *British Journal of Experimental Pathology*, **10**, 226–36.

FOSTER, T. J. (1983). Plasmid-determined resistance to antimicrobial drugs and toxic metal ions in bacteria. *Microbiological Reviews*, **47**, 361–409.

FRASCH, C. E. & ROBBINS, J. D. (1978). Protection against group B meningococcal disease. III. Immunogenicity of serotype 2 vaccines and specificity of protection in a guinea pig model. *Journal of Experimental Medicine*, **147**, 629–44.

FRASER, D. W. & McDADE, J. E. (1979). Legionellosis. *Scientific American*, **241**, 82–96.

FULTON, R. E. & MIDDLETON, P. J. (1974). Comparison of immunofluorescence and isolation techniques in diagnosis of respiratory viral infections of children. *Infection and Immunity*, **10**, 92–101.

GREENBERG, R. N. & GUERRANT, R. L. (1981). *E. coli* heat-stable enterotoxin. *Pharmacology and Therapeutics*, **13**, 507–31.

HADLEY, W. K. & SENYK, A. (1975). Early detection of microbial metabolism and growth by measurement of electrical impedance. In *Microbiology 1975*, ed. D. Schlessinger, pp. 12–21. Washington, DC: American Society for Microbiology.

HARDY, K. (1981). *Bacterial Plasmids*. London: Nelson.

HAVARD, C. W. H. (1982). The transfiguration of infection. In *The Medical Annual*, ed. R. B. Scott, pp. 189–92. Bristol, London & Boston: Wright PSG.

HINMAN, A. R. (1982). World eradication of measles. *Reviews of Infectious Diseases*, **4**, 933–6.

HOLMGREN, J. (1981). Actions of cholera toxin and the prevention and treatment of cholera. *Nature, London*, **292**, 413–17.

HONDA, T. & FINKELSTEIN, R. A. (1979). Selection and characteristics of a *Vibrio cholerae* mutant lacking the A(ADP-ribosylating) portion of the cholera enterotoxin. *Proceedings of the National Academy of Sciences, USA*, **76**, 2052–6.

HONE, R., CAFFERKEY, M., KEANE, C. T., HARTE-BARRY, M., MOORHOUSE, E., CARROLL, R., MARTIN, F. & RUDDY, R. (1981). Bacteraemia in Dublin due to gentamicin-resistant *Staphylococcus aureus. Journal of Hospital Infection*, **2**, 119–26.

HOWARTH, T. T., BROWN, A. G. & KING, T. J. (1976). Clavulanic acid, a novel β-lactam antibiotic isolated from *Streptomyces clavuligerus. Journal of the Chemical Society Chemical Communications*, (1976), 266–7.

HRUSKA, J. F. (1979). Yellow Fever virus. In *Principles and Practice of Infectious Disease*, ed. G. I. Mandell, R. G. Douglas & J. E. Bennett, pp. 1253–7. New York: Wiley.

JACOBS, M. R., KOORNHOF, H. J., ROBBINS-BROWNE, R. M., STEVENSON, C. M., VERMAAK, Z. A., FREIMAN, I., MILLER, G. B., WHITCOMB, M. A., ISAACSON, M., WARD, J. I. & AUSTRIAN, R. (1978). Emergence of multiply resistant pneumococci. *New England Journal of Medicine*, **299**, 735–40.

KLECKNER, N. (1981). Transposable elements in procaryotes. *Annual Reviews of Genetics*, **15**, 341–404.

LARSON, H. E. & PRICE, A. B. (1977). Pseudomembranous colitis; presence of clostridial toxin. *Lancet*, **II**, 1312–14.

LERNER, R. A. (1983). Synthetic vaccines. *Scientific American*, **245**, 48–56.

LERNER, R. A., GREEN, N., ALEXANDER, H., LIU, F.-T., SUTCLIFFE, J. G. & SHINNICK, T. M. (1981). Chemically synthesized peptides predicted from the nucleotide sequence of hepatitis B virus genome elicit antibodies reactive with the native envelope protein of Dane particles. *Proceedings of the National Academy of Sciences, USA*, **78**, 3403–7.

McMICHAEL, A. J., ACKONAS, B. A., WEBSTER, R. G. & LAVER, W. G. (1982). Vaccination against influenza: B-cell or T-cell immunity? *Immunology Today*, **3**, 256–60.

MAKELA, P. H., PELTOLA, H., KAYTHY, H., JOUSIMIES, H., PETTAY, O., RUOSLATHTI, E., SIVONEN, A. & RENKONEN, O. (1977). Polysaccharide vaccines of group A *Neisseria meningitidis* and *Haemophilus influenzae* type b: a field trial in Finland. *Journal of Infectious Diseases*, **136**, S43–S56.

MERIGAN, T. C. (1982). Interferon: the first quarter century. *Journal of the American Medical Association*, **248**, 2513–16.

MEYER, R. D. (1983). *Legionella* infections: a review of five years of research. *Reviews of Infectious Diseases*, **5**, 258–78.

MINOR, P. D., KEW, O. & SCHILD, G. C. (1982). Poliomyelitis: epidemiology, molecular biology and immunology. *Nature, London*, **299**, 109–10.

MOSELEY, S. L., ECHERERRIA, P., SERRINATAMA, J., TIRAPAT, C., CHJAOICUMPA, W., SAKULDAIPEARA, T. & FALKOW, S. (1982). Identification of enterotoxigenic

Escherichia coli by colony hybridization using three gene probes. *Journal of Infectious Diseases*, **145**, 863–9.

NATHANSON, N. (1982). Eradication of poliomyelitis in the United States. *Reviews of Infectious Diseases*, **4**, 940–5.

NEU, H. C. (1982). Mechanisms of bacterial resistance to antimicrobial agents with particular reference to cefotaxime and other β-lactam compounds. *Reviews of Infectious Diseases*, **4**, Supplement, S288–S299.

NEWMARK, P. (1983). What chance a malaria vaccine? *Nature, London*, **302**, 473.

O'BRIEN, W. (1982). Tropical diseases. In *The Medical Annual*, ed. R. B. Scott, pp. 320–30. Bristol, London & Boston: Wright PSG.

O'GRADY, F. & SMITH, H. (1981). *Microbial Perturbation of Host Defences*. New York & London: Academic Press.

PARKER, M. T. (1979). Perinatal and neonatal infections: infections with group B streptococci. *Journal of Antimicrobial Chemotherapy, Supplement A 5*, 27–37.

PELTOLA, H. *et al.* (1977). Clinical efficacy of meningococcus Group A capsular polysaccharide vaccine in children three months and five years of age. *New England Journal of Medicine*, **297**, 686–91.

PERCIVAL, A., ROWLANDS, J., CORKILL, J. E., ALERGARD, C. D., ARYA, O. P., REES, E. & ANELS, E. H. (1976). Penicillinase-producing gonococci in Liverpool. *Lancet*, **11**, 1379–82.

RAJAN, V. S., THIRUMOORTHY, T. & TAN, N. J. (1981). Epidemiology of penicillinase-producing *Neisseria gonorrhoeae* in Singapore. *British Journal of Venereal Diseases*, **57**, 158–61.

RANDALL, E. (1975). Long-term evaluation of a system for radiometric detection of bacteremia. In *Microbiology 1975*, ed. D. Schlesinger, pp. 39–44. Washington, DC: American Society for Microbiology.

REPORT (1981). National survey of infection in hospitals, 1980. *Journal of Hospital Infection*, **2**, 1–51.

ROBBINS, J. B. (1978). Vaccines for the prevention of encapsulated bacterial diseases: current status, problems and prospects for the future. *Immunochemistry*, **15**, 839–54.

SACK, D. A., CHOWDHURY, A. M. A. K., EUSOF, A., AKBAR, ALI, MERSON, M. H., ISLAM, S., BLACK, R. E. & BROWN, K. H. (1978). Oral hydration in rotavirus diarrhoea: a double blind comparison of sucrose with glucose electrolyte solutions. *Lancet*, **II**, 280–3.

SALTER, A. J. (1982). Overview of trimethoprim-sulfamethoxazole: an assessment of more than 12 years of use. *Reviews of Infectious Diseases*, **4**, 196–236.

SCHLIEVERT, P. M., SHANDS, K. N., DAN, B. D., SCHMID, G. P. & NISHIMURA, R. D. (1981). Identification and characterization of an exotoxin from *Staphylococcus aureus* associated with toxic shock syndrome. *Journal of Infectious Diseases*, **143**, 509–16.

SMITH, H. (1982). Infectious diseases. In *The Medical Annual*, ed. R. B. Scott, pp. 169–76. Bristol, London & Boston: Wright, PSG.

SMITH, H., ARBUTHNOTT, J. P. & MIMS, C. (1983). The determinants of bacterial and viral pathogenicity. *Proceedings of the Royal Society of London, Series B*, **303**, in press.

SOUSSY, C. J., DUBLANCHET, A., CROMIER, M., BISMUTH, R., MIZON, F., CHARDON, H., DUVAL, J. & FABIANI, G. (1976). Nouvelles resistances plasmidiques de *Staphylococcus aureus* aux aminosides (gentamicine, tobramycin, amikacine). *Nouvelle presse médicale*, **5**, 2599–602.

STUART-HARRIS, C. (1982). Report on the International Conference on the Eradication of Infectious Disease. Foreword: Can infectious diseases be eradicated? *Reviews of Infectious Diseases*, **4**, 913–14.

THIN, R. N. T. (1982). Sexually transmitted diseases. In *The Medical Annual*, ed. R. B. Scott, pp. 295–9. Bristol, London & Boston: Wright, PSG.

THIRUMOORTHI, M. C. & DAJANI, A. S. (1979). Comparison of staphylococcal coagglutination, latex agglutination and counter-immunoelectrophoresis for bacterial antigen detection. *Journal of Clinical Microbiology*, **9**, 28–32.

TILTON, R. C. (1982). The laboratory approach to the detection of bacteremia. *Annual Reviews of Microbiology*, **36**, 467–93.

TIMBURY, M. C. (1982). Acyclovir. *British Medical Journal*, **285**, 1223–4.

TODD, J., FISHAUT, M., KAPRAL, F. & WELCH, T. (1978). Toxic-shock syndrome associated with page-group I staphylococci. *Lancet*, **II**, 1116–18.

WAHDAN, M. H., SERIE, C., CERISIER, Y., SALLAM, S. & GERMANIER, R. (1982). A controlled field trial of life *Salmonella typhi* strain Ty21a oral vaccine against typhoid: three year results. *Journal of Infectious Diseases*, **145**, 292–5.

WALDVOGEL, F. A. (1982). The future of β-lactam antibiotics. *Reviews of Infectious Diseases*, **4**, *Supplement*, S491–S495.

WATERSON, A. P. (1983). Acquired immune deficiency syndrome. *British Medical Journal*, **286**, 743–6.

WEISS, A. A. & FALKOW, S. (1983). The application of molecular cloning to the study of bacterial virulence. *Proceedings of the Royal Society London, Series B*, **303**, in press.

WHO TECHNICAL REPORT (1976). Series 588: WHO Expert Committee on Biological Standardisation. Rome: WHO.

WHO TECHNICAL REPORT (1978). Series 624: Surveillance and control of health hazards due to antibiotic resistant enterobacteria. Rome: WHO.

WILLIAMS, J. D. (1978). Spread of R-factor outside the Enterobacteriaceae. *Journal of Antimicrobial Chemotherapy*, **4**, 6–8.

WILLIAMS, J. D. (1981). The needs for new antibiotics for infection in man. In *The Future of Antibiotherapy and Antibiotic Research*, ed. L. Ninet, P. E. Bost, Bouanchaud, D. H. & J. Florent, pp. 117–25. London & New York: Academic Press.

WILLIAMS, J. D., KATLAN, S. & CAVANAGH, P. (1974). Penicillinase production by *Haemophilus influenzae*. *Lancet*, **II**, 103.

YOLKEN, R. H. (1982). Enzyme immunoassays for the detection of infectious antigens in body fluids: current limitations and future prospects. *Reviews of Infectious Diseases*, **4**, 35–68.

YORKE, J. A., NATHANSON, N., PIANIGIANI, G. & MARTIN, J. (1979). Seasonality and the requirements for perpetuation and eradication of viruses in populations. *American Journal of Epidemiology*, **109**, 103–23.

MICROBES, MICROBIOLOGY AND THE FUTURE OF MAN

JOHN POSTGATE

ARC Unit of Nitrogen Fixation,
University of Sussex, Brighton BN1 9RQ, UK

Oubliez pour un moment le point que nous occupons dans l'espace et dans la durée, et étendons notre vue sur les siècles à venir, les régions les plus éloignées et les peuples à naitre. Songeons au bien de notre espèce . . .

Diderot, *Le Neveu de Rameau* (Billy, 1951)

I have been urged to assess the future, to move from real science to speculation. Confronted with a similar instruction from his philosophical companion, the extrovert nephew of Rameau sang a couple of songs, danced a few steps, mimed a little and feigned sleep, thus contributing rather little to the substance of their dialogue but making, nonetheless, a point with which I have sympathy. In a figurative sense I am obliged to do something similar, although the song-and-dance routine will have melancholy overtones and it will be the reader, not me, who may fall asleep.

Other contributions to this volume survey the present state of many aspects of microbiology; I take it that my brief is to seek indications of the directions in which the subject may go in the future. Naturally, I interpret my brief as including the applications of microbiology as well as the science itself.

The future of microbiology cannot be discussed without reference to the future of human society, and this is where the problems which I must face start immediately. Forecasting mankind's future has become a highly specialized affair and is the subject of numerous books and pamphlets, as well as more ephemeral presentations. Jahoda & Freeman (1978) collated many of the more serious forecasts (notably in Table 2.2 of S. Cole's contribution thereto) and the wide range of projections available, optimistic and pessimistic, is clear from a perusal of their work. This is an area where polemic and political posture can obscure the scientific detachment which most microbiologists expect, and I have no wish to tangle with such matters. There is a wide variety of human societies today – Western Capitalist, Communist, Islamic, nomadic and tribal, for example – and even analyses of their structures, as a prerequisite of a consideration of their futures, raise questions well beyond the provenance of even a relatively unspecialized topic such as is the subject of this volume. Yet it is necessary to have some sort of intellectual framework within which to work, if only to answer the question of whether there will be microbiologists available to further microbiology. Therefore, with some trepidation, I shall try to present a biological rather than political global background against which to discuss the future.

Any reasonable scenario must take into account the consequences of this century's population explosion. For it is now self-evident that, unless a global catastrophe intervenes, the world's population, at present just over 4×10^9, will rise to between 7.5×10^9 and 8×10^9 in the early decades of the next millennium: 30 or so years ahead. This is true even if all international plans for population control are successful, for the simple reason that the children who will mate and produce all these extra offspring are already here, and they will do so at a greater rate than that at which their parents will die (see, for example, Study Committee, 1971). Microbiology, through its contribution to the control of disease, made a substantial contribution to the net increases in human fertility and longevity which underlie the population explosion. A major consequence of this situation is already upon us: there is already evidence in many countries that population growth has outstripped the technology required to support it. Recognition of this fact informs all the contributions to Jahoda & Freeman (1978) and the other works commented upon therein as well as more recent discussions such as Faaland (1982). I could cite specific examples from many nations, but to choose would be invidious and would again risk raising political and/or patriotic prejudices. I leave such examples to the reader's familiarity with the social stresses in both developing and developed countries and, instead, I shall describe a more general, well-documented situation which illustrates my point.

It is axiomatic that all these new people will need to be fed. At present about a third of the world's population is alive and fed, albeit sometimes inadequately, by courtesy of the Haber process. For the fertilizer industry provides, from atmospheric nitrogen, the fixed nitrogen which underpins over 30% of the world's agricultural productivity. Support of a nearly doubled population will require expansion of both industrial and biological (i.e. microbial) nitrogen inputs by nearly two-fold. Appropriate expansion of the production and transport of industrial nitrogen fertilizer, together with appropriate changes in agricultural and food-processing practices so as to exploit biological nitrogen fixation more, are mandatory, yet they entrain economic, social, political and even natural obstacles (e.g. poor harvests) which have already led to localized starvation. The technology to support the prospective world's population exists already, but the logistic and economic problems of how to do so are daunting. I choose this example not only because it is familiar to me, in that it relates to my own major research area, but also because it is not disputed: fertilizer manufacturers and eco-freaks may argue about the relative merits of chemical nitrogen fertilizer versus microbial diazotrophy as a solution to the problem, but none denies that the problem exists and that it is serious (see Hardy, 1976; Wittwer, 1977; Postgate, 1980).

The matter of nitrogen inputs in world agriculture is an urgent and important problem, but it is one which has been perceived by responsible scientists and administrators. It is also in principle soluble, largely because it concerns a renewable resource which can be recycled more rapidly. However, numerous shortages exist of raw materials – sulphur, coal, oil, nickel, silver, copper, mercury, water – which seem to crowd in on us as demand expands and concentrated sources are depleted. They fall into two major classes: renewable resources (wood, hydroelectricity, soil-nitrogen, etc.) and non-renewable resources (copper, chromium, uranium). Strictly, though, this distinction is artificial: most renewable resources are materials which can be recycled rapidly, most non-renewable ones are those which can be recycled only very slowly or with great difficulty (radioactive or fissionable materials are probably the only truly non-renewable materials on this planet).

Renewable resources, in the conventional sense, depend on the thermodynamic flux in the biosphere, much of it engendered by living things in the carbon, nitrogen, oxygen, sulphur and other cycles of the biological elements. The transcendent importance of microbes in these cycles does not need spelling out to microbiologists; as an illustration, which continues my earlier example, some agronomists are encouraging a return to biological nitrogen inputs in agriculture as a means of avoiding some of the expenditure of fossil energy (the natural gas feedstock of the Haber process) to supplement the nitrogen cycle. Non-renewable resources such as coal, oil, natural gas, sulphur, copper, etc., etc., are materials which have been available in the past as concentrated deposits, sometimes formed by microbial action in earlier geological eras (e.g. sulphur, bog-iron, soda, probably coal and oil), but which are becoming exhausted at a rate greater than that at which new sources are being discovered. Forecasters differ remarkably in their estimates of the time for which such resources will last (some are collated in the contributions to Jahoda & Freeman, 1978) and recycling is becoming increasingly economic. However, despite the optimism of Kahn, Brown & Martel (1977), such recovery processes are never quantitative and new sources have to be exploited which are generally more dilute and less accessible than the older sources, thus requiring greater capital and technological investment for their exploitation.

In times of economic recession, some of these shortages may become masked – at the time of writing there is a world glut of oil – but in terms of the needs and expectations of mankind, the world recession is only a hiccup. In the absence of corrective action, the shortages of both renewable and non-renewable materials cannot but get worse as society expands, adjusts (one hopes) to the new employment situation by coming to terms with the micro-chip and generates a return to some expansion of prosperity.

Because these materials are essential to the continuation of society, of society as a whole and not just of high-technology society, increasing amounts of energy have to be expended to compensate for their shortages. The energy is expended in two ways. One is in supplementing renewable resources: when wood became inadequate as a self-renewing fuel many centuries ago, mankind accelerated the carbon cycle by burning fossil fuels (first coal and later natural gas and oil as well). The other way is in winning non-renewable materials from dilute, poor or difficult sources such as deeper mines, sea-bed wells or low-grade ores. Coal provides a particularly instructive example in this class, being an energy source itself: as the more accessible seams become exhausted, the energy cost of mining coal begins to approach the energy content of the coal itself.

The message here is that the winning of materials from dilute, poor or difficult sources, or the generation of supplements or substitutes for dwindling resources, costs energy. And energy costs are financial costs, as the world learned abruptly in the early 1970s, by courtesy of the OPEC nations.

Thus the world is approaching a situation in which the population increases uncontrollably, the expectations of all members of that population also increase, yet at the same time the economic (energy) input required to support each individual, and to satisfy those expectations, increases.

In a rational society this situation ought to be manageable: twentieth-century technology is perfectly capable of coping with a doubling of the world's population together with a general upgrading of living standards (see Kahn *et al.*, 1977; and Jahoda & Freeman, 1978). The two extreme scenarios on which I shall base my presentation differ principally according to their view of whether mankind is or is not capable of directing his technology towards the survival of a high-technology society.

THE DISASTER SCENARIO

The disaster scenario is representative of several more detailed and authoritative projections. It is based on a simple but regrettably compelling view of the social effects of overpopulation. The distribution of public-spirited versus anti-social behaviour within a human population, if it could be measured, would be Gaussian, like the distribution of a property such as doubling time or motility within a bacterial population. In effect, the vast majority would be close to average in their behaviour and the proportion of exceptionally anti-social or, conversely, exceptionally public-spirited people would be very small, with saintliness and extreme evil an asymptotically vanishing component. The population explosion does nothing to alter this distribution, but it has the following most important effect: *the absolute number* of social deviants, whether benevolent or evil, is larger today than it has ever been, in virtually any society one cares to consider. It is, of course, the social deviants who have a dominant effect on the behaviour of their fellows.

The disaster scenario arises from the observation, not restricted to mankind but also true of overcrowded mammals, that overcrowding and competition for resources favour destructive deviants. Unlike other mammals, mankind has highly effective means of communication, so the effect of these deviants is amplified. To express the point differently, population expansion increases the total number of foci of social breakdown within a population; population pressure also increases the range of their disruptive effect. We see this phenomenon in several forms today: in the major Western cities it takes the form of street crime, vandalism and violence; in several Western countries it takes the form of political extremism and terrorism; in many less developed, but sometimes quite rich, countries (usually but not always outside the Judeo-Christian ethic) it takes the form of a retreat from technological society into religious fundamentalism and holy war; between countries it takes the form of nationalism, xenophobia and bellicosity. It is reflected at an intellectual level in an increasingly anti-science posture adopted by the more demagogic politicians and religious leaders. The disaster scenario is based on the view that Western technological society (for these purposes Japan and probably India are 'honorary' Western societies) will shortly be precipitated into a global war, followed rapidly by the collapse of Western civilization and its disintegration into isolated pockets of relatively unsophisticated war-orientated technology.

This type of scenario is familiar to most thinking people, and is a rich source of material for science fiction writers. Yet it is important to re-emphasize that it is a social, not a technological matter. The social diseases – nationalism, terrorism, racism, fundamentalism and extremism – demoralize and undermine faith in scientific rationalism. Luria (1971), who unwittingly led me to my opening quotation, expressed the point with elegant melancholy:

An enormous amount of scientific progress has generated stupendous technologies; but the rationalist hopes of the past two centuries, that a free-wheeling technology would satisfy the biological needs of man, remove the causes of conflict among men, and open up a royal way to a golden era of humanity, have proved illusory. The brutalities of two world wars at a time when educated men considered war unthinkable; the rise of aggressive nationalisms just when the way seemed open to international brotherhood; the persistence and increase of poverty even in the most affluent societies; and the current twin threats of nuclear destruction and uncontrolled overpopulation – all these events have shattered confidence in the values of the rationalist-scientific revolution.

Microbiology in catastrophe

It is scarcely necessary to spell out the consequences of global war and/or social breakdown as far as the title of this paper is concerned. Microbes will indeed flourish, being relatively immune to radiation and thriving in conditions of squalor, disease and human deprivation. Microbiologists will cease to exist. Microbiology will regress, persisting only in such fragments of contemporary medicine as may survive and, possibly, in primitive forms of biological warfare. The one new point which the microbiologist can add to this gloomy scenario, and it is an important one, is that one can expect the collapse of our high-technology society to be irreversible. For we developed our present industry on the basis of raw materials – coal, sulphur, bog-iron, oil, several metal ores – many of which were concentrated by microbiological processes over preceding geological ages. Most of these concentrated sources are now exhausted or inaccessible. So the resources that came freely to hand over the last two or three centuries, and which supported the industrial revolution, are no longer there. It is difficult to envisage any kind of technological society re-emerging without them.

THE UTOPIAN SCENARIO

The Utopian scenario is based on the view that the social stresses consequent upon overpopulation can be contained sufficiently effectively for high technology to catch up with, and innovate sufficiently to satisfy, the demands and expectations of the impoverished billions. Since the Second World War, several countries in Asia, Central and South America have transformed their economies from 'developing' or 'underdeveloped' status to something closely approaching the Western 'developed' status. Some have had the advantage of newly discovered energy reserves; some have undergone political convulsions of a devastating kind; some have developed regimes of appalling harshness; some have announced their retreat from technology while importing it from elsewhere. As monuments to human folly, the patterns of man's adaptation to his own science and technology would defeat the pen of a Dean Swift. Yet in real terms the comfort and well-being of very substantial portions of mankind has improved, particularly during the second half of the twentieth century.

It is this evidence of a capacity to muddle through despite all threats of disaster which gives hope to optimists and provides the Utopian scenario: a peaceful society in which science and its applications flourish, for the furtherance of knowledge as well as the benefit of mankind and his biosphere. So I shall now close my mind to the news bulletins and consider the ways in which our subject might develop. I apologize in advance for the random and subjective nature of my speculations and for the fact that they echo to some extent an earlier consideration of this topic (in Postgate, 1969).

Microbiology in Utopia

Information

Like all sciences, microbiology is suffering from the information explosion: publication has outstripped the capacity of microbiologists to read and assimilate papers. The days of 'browsing' are over and most microbiologists rely heavily on contents lists and reviews. Computerized bibliographies (e.g. the Biosis and Agricola files) are proving their worth but represent only a beginning. I imagine that conventional publication will be replaced by deposition at some centre, with access for enquiry (via satellite or land-line) on video-display at the desk; video-cassette recording or print-out would supply information for continued local use. Editors, referees and editorial boards will nevertheless have as important a job as today, since the ability to store information is a temptation to store rubbish . . .

Taxonomy

With the discovery of plasmids and R-factors, the concept of the bacterial genome began to crumble over a decade ago. Few of us then realized how literally it would fragment: today most microorganisms are known to possess plasmids, some so large as to be mini-chromosomes (e.g. the megaplasmids of *Rhizobium meliloti*); among bacteria, plasmid-free organisms seem to be the exception rather than the rule. Putatively definitive characters (e.g. tumorogenicity in *Agrobacterium*) are plasmid-borne and transferable intergenerically. At another structural level the transposons, originally discovered as sub-units of plasmids, are now proving universal. The microbiologist has long since faced the genetic fluidity of bacteria with its attendant problems of discontinuous and reticulate evolution (see Sneath, 1974). Now these problems trouble macrobiologists.

The classical criteria of microbial genera and species are readily breached by deliberate or fortuitous gene exchange and even the Adansonian approach cannot accommodate a substantial intergeneric exchange of properties. Molecular taxonomy using rRNA sequencing has proved remarkably fruitful as a coarse tool, notably in substantiating the reality of the Archaebacteria and in rationalizing a variety of putative genera and species (see Stackebrandt & Woese, 1981). No doubt the molecular approach will continue to be refined towards a taxonomy related to molecular phylogeny. Will it ever accommodate gross intergeneric DNA transfers and provide an organismic phylogeny? If so, perhaps we shall learn the answer to the microbiologist's current variant of the hen/egg paradox: which came first, the chromosome or the plasmid? Until then, working bacteriologists will make do with working classifications, as they always have done.

Medicine
The hypothesis of the prokaryotic ancestry of organelles is attractive and has considerable experimental support, despite the presence of such un-prokaryotic genomic elements as introns in chloroplast DNA. If this view is correct, the organelle is probably a specialized example of a more general phenomenon whereby prokaryotic material has formed part of the eukaryotic genome, not in the sense of sharing a common evolutionary origin but by positive accretion into eukaryotic material long after the two classes separated. Precedents exist, such as the integration of virus DNA into animal chromosomes, or integration of agrobacterial plasmid DNA into the plant chromosome. The consequences of such events can be deleterious, and the study of prokaryotic genomes in such backgrounds may well preoccupy medical microbiologists to a substantial extent, particularly in view of its potential for the correction of genetic defects. Virus action and its control will remain a burning problem, perhaps with emphasis turning to the slow viruses, but some microbiologists will have to keep an eye open for new bacterial infections, lest an episode comparable to the *Legionella* story should take us by surprise again. Although drug-resistant coliforms, neisserias and other pathogens are now commonplace and usually assignable to plasmid transfer, the fact that much other genetic information is in principle transferable means that taxonomically totally new pathogens might emerge through the spontaneous interaction of prokaryotic genomes which are only distantly related. The deliberate exploitation of this information for biological warfare must not escape the attention of the thinking microbiologist, though it belongs in my other scenario.

Of course, medicine will provide the background for much microbiological research because we are all deeply, and rightly, concerned with health. New antibiotics, drugs and therapeutic agents will emerge, by conventional as well as recombinant DNA techniques, and a proportion will duly be rendered useless by uncontrolled or incompetent application. Clinical medicine will continue to require a substantial input from microbiology.

Preventive medicine will also need microbiology. The triumphant cry of the World Health Organization on 8 May 1980, announcing the elimination of smallpox from the world and its peoples (see *The Times* (London), 9 May 1980, p. 9), represented a truly magnificent feat of social medicine. I regret that laboratory stocks of the virus must still be held, happily under tight security, yet I am sufficient of a conservationist to agree that all trace of a life-form should never be deliberately eliminated. My own solution, based on a superficial assessment of the situation, is that the total DNA sequence should be obtained and recorded and that all viable stocks should then be destroyed. Future generations could then, at least in principle, reconstruct the organism should any good reason for doing so arise. The elimination of other serious pathogens has become possible and will doubtless proceed, sometimes spontaneously as North-Western standards of hygiene become universal. A different problem is that increased freedom to travel has already brought diseases such as tuberculosis back to areas from where it had been all but eliminated. Some may fear that universally high levels of hygiene will lead to a population unduly susceptible to epidemic, as were the inhabitants of Tristan da Cunha when, in 1961 (Roberts, 1971), their volcano imposed upon them an unwelcome few months' holiday in Britain. Personally, I do not doubt that small children, pets, persons averse to 'washing their hands before leaving' and other foci of infection will provide sufficient inocula for reasonable levels of general immunity to develop locally, but in principle the microbiologist and immunologist will need to be increasingly on the alert for epidemic situations as standards of hygiene increase.

Veterinary microbiology
Meat is still an essential component of the diet for maximum growth and health of most of humanity. Therefore the control of disease in the production of livestock will remain a problem, not only in the developing world but also under intensive farming conditions, where diseases or drug resistances can spread very rapidly. Yet the production and processing of livestock for food is becoming increasingly repugnant and, in the Utopian scenario, it may well die out, albeit over many generations. Microbiology may well bring forward the age of the vegetarian, not so much through the use of microbes directly as food, as through their use as vehicles for gene transfer to upgrade vegetable (and microbial) protein to a quality equivalent to that of fresh meat. Will there then be a murky trade for illicit carnivores, with police raids on back-street abattoirs or

broiler houses? The science fiction writer, as so often, has been here before us.

Biotechnology

It was recombinant DNA technology that made the non-microbiological world conscious of biotechnology (earlier economic microbiology, incorporating industrial microbiology and applied microbiology. What's in a name? Answer: substantial research grants . . .). Yet it has been with us for a long time and older microbiologists will have to suppress their irritation and adjust to the term. For biotechnological processes have two economic virtues which will become increasingly valuable as energy costs escalate: they consume low-grade energy (high pressures and temperatures are not required) and, generally speaking, they exploit renewable resources. (A few, such as the microbial desulphuring of coal, or leaching of copper and uranium, upgrade non-renewable resources, but these instances do not seriously affect my generalization.) Therefore, until cheap fusion power becomes available, the major disadvantage of biotechnological processes – that the product usually has to be concentrated from relatively dilute solution – need not prevent such processes from being cost-effective.

It would not be appropriate to review here the prospects which biotechnology holds for the future; this has been done many times (e.g. a series of introductory articles by Dunnill, 1981; Kirsop, 1981; Coombs, 1981; Fowler, 1981; Bucke & Wiseman, 1981; Sherwood & Atkinson, 1981, which present a reasonably down-to-earth survey of the topic). But some general principles may be worth recalling. Microbes are economical consumers or transformers of energy. They may well, therefore, re-appear in industry as producers of fuel (ethanol, methane), solvents (acetone, butanol) and even relatively 'heavy' chemicals (sulphur, ammonia), but this is only likely to occur where high-grade energy sources are particularly short. As direct sources of energy, microbes have rather little promise, but as components of animal and human food, or as agents in food processing, their value is already established. Their role in packaging, storage and transport of food is already so important that it seems unnecessary to remind microbiologists that the scale of their processes, and their associated problems, cannot but increase.

Recombinant DNA technology has opened vistas of new and unusual ways in which genetically manipulated microbes might be exploited. We are today in a curious phase recalling the South Sea

Bubble: many molecular biologists and microbial geneticists are going private, joining newly formed companies for the commercial exploitation of new (and old) biotechnological processes. The pessimist is confident that the progress of microbiology will suffer, because industrial secrecy is already inhibiting scientific communication and requests for published materials are already being refused among 'rival' laboratories. The struggle for priority and credit, reprehensible at any time, becomes truly counter-productive when large sums of money are involved. The optimist points out that the surge in basic organic chemistry in the late nineteenth century owed much to the dyestuffs industry, that the dead hand of the patent application can be overcome and that industrial involvement need not always be secretive. The pessimist merely hopes that, when the Biotechnology Bubble bursts, the subject itself will emerge reasonably unscathed. For the Utopian scenario we must determinedly assume that the commercialization of biotechnology does not inhibit fundamental advance, for the prospects of generating entirely new species are before us: the imagination can range from oil-slick-consuming bacteria through nitrogen-fixing plants to giant mammals; in many instances the strategy seems reasonably straightforward even if much new fundamental information is still needed.

Environment
The biosphere approximates to a steady-state system in which the activities of microbes are dominant. Recognition of this fact has prompted the surge of interest in microbial ecology of the last two decades and international interest in its applications, represented, for example, by the activities and publications of SCOPE (Scientific Committee on Problems of the Environment). In the Utopian scenario, this trend will continue, the cycles of the biological elements will be increasingly understood, not only at the level of biological processes but at the level of interactive microbiology: already the involvement of microbial consortia is becoming evident and obligate syntrophs have been discovered. The detritus of a world populated by 7×10^9 to 8×10^9 people will require more elaborate and efficient recycling processes than we have today, with our overloaded sewage and effluent plants which pour most of the useful products into the sea; the microbial degradation of anthropogenic pollutants – agrochemicals, plastics, hydrocarbons, etc. – will become systematized and controlled. Recalcitrant materials (e.g. polyfluorocarbons) and radioactive residues suggest no obvious role for microbiological disposal.

The environmental impact of genetically manipulated microbes, or other systems such as plants carrying microbial genes, will need careful consideration. I doubt whether hydrocarbon-oxidizing super-bacteria will corrode our roads and I do not fear a takeover of this planet by genetically constructed nitrogen-fixing weeds. But I should certainly share the anxieties implicit in Clement's (1980) novel if someone successfully generated an aerobic nitrogen-oxidizing bacterium.

The further future
One must assume that, in due time, the biosphere, and in particular its human component, will cease to depend on fossil energy (by fossil energy I include geomicrobiologically produced resources such as sulphur, soda and certain metal ores, which microbes concentrated by energy-consuming processes during the early history of this planet). Our major energy sources will be solar energy (directly or via biological photosynthesis) and fusion. By then we ought to understand the physiology of exotic ecosystems, such as the sulphuretum-based fauna of the hydrothermal vents in the Galapagos reef of the deep Pacific Ocean, or the acidophilic sulphur-reducing Archaebacteria of certain hot springs. Sooner or later mankind will have to cool down Venus and render it habitable, for Mars is just too dry (and chilly) for all but elaborately climatized settlements. The use of microbes as recycling agents in space vehicles is already familiar, at least as an idea; their use as primary colonists of Venus is attractive and on the horizon of possibility.

REALITY

I have chosen a somewhat Hegelian presentation – disaster versus Utopia – fully realizing that my scenarios represent extreme futures of roughly equal improbability. The actual future will lie somewhere between them and will be geographically non-uniform, much like the position today. Obviously the basic biology of microbes will not change significantly in the foreseeable future, but our knowledge of it will develop in ways which will be determined, at least in part, by the pressures and requirements of society. To the working microbiologist this will probably seem like a prolongation of the *status quo*, but there is no *status quo*. In the last 40 years spectacular developments in microbiology have taken place – antibiotics, structural analogues, metabolic pathways, genetic regulation and gene cloning are a few illustrative key words – and the microbiologist has had constantly to readjust his thinking to accommodate them. Simultaneously, the requirements of applied microbiology – medicine, food and, recently, biotechnology – have imposed directions and outlooks which were unfamiliar to the pioneers of this Society. All these changes and counterchanges will continue, unless a genuine catastrophe intervenes. Both microbiology and microbiologists will survive and, one hopes, prosper.

It would be idle for me to try to predict actual changes in detail, so I shall conclude with a single message. The trend towards research orientated towards the good of mankind is admirable, but it is axiomatic that it must never be allowed to exclude or suppress the advance of fundamental knowledge. Although grant-giving agencies, Research Councils, industrialists and administrators often pay lip-service to this axiom, they rarely understand its implications. In the near future, the thinking microbiologist will be called upon increasingly to defend fundamental research. It is difficult for non-scientists to understand that in the search for pure knowledge one does not really blaze trails, but sneaks round obstacles by the easiest route, only to encounter new ones. The most important point, which all scientists have to make incessantly, is that the sorts of practical benefits I touched upon in the Utopian scenario will not come about without substantial and continuing advances in fundamental knowledge. The need for basic, exploratory, innovative – call it what you will – research must be recognized and explained. Then we may expect to know more about everything and the new questions will be more complex than ever, costing more in time, equipment and, above all, in patience for their answers. By patience I mean patience with one's fellow scientists, for collaboration, particularly interdisciplinary collaboration, will become an increasingly important component of scientific advance. Microbiology grew out of chemistry, for it was the chemical activities of microbes which attracted early microbiologists such as Pasteur and Beijerinck; it hybridized with bacteriology at an early stage and has effortlessly assimilated elements of physics and botany; it still underpins the whole of molecular biology. We must retain this ability to draw from and interact with the rest of the sciences; for the really good microbiologists are not really microbiologists at all, they are scientists who happen to be very interested in microbes.

I am grateful to Professor Mark Richmond, FRS, and Professor Harry Smith, FRS, who read and commented upon an early draft of this paper most constructively. I thank Professor Douglas Watson for confirming some virological information. They bear no responsibility for the views I have expressed.

REFERENCES

BILLY, A. (ed.) (1981). *Diderot: Oeuvres*, p. 403. Paris: Gallimard.

BUCKE, C. & WISEMAN, A. (1981). Immobilized enzymes and cells. *Chemistry and Industry*, 234–42.

CLEMENT, H. (1980). *The Nitrogen Fix*. New York: Ace Books.

COOMBS, J. (1981). Biogas and power alcohol. *Chemistry and Industry*, 223–9.

DUNNILL, P. (1981). Biotechnology and industry. *Chemistry and Industry*, 204–17.

FAALAND, J. (ed.) (1982). *Population and the World Economy in the Twenty-first Century*. Oxford: Blackwell Scientific.

FOWLER, M. W. (1981). Plant cell biotechnology to produce desirable substances. *Chemistry and Industry*, 229–33.

HARDY, R. W. F. (1976). Potential impact of current abiological and biological research on the problem of providing fixed nitrogen. In *Nitrogen Fixation*, vol. 2, ed. W. E. Newton & C. J. Nyman, pp. 693–717. Pullman, WA: Washington State University Press.

JAHODA, M. & FREEMAN, C. (1978). *World Futures. The Great Debate*. London: Martin Robertson.

KAHN, H., BROWN, W. & MARTEL, L. (1977). *The Next 200 Years*. London: Associated Business Programmes.

KIRSOP, B. H. (1981). Biotechnology in the food processing industry. *Chemistry and Industry*, 218–22.

LURIA, S. E. (1971). A latter-day rationalist's lament. In *Of Microbes and Life*, ed. J. Monod & E. Borek, pp. 56–61. New York: Columbia University Press.

POSTGATE, J. R. (1969). *Microbes and Man*. Harmondsworth, Middx: Penguin Books.

POSTGATE, J. R. (1980). The nitrogen economy of marine and land environments. In *Food Chains and Human Nutrition*, ed. Sir K. Blaxter, pp. 161–85. London: Applied Science Publishers.

ROBERTS, D. F. (1971). The demography of Tristan da Cunha. *Population Studies*, **25**, 465–79.

SHERWOOD, R. & ATKINSON, T. (1981). Genetic manipulation in biotechnology. *Chemistry and Industry*, 243–7.

SNEATH, P. H. A. (1974). Phylogeny of micro-organisms. In *Evolution in the Microbial World*, ed. M. J. Carlile & J. J. Skehel, pp. 1–39. *Symposia of the Society for General Microbiology*, **24**. Cambridge University Press.

STACKEBRANDT, E. & WOESE, C. (1981). The evolution of prokaryotes. In *Molecular and Cellular Aspects of Microbial Evolution*, ed. M. J. Carlile, J. F. Collins & B. E. B. Moseley, pp. 1–32. *Symposia of the Society for General Microbiology*, **32**. Cambridge University Press.

STUDY COMMITTEE (1971). *Rapid Population Growth* (Study Committee of the US Academy of Sciences). Baltimore: Johns Hopkins Press.

WITTWER, S. H. (1977). Agricultural productivity and biological nitrogen fixation: an international view. In *Genetic Engineering for Nitrogen Fixation*, ed. A. Hollaender, pp. 515–19. New York & London: Plenum Press.

INDEX

acetate: metabolism of, by deep-sea bacteria, 104, by methanogens, 139–41, by sulphur- and sulphate-using bacteria, 141–2; production of, in anaerobic degradation of organic compounds, 39, including aromatic compounds, 143–4

acetic acid, bacterial cell membranes permeable to, 138

Acetobacter woodii: carbon monoxide dehydrogenase of, containing nickel, 150; in cooperative breakdown of aromatic compounds, 40; fixation of carbon dioxiode by, 145–6

acetyl CoA, fixation of carbon dioxide into, 145, 146, 147

Achromobacter: medium, and growth characteristics of, 44

Acidaminococcus, production of glutaconate by, 148

Acidaminococcus fermentans, glutaconyl-CoA decarboxylase of, 132

acidophiles: internal pH of, 76; media for isolation of, 233

Acidiphilium cryptum (new), from coal mine, 226

acquired immune deficiency syndrome, new disease, 308

acrylate, produced from lactate by *Cl. propionicum*, 148

actin, in intracytoplasmic transport, characteristic of eukaryotes, 8

Actinomadura kijania, antibiotic-producer, 238

Actinomadura luzonensis, producer of anti-tumour complex, 238

actinomycetes: antibiotics from, 237, 287; clustering of genes for antibiotics in, 261; convergent evolution of fungi and, 28; heat treatment in isolation of, 231

Actinoplanes, isolated by flotation, 231

Actinopolyspora halophila, unintentional isolation of, 224

adaptation to environment, 51–2; to changes in ionic strength, 76–8, in nutrient supply, 62–6, in pH, 75–6, in temperature, 71–5; by changes in microbe, in chemical composition, 53, in enzymes, *see* enzymes; in morphology, 53–4, in ultrastructure, 53; by phenotypic change of all cells in population, or by mutation and selection, 87

adenine nucleotides: content of, in sea water, at surface, at 2400 m, and at thermal vent, 110; pool of, in *E. coli*, 81

adenosine phosphosulphate reductase, of bacteria in trophosome of *Riftia*, 116

adenylate cyclase: activated in induction of enzymes in *E. coli*, 59; inactivated in substrate-accelerated death? 86

ADP sulphurylase, of bacteria in trophosome of *Riftia*, 116

aeration during incubation, in isolation of new microbes, 234

Aerobacter aerogenes, in starvation conditions, 30

aerotaxis: negative, of purple sulphur bacteria, 36

agar: acid, for isolation of acidophiles, 233; containing chitin, for streptomycetes, 232–3; containing chromogen to indicate presence of galactosidase, 171; isolation of *Spirochaeta* by migration through, 231

agarose, degraded by extracellular and membrane-bound agarases of *Cytophaga flevensis*, 70–1

Agilococcus lubricus (new), from fresh water, 226

Agrobacterium tumefaciens: integration of plasmid DNA from, into plant chromosome, 326, 327; strains of, with galactosidase introduced by phage, 175; transfer of *nif* genes into, 272

alcohol oxidase, in metabolism of methanol by yeasts, 61, 65–6

algae, planktonic: as source of organic carbon in sea water, 98–9

alkalophiles: internal pH of, 76; in soda lakes and deserts, 228

Alteromonas aurantis, from sea water, antibiotic-producer, 238, 240

amino acids: in medium, and secretion of extracellular proteases by *Vibrio* SAI, 70; as osmoregulators, 77; pool of, in starvation, disappears in *Staphylococcus epidermidis*, 85, is retained in *E. coli*, 81

aminoglycoside antibiotics (streptomycin, neomycin, gentamycin, etc.): inhibit translation, 288; plasmid-coded enzymes inactivating, 293

ammonia: emitted at thermal vent, Gulf of California, 114; enzymes involved in assimilation of, 60, 66, 263; nitrogen fixation inhibited by, 270–1; oxidation of, to nitrite and nitrate, 11; in primeval atmosphere, 2; released by mutant of *K. aerogenes*, 273

ammonium sulphate, preferred nitrogen source for enterobacteria, 59